Lecture Notes in Computer Scien

T0237842

Commenced Publication in 1973
Founding and Former Series Editors:
Gerhard Goos, Juris Hartmanis, and Jan van Leeuwen

Gerd Kortuem   Joe Finney   Rodger Lea
Vasughi Sundramoorthy (Eds.)

Smart Sensing
and Context

Second European Conference, EuroSSC 2007
Kendal, England, October 23-25, 2007
Proceedings

 Springer

Volume Editors

Gerd Kortuem
Lancaster University
InfoLab21, Computing Department
Lancaster, LA1 4W4, UK
E-mail: kortuem@comp.lancs.ac.uk

Joe Finney
Lancaster University
InfoLab21, Computing Department
Lancaster, LA1 4WA, UK
E-mail: joe@comp.lancs.ac.uk

Rodger Lea
University of British Columbia
Media and Graphics Interdisciplinary Centre
FSC 3640 - 2424 Main Mall, Vancouver, B.C.,Canada
E-mail: rodgerl@ece.ubc.ca

Vasughi Sundramoorthy
Lancaster University
InfoLab21, Computing Department
Lancaster, LA1 4W4, UK
E-mail: v.sundramoorthy@comp.lancs.ac.uk

Library of Congress Control Number: 2007936932

CR Subject Classification (1998): H.3, H.4, C.2, H.5, F.2

LNCS Sublibrary: SL 5 – Computer Communication Networks
and Telecommunications

ISSN   0302-9743
ISBN-10   3-540-75695-7 Springer Berlin Heidelberg New York
ISBN-13   978-3-540-75695-8 Springer Berlin Heidelberg New York

Springer is a part of Springer Science+Business Media

springer.com

© Springer-Verlag Berlin Heidelberg 2007
Printed in Germany

Typesetting: Camera-ready by author, data conversion by Scientific Publishing Services, Chennai, India
Printed on acid-free paper   SPIN: 12175093   06/3180   5 4 3 2 1 0

Preface

On behalf of the organizing committee for EuroSSC 2007, we would like to welcome you to the proceedings of the second European Conference on Smart Sensing and Context. Although only in its second year, EuroSSC has already begun to attract significant interest from researchers in the rapidly evolving area of embedded sensing and intelligent objects in support of smart surroundings. Building on the success of the previous conference held in Enschede, The Netherlands, EuroSSC 2007 aimed to reflect two complementary viewpoints: a technology-driven viewpoint which encompasses the diversity of intelligent sensing and associated information processing and a user-driven viewpoint exploring scenarios, applications and interaction methods as they relate to smart surroundings.

This year the conference received a total of 51 paper submissions. These represented contributions from 21 countries from Europe, the Middle East, and Africa (69%); Asia (21%); and North America (10%). This obviously reflects the European origins of the conference. However we were particularly pleased by the high number of submissions from Asia reflecting the range and diversity of relevant work in that region. From these the program committee selected 17 papers (33%) for presentation at the conference after a rigorous review process. Every paper received at least three independent reviews, most received four and some five reviews.

Our main concern during the review process was not only to select the highest quality papers, but also to choose a subset of the best papers that reflected the two complementary viewpoints of the conference and that collectively represented the diversity of the research work in the community. As such, the 17 papers chosen cover 4 core areas: sensing and sensor data interpretation; context inference and management; context aware models and platforms; and high level applications and scenarios. These were divided into 4 sessions: context frameworks and platforms; spatial and motion context; human behavior as context; and sensing technologies and case studies. In addition we invited Andrew T. Campbell, Associate Professor of Computer Science at Dartmouth College, to give an invited paper presentation on sensing systems for social network applications. Ours is a diverse field, and we hope that these papers will show the diversity and range of our community, but also allow us to explore the connections and interactions between different aspects of the field.

Given the broad range of this year's submissions the reviewing process was a challenging one. The members of the program committee and the reviewers worked hard under an extremely tight deadline to complete all reviews to ensure a balanced and fair assessment of each paper. Once the papers had been reviewed, we had four days of online discussion to decide on the final selection and to ensure a balanced program. We also worked hard to ensure that all of the authors received useful feedback, so that the papers we could not include in the program could be revised and perhaps submitted elsewhere. Conferences like these are sustained by the community, and we are extremely grateful to all those who volunteered their valuable time to help in the review process and in the creation of such a strong program for EuroSSC 2007.

Equally we extend our thanks to the 106 authors of the submitted papers for their efforts in writing and submitting their work.

Finally, we would like to extend our thanks to the other conference chairs and everyone involved in the organization of EuroSSC 2007, among them Christos Efstratiou, Mark Lowton, Rene Mayrhofer, and Utz Roedig.

August 2007

Gerd Kortuem
Joe Finney
Rodger Lea
Vasughi Sundramoorthy

Conference Organization

General Chair

Gerd Kortuem, Lancaster University, UK

Program Chairs

Joe Finney, Lancaster University, UK
Rodger Lea, University of British Columbia, Canada
Gerd Kortuem, Lancaster University, UK
Vasughi Sundramoorthy, Lancaster University, UK

Demonstrations and Poster Chair

Urs Bischoff, Lancaster University, UK

Panels Chair

Rodger Lea, University of British Columbia, Canada

Publicity Chairs

Rene Mayrhofer, Lancaster University, UK
Vasughi Sundramoorthy, Lancaster University, UK

Local Arrangements

Christos Efstratiou, Lancaster University, UK
Joe Finney, Lancaster University, UK
Mark Lowton, Lancaster University, UK
Utz Roedig, Lancaster University, UK

Program Committee

Stefan Arbanowski, Fraunhofer Institut FOKUS, Germany
Honary Bahram, Lancaster University, UK
Michel Banatre, IRISA, Rennes, France
George Bilchev, British Telecommunications, UK
Azzedine Boukerche, Univ. of Ottawa, Canada
Götz-Philip Brasche, European Microsoft Innovation Center, Germany
Vic Callaghan, University of Essex, UK
Jessie Dedecker, Vrije Universiteit Brussel, Belgium
Hakan Duman, University of Essex, UK

Schahram Dustdar, Vienna University of Technology, Austria
Martin Elixmann, Philips Research, Germany
Ling Feng, Tsinghua University China, P.R. China
Alois Ferscha, University of Linz, Austria
Joe Finney, Lancaster University, UK
Patrik Floréen, University of Helsinki, Finland
Kaori Fujinami, Tokyo Univ. of Agriculture and Technology, Japan
Manfred Hauswirth, DERI Galway, Ireland
Paul Havinga, University of Twente, The Netherlands
Markus Huebscher, Imperial College London, UK
Maddy Janse, Philips Research, The Netherlands
Gerd Kortuem, Lancaster University, UK
Koen Langendoen, Delft University of Technology, The Netherlands
Marc Langheinrich, ETH Zurich, Switzerland
Rodger Lea, University of British Columbia, Canada
Maria Lijding, University of Twente, The Netherlands
Paul Lukowicz, University of Passau, Germany
Robin Mannings, BT Labs, UK
Pedro Marron, University of Bonn and Fraunhofer IAIS, Germany
Rene Mayrhofer, Lancaster University, UK
Nirvana Meratnia, University of Twente, The Netherlands
Kevin Mills, National Institute of Standards and Technology, USA
Tatsuo Nakajima, Waseda University, Japan
Thomas Plagemann, University of Oslo, Norway
Utz Roedig, InfoLab21, Lancaster University, UK
Kay Roemer, ETH Zurich, Switzerland
George Roussos, Birkbeck College, University of London, UK
Gregor Schiele, University of Mannheim, Germany
Klaus Schmid, University of Hildesheim, Germany
Albrecht Schmidt, Fraunhofer IAIS and b-it University of Bonn, Germany
Aruna Seneviratne, University of New South Wales, Australia
Thomas Strang, DLR Oberpfaffenhofen, Germany
Vasughi Sundramoorthy, Lancaster University, UK
Nicolae Tapus, Politehnica University of Bucharest, Romania
Khai Truong, University of Toronto, Canada
Kristof Van Laerhoven, Darmstadt University of Technology, Germany
Harald Vogt, SAP AG, Germany
Matthias Wagner, DoCoMo Communications Labs Europe, Germany
Maarten Wegdam, Alcatel-Lucent and University of Twente, The Netherlands
Hee Yong Youn, Sungkyunkwan University, Korea
Daqing Zhang, National Institute of Telecommunications (INT/GET), France

Table of Contents

Invited Paper

CenceMe – Injecting Sensing Presence into Social Networking
Applications.. 1
  *Emiliano Miluzzo, Nicholas D. Lane, Shane B. Eisenman, and
  Andrew T. Campbell*

Spatial and Motion Context

Mapping by Seeing – Wearable Vision-Based Dead-Reckoning, and
Closing the Loop.. 29
  *Daniel Roggen, Reto Jenny, Patrick de la Hamette, and
  Gerhard Tröster*

The Design of a Pressure Sensing Floor for Movement-Based Human
Computer Interaction 46
  *Sankar Rangarajan, Assegid Kidane, Gang Qian,
  Stjepan Rajko, and David Birchfield*

Sensing Motion Using Spectral and Spatial Analysis of WLAN RSSI.... 62
  *Kavitha Muthukrishnan, Maria Lijding, Nirvana Meratnia, and
  Paul Havinga*

Inferring and Distributing Spatial Context 77
  Clemens Holzmann

Context Sensitive Adaptive Authentication........................ 93
  *R.J. Hulsebosch, M.S. Bargh, G. Lenzini, P.W.G. Ebben, and
  S.M. Iacob*

Human Behavior as Context

A Sensor Placement Approach for the Monitoring of Indoor Scenes 110
  Pierre David, Vincent Idasiak, and Frédéric Kratz

Recognition of User Activity Sequences Using Distributed Event
Detection 126
  *Oliver Amft, Clemens Lombriser, Thomas Stiefmeier, and
  Gerhard Tröster*

Behavior Detection Based on Touched Objects with Dynamic Threshold
Determination Model.. 142
  Hiroyuki Yamahara, Hideyuki Takada, and Hiromitsu Shimakawa

Towards Mood Based Mobile Services and Applications 159
  Alexander Gluhak, Mirko Presser, Ling Zhu,
  Sohail Esfandiyari, and Stefan Kupschick

Recognising Activities of Daily Life Using Hierarchical Plans 175
  Usman Naeem, John Bigham, and Jinfu Wang

Context Frameworks and Platforms

GlobeCon – A Scalable Framework for Context Aware Computing 190
  Kaiyuan Lu, Doron Nussbaum, and Jörg-Rüdiger Sack

ESCAPE – An Adaptive Framework for Managing and Providing
Context Information in Emergency Situations 207
  Hong-Linh Truong, Lukasz Juszczyk, Atif Manzoor, and
  Schahram Dustdar

Capturing Context Requirements 223
  Tom Broens, Dick Quartel, and Marten van Sinderen

Deployment Experience Toward Core Abstractions for Context Aware
Applications... 239
  Matthias Finke, Michael Blackstock, and Rodger Lea

Sensing Technologies and Case Studies

Ambient Energy Scavenging for Sensor-Equipped RFID Tags in the
Cold Chain 255
  Christian Metzger, Florian Michahelles, and Elgar Fleisch

Escalation: Complex Event Detection in Wireless Sensor Networks 270
  Michael Zoumboulakis and George Roussos

Multi-sensor Cross Correlation for Alarm Generation in a Deployed
Sensor Network 286
  Ian. W. Marshall, Mark Price, Hai Li, Nathan Boyd, and Steve Boult

Author Index 301

CenceMe – Injecting Sensing Presence into Social Networking Applications

Emiliano Miluzzo[1], Nicholas D. Lane[1],
Shane B. Eisenman[2], and Andrew T. Campbell[1]

[1] Dartmouth College, Hanover NH 03755, USA
{miluzzo,niclane,campbell}@cs.dartmouth.edu
[2] Columbia University, New York NY 10027, USA
shane@ee.columbia.edu

Abstract. We present the design, prototype implementation, and evaluation of CenceMe, a personal sensing system that enables members of social networks to share their *sensing presence* with their buddies in a secure manner. Sensing presence captures a user's status in terms of his activity (e.g., sitting, walking, meeting friends), disposition (e.g., happy, sad, doing OK), habits (e.g., at the gym, coffee shop today, at work) and surroundings (e.g., noisy, hot, bright, high ozone). CenceMe injects sensing presence into popular social networking applications such as Facebook, MySpace, and IM (Skype, Pidgin) allowing for new levels of "connection" and implicit communication (albeit non-verbal) between friends in social networks. The CenceMe system is implemented, in part, as a thin-client on a number of standard and sensor-enabled cell phones and offers a number of services, which can be activated on a per-buddy basis to expose different degrees of a user's sensing presence; these services include, life patterns, my presence, friend feeds, social interaction, significant places, buddy search, buddy beacon, and "above average?"

1 Introduction

The growing ubiquity of the Internet provides the opportunity for an unprecedented exchange of information on a global scale. Those with access to this communication substrate, and among these especially the youth, increasingly incorporate information exchange via technologies such as email, blog, instant message, SMS, social network software, and VOIP into their daily routines. For some, the electronic exchange of personal information (e.g., availability, mood) has become a primary means for social interaction [2]. Yet, the question of how to incorporate personal sensing information such as human activity inferencing into these applications has remained largely unexplored. While existing communication forums allow the exchange of text, photos, and video clips, we believe a user experience with a richer texture can be provided in a more natural way by integrating automatic sensing into the various software clients used on mobile

G. Kortuem et al. (Eds.): EuroSSC 2007, LNCS 4793, pp. 1–28, 2007.

communication devices (e.g., cellular phone, PDA, laptop) and popular Internet applications.

It seems we are well situated to realize this vision now. The technology push driving the integration of sensors into everyday consumer devices such as sensor-enabled cell phones is one important enabler, setting the scene for the deployment of large-scale people-centric sensing applications [41] over the next decade. We can observe change in the marketplace: the commonly carried cell phone of today with its microphone and camera sensors is being superseded by smart-phones and PDA devices augmented with accelerometers capable of human activity inferencing, potentially enabling a myriad of new applications in the healthcare, recreational sports, and gaming markets. We imagine people carrying sensor-enabled cell phones or even standard cell phones for that matter will also freely interact over short range radio with other sensors not integrated on the phone but attached to different parts of the body (e.g., running shoes, BlueCel dongle, as discussed in Section 5), carried by someone else (e.g., another user), attached to personal property (e.g., bike, car, ski boot), or embedded in the ecosystem of a town or city (e.g., specialized CO_2, pollen sensors).

In this paper, we present the design, prototype implementation, and evaluation of CenceMe, a personal sensing system that enables members of social networks to share their *sensing presence* with their buddies in a secure manner. CenceMe allows for the collection of physical and virtual sensor samples, and the storage, presentation and controlled sharing of inferred human sensing presence. When CenceMe users engage in direct communication (e.g., instant messaging), we aim to allow the conveyance of non-verbal communication that is often lost (or must be actively typed - an unnatural solution) when human interaction is not face to face. When indirect communication is used (e.g., Facebook profile), we aim to make people's personal status and surroundings information - i.e., their sensing presence - available. Similarly, through mining of longer term traces of a user's sensed data CenceMe can extract patterns and features of importance in one's life routine.

The concept of "sensing presence" is fundamental to the CenceMe system, capturing a user's status in terms of his activity (e.g., sitting, walking, meeting friends), disposition (e.g., happy, sad, doing OK), habits (e.g., at the gym, coffee shop today, at work) and surroundings (e.g., noisy, hot, bright, high ozone). CenceMe injects sensing presence into popular social networking applications allowing for new levels of "connection" between friends in social networks. We believe that providing a framework allowing the collection, organization, presentation, and sharing of personal sensing presence and life pattern information will serve a broad spectrum of people, and represents the core challenge of making people-centric sensing [41] a reality. One can imagine many situations where the availability of sensing presence would provide great utility. We have, for example, on the one end a mum who wants to simply know where her kid is, to the youth interested in knowing and catching up with what is hot, where his friends are and what they are doing, what are the trendy hang outs, where a party is taking place, and comparing himself to his peer group and the broader CenceMe community.

The CenceMe architecture includes the following components:

– thin sensing clients focus on gathering information from mobile user communication computing devices including a number of standard and sensor-enabled cell phones. We leverage physical sensors (e.g., accelerometer, camera, microphone) embedded in off-the-shelf mobile user devices (e.g., Nike+ [5], Nokia 5500 Sport [7], Nokia N95 [7]), and virtual software sensors that aim to capture the online life of the user. The thin-client also supports interaction between the phone and external sensors over short-range radio such as the BlueCel dongle (discussed in Section 5), which integrates a 3-axis accelerometer with a Bluetooth radio that can be attached to the body (e.g., as a badge) or other entities (e.g., bike).
– a sensor data analysis engine that infers sensing presence from data.
– a sensor data storage repository supporting any-time access via a per-user web portal (a la Nike+ [5]).
– a services layer that facilitates sharing of sensed data between buddies and and more globally, subject to user-configured privacy policy. CenceMe services can be activated on a per-buddy basis to expose different degrees of a user's sensing presence to different buddies, as needed. These services include *(i) life patterns*, which maintains current and historical data of interest to the user - consider this a sensor version of MyLifeBits [46]; *(ii) my presence*, which reports the current sensing presence of a user including activity, disposition, habits, and surroundings, if available; *(iii) friend feeds*, which provides an event driven feed about selected buddies; *(iv) social interaction*, which uses sensing presence data from all buddies in a group to answer questions such as "who in the buddy list is meeting whom and who is not?"; *(v) significant places*, which represents important places to users that are automatically logged and classified, allowing users to attach labels if needed - in addition, users can "tag" a place as they move around in a user-driven manner; *(vi) health monitoring*, which uses gathered sensing presence data to derive meaningful health information of interest to the user; *(vii) buddy search*, which provides a search service to match users with similar sensing presence profiles, as a means to identify new buddies; *(viii) buddy beacon*; which adapts the buddy search for real-time locally scoped interactions (e.g., in the coffee shop); and finally *(ix) "above average?"*, which compares how a user is doing against statistical data from a user's buddy group or against broader groups of interest (e.g., people that go to the gym, live in Hanover) - in the latter case the presentation of data is always strictly anonymous.
– Consumers of sensing presence run as plugins to popular social networks software (e.g., Skype [9], Gaim/Pidgin [12], Facebook [36], MySpace [37]), rendering a customizable interpretation of the user information.

The rest of the paper is organized as follows. In Section 2, we discuss the CenceMe architecture in more detail, including the types of physical and virtual

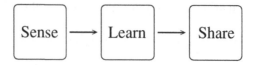

Fig. 1. Information and process flow in the CenceMe system

sensors we design for, the inference and analysis techniques we use, and the application plugin infrastructure and the per-user web portal that exist to present the distilled sensing presence of the user. We describe the set of services CenceMe provides to applications in Section 3. The CenceMe privacy strategy is discussed in Section 4. Section 5 describes our current prototype implementation; results from the implementation are shown in Section 6. Related work is discussed in Section 7 before we conclude.

2   CenceMe Architecture

With the CenceMe system, we aim to leverage the growing integration of sensors into off-the-shelf consumer devices (e.g., cell phones, laptops) to unobtrusively collect information about device users. The goal of the information collection is to allow the system to learn (via data fusion and analysis) characteristics and life patterns of individuals and groups, and to feed this learned information back to users in the form of application services. Noting the increasing popularity of social network applications such as Facebook and MySpace, along with the increasing usage of instant messaging as a replacement to email in both the business world and otherwise, it is clear there is a strong market for the *sharing* of information learned from sensed data as well. We apply this sense/learn/share model in the design of the CenceMe architecture described in the following.

Conceptually, the core of the CenceMe architecture resides on a set of servers that hold a database of users and their sensing presence data, implement a web portal that provides access to processed user data via per-user accounts, and contain algorithms to draw inferences about many objective and subjective aspects of users. APIs to the CenceMe core are used by thin clients running on consumer computing and communication devices such as cell phones, PDAs and laptop computers to push to the core information about the user and his life patterns based on sensed data. While this processed user information is available (both for individual review and group sharing) via the CenceMe web portal, APIs for the retrieval and presentation of (a subset of) this information are used by plugins to popular social network applications (e.g., Skype, Pidgin, Facebook, MySpace) to pull from the core. The CenceMe core in concert with the implemented APIs provide the services discussed in Section 3. A diagram of the relative positioning of the CenceMe core is shown in Figure 2.

In terms of the physical separation of functionality, the CenceMe architecture can be separated into two device classes: back end servers implementing the

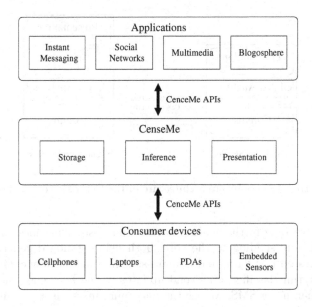

Fig. 2. The relative positioning of the CenceMe core between sensors on devices producing data applications consuming information gleaned from sensed data

CenceMe core, and off-the-shelf consumer computing and communications devices that are integrated into the system via APIs to the core. Figure 3 shows the high-level software components that reside on each of these two device classes. Communication between consumer devices takes place according to the availability of the 802.11 and cellular data channels, which is impacted both by the device feature set and by radio coverage. For devices that support multiple communication modes, communication is attempted first using a TCP/IP connection over open 802.11 channels, second using GPRS-enabled bulk or stream transfer, and finally SMS/MMS is used as a fallback. In the following, we describe the sensing, analysis and presentation components in more detail.

2.1 Sensing

Conceptually, the thin sensing client installed on the user device periodically polls on-board sensors (both hardware and software) and pushes the collected data samples via an available network connection (wired or wireless) to the CenceMe servers for analysis and storage. For sensing modalities that are particularly resource taxing (especially for mobile devices), sensor sampling may be done on demand via an explicit query. Sampling rates and durations for each of the sensors discussed in this section are set in accordance with the needs of our inferencing engine. Typically, the sensing clients use low rate sampling to save energy and switch to a higher rate sensing upon detection of an interesting *event* (i.e., set of circumstances) to improve sampling resolution. Given the pricing schemes of MMS/SMS and the battery drain implied by 802.11 or cellular

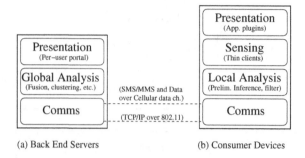

(a) Back End Servers   (b) Consumer Devices

Fig. 3. High level Software architecture of the CenceMe core and clients

radio usage, we take further energy-saving and cost-saving measures. Data is compressed before sending to the core, both using standard generic compression techniques on raw data, and domain-specific run-length encoding (e.g., for a stand/walk/run classifier, only send updates to the back end when the state changes). When using SMS, we use the maximum message size to minimize the price per bit. Also, we migrate preliminary data analysis (e.g., filtering, inference) when it makes sense to do so. Given the computational power of most new cellular phones, significant processing can be done on the mobile device to save on communication costs. However, all aggregate (trans-users) analysis is done on the back end. A discussion of the CenceMe hardware and software sensors follows.

Hardware Sensors. In the CenceMe architecture, we focus on the following classes of computer communication devices: cell phones like the Nokia N80 and N95 [7]; PDAs like the Nokia N800 [7]; phone/PDA hybrids like Apple iPhone [38]; embedded sensor platforms like Nike+ [5]; recreational sensor platforms like Garmin Edge [40], SkiScape [51] and BikeNet [43]; and laptop/desktop computers. Through a survey of the commonly available commercial hardware, including the examples just mentioned, the following hardware sensors are currently available on one or more COTS devices: embedded cameras, laptop/desktop web cameras, microphone, accelerometer, GPS, radio (e.g., BlueTooth device contact logs, 802.15.4 ranging, 802.11 localization [45] [48]), temperature, light, humidity, magnetometer, button clicks, and device state (e.g., ringer off). CenceMe exploits the availability of these sensors.

Virtual Software Sensors. Software sensors are those that measure artifacts of other software that runs on the computing platform in an effort to understand the context of the human's behaviour, mood, etc. They are "virtual" in that they do not sense physical phenomena but rather sense electronic breadcrumbs left as the human goes about his daily routine. Examples of virtual software sensors include, a trace of recent/current URLs loaded by the web browser, a trace of recent songs played on the music player to infer mood or activity, and mobile phone call log mining for structure beyond what your cell phone bill provides.

As an example of how hardware and software sensor samples can be combined to infer activity or status, based on recent web searches (e.g., moviefone.com), and a call to a particular friend, and the time of day, and the day of week, and the phone ringer turned off, we can be nearly sure the human is at the theatre.

2.2 Analysis

Sensed data pushed from CenceMe device clients are processed by the analysis component resident on the back end CenceMe servers. Generally, this analysis component combines historical per-user information, with inferences derived from combinations of the current data from multiple sensors to try to reconstruct the status of the individual, i.e., their personal sensing presence. Here we use sensing presence as a broad term meant to cover objective items such as location and activity, and subjective items like mood and preference. While a number of data fusion, aggregation, and data processing methods are possible, the following are examples of analysis/inference outputs we use to generate the sensing presence used by the CenceMe services discussed in Section 3.

Location is a key primitive in any sensing system, to be able to provide geographical context to raw sensor readings. When explicit localization services like GPS are not available either due to hardware limitation or issues with satellite coverage, we infer location of the client devices based on observed WiFi (e.g., access point identifiers), Skyhook service [48], BlueTooth (e.g., static devices) and cellular base station neighborhoods, and other unique sets of sensed data in a manner similar to ambient beacon localization [53].

We incorporate human activity inferring algorithms to log and predict users' behaviour. A simple classifier to determine whether a user is stationary or mobile can be built from several different data inputs, alone or in combination (e.g., changes in location by any possible means, accelerometer data). We evaluate accelerometer data to identify a number of physical activities, including sitting, standing, using mobile phone, walking, running, stair climbing, and others.

Human behaviour is often a product of the environment. To better understand people's behaviour then, it is useful to quantify the environmental context. We gather and process image and sound data to derive the noisiness/brightness of the environment. Further, we run conversation detection and voice detection algorithms to identify the people in a given user's vicinity that may impact behaviour and mood.

Part of a person's daily experience is the environment where the person lives and spends most of the time. For example, health related issues of interest may include the level of an individual's exposure to particulates (e.g., pollen) and pollution. We incorporate mechanisms that enable air quality monitoring around the individual through opportunistic interaction with mobile sensors [43] or static pre-deployed infrastructure [41] [44].

2.3 Presentation

Since communication devices, and in particular mobile communication devices, provide varying amounts of application support (e.g., web browser, Skype, and

Rhythmbox on a laptop; web browser and Skype on the N800, SMS only on the Motorola L2 [52]), we provide a variety of means for pulling the human sensing presence distilled from the sensed data from the CenceMe servers and displaying this status on the end user device.

Text only: Email/SMS. More limited platforms, such as older/low-end cell phones and PDAs, likely do not have the capability to browse the Internet and have a limited application suite. These platforms can still participate as information consumers in the CenceMe architecture via simple text-based updates via SMS or email, rather than graphical indicators of status embedded in other applications.

CenceMe Web Portal. Platforms that support at least general Internet browsing allow users to access their personal CenceMe web portal whose content is served from the CenceMe data repositories. The particular visualizations are customizable to a degree in a manner similar to Google Gadget [59] development/configuration on personalized iGoogle [56] pages. This web portal allows for the most flexible and complete presentation of one's own collected data log, and data shared by others (e.g., via buddy list). Through this portal the user can configure all aspects of his account, including fine grained sharing preferences for his buddies.

Application-specific Plugins. Depending on the application support on the device, any of the following plugins are possible. In each case, in addition to status information rendered by the plugin in the applications' GUI, the plugin provides click-through access to the CenceMe portal - both to the user's pages and the shared section of any friends' pages.

- Instant messaging client buddy list shows an icon with a particular status item for the buddy.
- Facebook and MySpace pages have plugins to show your status and that of your friends.
- iGoogle gadgets show various status items from a device user and his buddies. The iGoogle page periodically refreshes itself, so it follows the data pull model from the CenceMe servers.
- Photography applications have plugins to allow pictures to be stamped with metadata like location (minimally) and other environmental (light, temperature) and human status elements.

3 CenceMe Services

The goal of the CenceMe system is twofold: (i) to provide information to individuals about their life patterns; and (ii) to provide more texture to interpersonal communication (both direct and indirect) using information derived from hardware and software sensors on user devices. In the following, we describe a number of services built on the CenceMe architecture that aim to meet these goals.

3.1   Life Patterns

Enriching the concept put forward in MyLifeBits project [46], we automatically sense and store location traces, inferred activity history [3], history of sensed environment (e.g., sound and light levels), rendezvous with friends and enemies, web search history, phone call history, and VOIP and text messaging history. In this way, we can provide context in the form of sensed data to the myriad other digital observations being collected. Such information may be of archival interest to the individual as a curiosity, and may also be used to help understand behaviour, mood, and health. Sections 5.2, 5.3, and 6 describe our current prototype implementation of the sensing, inferring and display of human activity and environment.

3.2   My Presence

As indicated by the increasing popularity of social networking sites like Facebook and MySpace, people (especially the youth) are interested both in actively updating aspects of their own status (i.e., personal sensing presence), and surfing the online profiles of their friends and acquaintances for status updates. However, it is troublesome to have each user manually update more than one or two aspects of his or her sensing presence on a regular basis. We add texture and ease of use to these electronic avatars, by automatically updating each user's social networking profile with information (e.g., "on the phone", "drinking coffee", "jogging at the gym", "at the movies") gleaned from hardware and software sensors.

3.3   Friends Feeds

In the same way people subscribe to news feeds or blog updates, and given the regularity with which users of social networking sites browse their friends' profiles, there is clearly a need for a profile subscription service a la RSS (Facebook has a similar service for the data and web interface it maintains). Under this model, friend status updates might be event driven; a user asks to be informed of a particular friends state (e.g., walking, biking, lonely, with people at the coffee shop) at, for example, his cell phone.

3.4   Social Interactions

Using voice detection, known device detection (e.g., cell phone Bluetooth MAC address), and life patterns, group meetings and other events that involve groupings of people can be detected. In social group internetworking, friends are often interested in who is spending time with whom. This CenceMe service allows individuals to detect when groups of their friends are meeting, or when illicit rendezvous are happening. A further level of analysis can determine whether a conversation is ongoing (we report results on this in Section 6) and further group dynamics [60] (e.g., who is the dominant speaker).

3.5 Significant Places

Have you ever found yourself standing in front of a new restaurant, or wandering in an unfamiliar neighborhood, wanting to know more? A call to 411 is one option, but what you really want are the opinions of your friends. Phone calls to survey each of them are too much of a hassle. Or alternatively, maybe you just want to analyze your own routine to find out where you spend the most time. To satisfy both aims, CenceMe supports the identification and sharing of significant places in people's life patterns.

Significant places are derived through a continuously evolving clustering, classification, and labelling approach. In the first step, we collect location traces from available sources (e.g., wifi association, GPS, etc.) for the given user. Since location traces always have some level of inaccuracy, we cluster the sensed locations according to their geographical proximity. The importance of a cluster is identified by considering time-based inputs such as visitation frequency, dwell time, and regularity. Once significant clusters are identified, a similarity measure is applied to determine how "close" the new cluster is to other significant clusters already identified (across a user's buddies) in the system. If the similarity is greater than a threshold then the system automatically labels (e.g., "Home", "Coffee shop", etc.) the new cluster with the best match. The user has the option to apply a label of his own choosing, if the automatic label is deemed insufficient. Finally, the user has the option of forcing the system to label places considered "insignificant" by the system (e.g., due to not enough visitations yet).

As implied above, the CenceMe system keeps the labels and the cluster information of important clusters for all users, applying them to subsequent cluster learning stages and offering to users a list of possible labels for given clusters. In addition to this "behind the scenes" type of place label sharing, users may also explicitly expose their significant places with their buddies or globally, using the normal methods (e.g., portal, plugins) previously described. In particular, this means that for a user that is visiting a location that is currently not a (significant) cluster to him based on his own location/time traces, the point can be matched against buddies' clusters as well to share. We report on the implementation and performance of this service in Sections 5.2 and 6, respectively.

Once the significant places of users have been automatically identified and either automatically or manually tagged, users may annotate their significant places. The annotation may include identifying the cafe that has good coffee or classifying a neighborhood as either dangerous, safe, hip or dull.

3.6 Health Monitoring

As many people are becoming more health-conscious in terms of diet and lifestyle, the CenceMe system also provides individuals with health aspects [1] [34] [21] [33] of their daily routines. CenceMe is able to estimate exposure to ultraviolet light, sunlight (for SAD afflictees) and noise; along with number of steps taken (distance traveled) and number of calories burned. These estimates are derived by combining inference of location and activity [3] of the users with weather information (e.g., UV index, pollen and particulate levels) captured by the CenceMe backend from the web. Inference techniques and results for activity classification and exposure to the weather environment are discussed in Sections 5.2 and 6.

3.7   Buddy Search

The past ten years have seen the growth in popularity of online social networks, including chat groups, weblogs, friend networks, and dating websites. However, one hurdle to using such sites is the requirement that users manually input their preferences, characteristics, and the like into the site databases. With CenceMe we provide the means for the automatic collection and sharing of this type of profile information. CenceMe automatically learns and allows users to export information about their favorite haunts, what recreational activities they enjoy, and what kind of lifestyle they are familiar with, along with near real-time personal presence updates sharable via application (e.g., Skype, MySpace) plugins and the CenceMe portal. Further, as many popular IM clients allow to search people by name, location, age, etc., CenceMe enables the search of users through a data mining process that involves also interests (like preferred listened music, significant places, preferred sport, etc).

3.8   Buddy Beacon

The buddy search service is adapted to facilitate local interaction as well. In this mode, a user configures the service to provide instant notification to his mobile device if a fellow CenceMe user has a profile with a certain degree of matching attributes (e.g., significant place for both is "Dirt Cowboy coffee shop", both have primarily nocturnal life patterns, similar music or sports interests). All this information is automatically mined via CenceMe sensing clients running on user devices; the user does not have to manually configure his profile information. Devices with this CenceMe service installed periodically broadcast the profile aspects the user is willing to advertise - a Buddy Beacon - via an available short range radio interface (e.g., Bluetooth, 802.15.4, 802.11). When a profile advertisement is received that matches, the user is notified via his mobile device.

3.9   "Above Average?"

Everybody is interested in statistics these days. What is popular? How do I measure up? Do I have a comparatively outgoing personality? By analyzing aggregate sensor data collected by its members, CenceMe provides such statistical information on items such as the top ten most common places to visit in a neighborhood, the average time spent at work, and many others. CenceMe makes this aggregate information available to users; each user can configure their portal page to display this system information as desired. Comparisons are available both against global averages and group averages (e.g., a user's friends). Tying in with the Life Patterns service, users can also see how their comparative behaviour attributes change over time (i.e., with the season, semester). The normal

CenceMe privacy model (see Section 4) relies on buddy lists. Therefore, the user must manually opt in to this global sharing of information, even though the data is anonymized through aggregation and averaging before being made available. On the other hand, access to the global average information is only made available to users on a quid pro quo basis (i.e., no free loaders).

4   Privacy Protection

Users' raw sensor feeds and inferred information (collectively considered as the user's sensing presence) are securely stored in the CenceMe back end database, but can be shared by CenceMe users according to group membership policies. For example, the data becomes available only to users that are already part of a CenceMe buddy list. CenceMe buddies are defined by the combination of buddy lists imported by registered services (Pidgin, Facebook, etc.), and CenceMe-only buddies can be added based on profile matching, as discussed in Sections 3.7 and 3.8. Thus, we inherit and leverage the work already undertaken by a user when creating his buddy lists and sublists (e.g., in Pidgin, Skype, Facebook) in defining access policies to a user's CenceMe data. Investigation of stronger techniques for the protection of people-centric data is currently underway [11].

Users can decide whether to be visible to other users via the buddy search service (Section 3.7) or via the buddy beacon service (Section 3.8). CenceMe users are given the ability to further apply per-buddy policies to determine the level of data disclosure on per-user, per-group, or global level. We follow the Virtual Walls model [57] which provides different levels of disclosure based on context, enabling access to the complete sensed/inferred data set, a subset of it, or no access at all. For example, a CenceMe user A might allow her buddy B to take pictures from her cell phone while denying camera access to buddy C; user A might make her location trace available to both buddies B and C. The disclosure policies are set from the user's account control page.

In addition to user-specific data sharing policies, the system computes and shares aggregate statistics across the global CenceMe population. For this service (Section 3.9), shared information is anonymized and averaged, and access to the information is further controlled by a quid pro quo requirement.

5   Prototype Implementation

As a proof on concept, we implement a prototype of the CenceMe architecture. Sensing software modules, written both as applications plugins and standalone clients, are installed on commodity hardware, and integrate user devices with the CenceMe core. A sample of the analysis and inference algorithms discussed in the previous sections are implemented as part of sensing clients (preliminary processing) and as back end processes. These automatically process incoming data pushed by the sensing clients. A number of presentation modules are implemented to display information both for individual viewing and sharing between CenceMe users. In the following, we describe the hardware and software details of our prototype implementation.

5.1   Sensing

To demonstrate the types of information we can collect from commodity devices and popular applications, we implement a number of sensing clients on a selection of COTS hardware. We use the Nokia 5500 Sport (Symbian OS [42], 3D accelerometer, BlueTooth), the Nokia N80 (Symbian OS, 802.11b/g, Blue-Tooth), the Nokia N95 (Symbian OS, 802.11b/g, BlueTooth, GPS), the Nokia N800 (Linux OS, 802.11b/g, BlueTooth) and Linux laptop computers. Each sensing client is configured to periodically push its sensed data to the CenceMe core. We provide a short description of each implemented sensing client in the following list.

- Rhythmbox is an open source audio player patterned after Apple iTunes. We write a Perl plugin to Rhythmbox to push the current song to the core. The plugin works on the Linux laptop and the Nokia N800.
- We write a Python script to sample the 3D accelerometer on the Nokia 5500 Sport at a rate that supports accurate activity inference.
- The BlueTooth and 802.11 neighborhoods (MAC addresses) are periodically collected using a Python script. CenceMe users have the option to register the BlueTooth and 802.11 MAC address of their devices with the system. In this way the CenceMe backend can convert MAC addresses into human-friendly neighbor lists.
- We write a Python script to capture camera and microphone samples on the Nokia N80 and Nokia N95 platforms. In addition to the binary image and audio data, we capture and analyze the EXIF image metadata.
- Pidgin is an instant messaging client that supports many commonly used instant messaging protocols (e.g., .NET, Oscar, IRC, XMPP), allowing users to access accounts from many popular IM services (e.g., Jabber, MSN Messenger, AOL IM) via a single interface. We write a Perl plugin to Pidgin to push IM buddy lists and status to the CenceMe core.
- Facebook is a popular web-based social networking application. We write a Perl plugin to Facebook to push Facebook friend lists to the core.
- We write a Python script to periodically sample the GPS location from the Nokia N95.
- Skyhook [48] is a localization system based on 802.11 radio associations. We use Linux libraries compiled for x86 devices and the Nokia N800 to periodically sample the WiFi-derived location and push to the CenceMe core.

In addition to the sensing just described based strictly on commodity hardware, we extend the capability of any BlueTooth enabled device (this includes all the commodity devices we mention above) by allowing for the connection of an external 3D accelerometer. We envision that such a gadget (i.e., a small form-factor BlueTooth/accelerometer accessory) may become popular due to its application flexibility. We implement a prototype BlueCel accessory by integrating a Sparkfun WiTilt module, a Sparkfun LiPo battery charger, and a LiPo battery. The size is 1.5in x 2.0in x 0.5in (see Figure 4). We write a python script to read accelerometer readings from the device over the BlueTooth interface. The placement of the

Fig. 4. Mobile devices currently integrated into the CenceMe system: the Nokia N800 Internet Tablet, Nokia N95, Nokia 5500 Sport, Moteiv Tmote Mini (above the N95), and prototype BlueCel accessory (above the 5500)

accessory (e.g., on weight stack, on bike pedal) defines the application. A sensing client menu allows the user to tell the system what the application is, allowing the client to set the appropriate sampling rate of the accelerometer. The data is tagged with the application so that the CenceMe back end can properly interpret the data. Further, we leverage the use of existing embedded sensing systems accessible via IEEE 802.15.4 radio [41] [43] by integrating the SDIO-compatible Moteiv Tmote Mini [10] into the Nokia N800 device.

5.2 Analysis

In the CenceMe architecture, unprocessed or semi-processed data is pushed by sensing clients running on user device to the CenceMe core. We implement a MySQL database to store and organize the incoming data, accessible via an API instantiated as a collection of PHP, Perl, and Bash scripts. To extract useful information about CenceMe users from the sensed data, we apply a number of data processing and inferring techniques. We use the WEKA workbench [39] for our clustering and classification needs. In the following, we provide a short description of the data analysis tools we implement in support of the CenceMe services discussed in Section 3. Results on the output of these tools are presented in Section 6.

Based on data from either the Nokia 5500 Sport or the BlueCel accelerometer, we implement an activity classifier (stand/walk/run). The classifier is based on learned features in the raw data trace such as peak and rms frequency, and peak and rms magnitude. This approach applies similar ideas to those found in other accelerometer-based activity inferencing papers (e.g., [62] [3] [14]). This classifier is an example of processing that occurs on the mobile device to avoid the cost (energy and monetary) of sending complete raw accelerometer data via SMS to the back end. Performance results of this classifier are shown in Section 6.1.

We construct a classifier for determining whether a user is indoors or outdoors. We combine a number of elements into the feature vector to be robust to different types of indoor and outdoor environments. The features we consider are: the ability of the mobile device to acquire a GPS estimate, number of satellites seen by GPS, number of WiFi access points and BlueTooth devices seen and their signal strengths, the frequency of the light (looking for the AC-induced flicker), and differential between the temperature measured by the device and the temperature read via a weather information feed (to detect air conditioning). Performance results of this classifier are shown in Section 6.1.

We construct a mobility classifier (stationary/walking/driving) based on changes to the radio neighbor set and the relative signal strengths (both for individual neighbors and the aggregate across all neighbors), for BlueTooth, WiFi, and GSM radios, respectively, of the mobile devices. The idea is to map changes in the radio environment (i.e., neighbors, received signal strength) to speed of movement. The classifier uses techniques similar to those used in existing work [54] [55] [50]. The result of the aforementioned indoor/outdoor classifier is also included in the feature vector. Locations traces are omitted due to their relatively high error with respect to the speed of human motion. Performance results of this classifier are shown in Section 6.1.

Using Matlab processing on the back end, we generate a noise index (expressed in decibels) from audio samples captured from N80 and N95 microphones. Similarly, using Matlab we generate a brightness index (ranging from 0 to 1) from images captures from N80 and N95 cameras. The sounds and brightness indices help us to infer information about a person's surroundings. In particular, we combine the noise index to estimate the cumulative effect of the sound environment on a user's hearing, and the positive effect of sunlight (when combined with an indoor/outdoor classifier) on those afflicted with seasonal affective disorder. Finally, we implement a classifier based on a voice detection algorithm [19] to determine if a user is engaged in a conversation or not. Performance results of this classifier are shown in Section 6.1.

By analyzing location traces, time statistics of mobility, and other data inputs, as described in Section 3.5 we derive a user's significant places. Raw location data is first clustered using the EM algorithm, then clusters are mapped against time statistics (viz., visitation frequency, dwell time, regularity, time of day, weekday/weekend, AM/PM) and other information (viz., indoor/outdoor, current and previous mobility class, number and composition of people groups visible in a location) to determine importance. Also, WiFi and Bluetooth MAC address of neighbors are used to differentiate between overlapping clusters. Finally, a similarity measure is computed between the new cluster and existing clusters known by the system. The system maintains generic labels for these significant clusters, but users may alias them as well to give more personally meaningful or group-oriented names. The clustering is adaptive since the model changes over time depending on how the mobility trace of the user (and other system users) evolves (the significance of a place may evolve over time). The algorithm to recognize significant locations by combining location trace data with other indicators shares concepts with prior work [49] [47] [66] [65] [63]. The CenceMe approach is distinguished by the way it clusters according to per-user models (rather than globally), and then shares models based on social connections (e.g., presence in a buddy list). This provides a more personal/group-oriented set of labeled significant places, at the expense of general applicability of the training data. While it is often advantageous to relate recognized significant clusters to physical locations (i.e., coordinates), we also enable the recognition of significant places for devices that do not have access to absolute localization capabilities with the use of local region recognition based on what is available to the device (e.g., WiFi, Bluetooth, GSM) [61]. In this way, a location cluster does not solely have to be an aggregated set of true coordinate estimates, but can comprise a set of location recognition estimates, a la ABL [53].

In terms of health and fitness, we estimate the number of calories burned by combining the inference of walking from the standing/walking/running classifier, time spent walking, and an average factor for calories burned per unit time when walking at a moderate pace [20]. Further, we estimate exposure to ultraviolet light by combining the inference of walking or running or standing, the inference of being outdoors, the time spent, and a feed to a web-based weather service to learn the current UV dose rate [24]. A similar technique is applied to estimate pollen exposure (tree, grass, weed) and particulate exposure.

As discussed in Section 5.1, the BlueCel facilitates a number of application-specific data collection possibilities. We implement commensurate application-specific data analysis tools for bicycle rides (BlueCel placed on the pedal), golf swing analysis (BlueCel affixed to the club head), and analysis of weight lifting activity for exercise motion correctness (injury avoidance) and workout logging (BlueCel affixed to the wrist).

5.3 Presentation

All of a user's processed sensor data can be viewed via a web browser by logging into the user's account on the CenceMe portal. Additionally, a subset of the user's status information is made available (via both data push and data pull mechanisms) to the user's buddies (subject to his configured sharing policies) through their CenceMe portal pages, and through plugins to popular social networking applications.

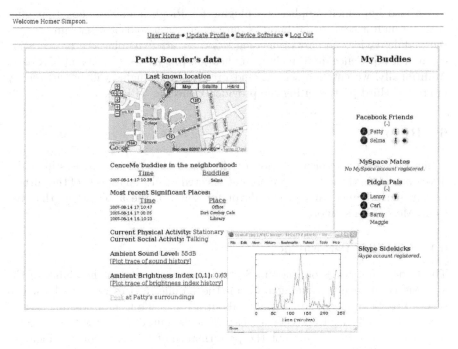

Fig. 5. Portal snaphot

Typically, the data a user shares with his buddies is rendered via a number of simple icons that distill the current sensing presence of the user. Figure 5 shows a snapshot of the data page of user on the CenceMe portal we implement. Buddy lists loaded on the right pane from registered Pidgin and Facebook accounts are annotated with icons representing the shared data. The icons offer click-through access to a fuller representation of the shared user data. In the Figure, buddies Patty and Selma are inferred to be standing and in a conversation, while buddy Lenny is inferred to be at the coffee shop, as indicated by the yellow activity icons next the each buddy's name. On login, the left pane shows the data of the logged in user, but shows a buddy's data if any one of the icons next to that buddy's name is clicked. In this case, the logged in user Homer Simpson has clicked on the icon for his buddy Patty. Patty has a sharing policy that allows the display of the data shown in the left pane: Patty's buddies in her vicinity (via BlueTooth and WiFi MAC address recognition), Patty's trace of her last significant places visited, etc. In sum, this is the detailed data behind the iconic representation of "standing" and "talking". Note that the link at the bottom of the page to take a picture ("peek") from Patty's cell phone is disabled; Patty has disabled access for Homer Simpson to image data in her privacy profile. Instead, Homer has clicked the link to view the sound level history plot for Patty, ostensibly to see how noisy Selma is. The black icons denote that a buddy imported from the user's Facebook account (using the Facebook developer API) or Pidgin account (using the Pidgin developer API) is also a registered CenceMe user.

We also implement application plugins for Pidgin and Facebook that offer click-through access to the buddies data page via a web browser launched on the mobile device. As part of ongoing work, we are extending the presentation capabilities of CenceMe to include application plugins for Skype and MySpace. Additionally, we envision sensing and presentation plugins will be written by interested third parties, using the provided CenceMe APIs.

6   Results

While our system implementation of the CenceMe architecture is still under development, in the following we present selected results on aspects of the current system performance, and give a flavor for some of the applications that the CenceMe system supports.

6.1   Classifier Performance

The CenceMe services outlined in Section 3 strongly rely on the ability of the analysis components (running both on the mobile devices and the back end servers) to glean meaningful insights about a user's life from a possible myriad of sensed data feeds. To provide a measure of the initial quality of the CenceMe services we provide, we present the performance of a number of classifiers we implement. Implementation descriptions of the presented classifiers are given in Section 5.2.

Activity: Standing/Walking/Running. Our activity classifier is currently based exclusively on accelerometer data. Since this sensor may not always be available on all mobile devices, beyond this initial evaluation the classifier will incorporate a broader array of sensors to improve the robustness of the technique. Additionally, we are in the process of enhancing the activity classifier to detect a broader range of activities.

We run experiments to evaluate the accuracy of the mobility classifier described in Section 5.2. The accelerometer thresholds of the classifier (i.e., between stationary/walking and walking/running) are learned based on mobility traces from three different people. The results shown in Figure 6(a), give the average test results for two others. The accelerometer is tested in two mounting positions (viz., belt, pocket) and the values represent the average of 4 one hour experiments. The scenarios include a normal office setting behaviour and sports (walking/running to the campus Green and back). From the matrix in Figure 6(a) we see that activities are classified correctly approximately 90% of the time for each of the three activities.

Mobility: Stationary/Walking/Driving. Our evaluation of the mobility classifier (described in Section 5.2) results from a week long controlled experiment with four members in our lab. All members carry Nokia N95 cell phones that execute the classifier. Participants manually label any state changes, which provide both training data for the classifier and a data set from which classifier

	Standing	Walking	Running
Standing	0.9844	0.0141	0.0014
Walking	0.0558	0.8603	0.0837
Running	0.0363	0.0545	0.9090

(a) Activity classifier.

	Stationary	Walking	Driving
Stationary	0.8563	0.3274	0.1083
Walking	0.1201	0.6112	0.2167
Driving	0.0236	0.0614	0.6750

(b) Mobility classifier.

	Indoors	Outdoors
Indoors	0.9029	0.2165
Outdoors	0.0971	0.7835

(c) Indoor/outdoor classifier.

	Background noise	Conversation
Background noise	0.7813	0.1562
Conversation	0.2187	0.8438

(d) Conversation classifier.

Fig. 6. Confusion matrices for the implemented classifiers

performance is derived. Figure 6(b) shows the confusion matrix that results with one third of the data used for training and the remainder used for evaluation. From this figure we observe that walking and driving are less accurately labeled. This is because the difference between these two states is less distinct. We intend to refine the classifier to incorporate additional modalities (such as audio) to address this weakness. We perform ten fold cross validation on our experiment data and find that 81% of instances still have correct labels. This allays some concerns about potential over-fitting.

Indoor/Outdoor. The evaluation of the indoor/outdoor classifier (described in Section 5.2) follows a similar methodology to that of the mobility classifier, with participants manually labeling their state changes. Since the results of the classification trials that we present here are performed only on the N95 hardware, many of the vector features that are part of the process have no effect (e.g., detection of the light flicker is not possible with this hardware). We perform independent tests with alternative sensing hardware, the Tmote Sky [10], using only temperature, light and humidity sensors available and find approximately 83% classification accuracy is achievable. As the sensing capabilities of cell phones mature, we will evolve our classification techniques to take advantage of the commercial technology. The results shown in Figure 6(c) are very promising given the initial stages of the development of this classifier. We observe that an accuracy of 86% results when we perform 10 fold cross validation.

We must qualify these results by saying these were performed in a college campus environment with a rather dense WiFi AP deployment. Further there are no significant high rise buildings that may induce urban canyoning effects with the GPS signal. On the other hand, due to the low population density of the area there is also a lower density of cell phone towers. As a result, some feature vector elements that are important within our test region will not be important for all regions. We plan to perform tests as part of a broader system study to evaluate the classifier performance in substantively different environments from that used for these initial tests.

Conversation detection. We conduct experiments to evaluate the accuracy of the conversation detection algorithm [19]. The experiment consists of taking samples from the cell phone's microphone from inside a pocket (to reproduce the common scenario of cell phones carried by a person) during conversations between a group of people in an office setting, sidewalk, and a restaurant. We annotate the time of the conversation for ground truth purposes. The microphone samples are also time stamped for comparison with the ground truth data. The result of the conversation detection algorithm is reported in Figure 6(d). The background noise and an ongoing conversation are correctly detected with an accuracy of almost 80% and 84% respectively. On average, 15% and 21% of the samples are mistaken, respectively, for conversation when they are noise and for noise when they are conversation. We conjecture that we can decrease the error by further processing the sound samples by applying low pass filters. In this way, sounds characterized by frequencies higher than 4KH (e.g, background noise) could be filtered out allowing a more accurate voice detection. As part of future work, we will augment our system with stronger voice processing capabilities. With these new capabilities, CenceMe users will have the ability to know who is involved in a conversation (subject to user privacy policy), rather than just a binary classification.

6.2 Significant Places

For the purpose of evaluating the CenceMe significant places service, four members of our lab execute data collection client code on Nokia N95 cell phones that sample and construct the types of data features discussed in Section 3.5. In the following, we demonstrate the sharing of significant place models (as described in Section 5.2) between buddies in the context of the general operational flow of the service.

For the purposes of this scenario we examine the interaction between two of the four system users. In Figure 7(a), we observe the location trace for a single user, Homer, with his location trace collected over the course of a week around Hanover, NH. Figure 7(b) provides the result of basic EM clustering of these raw location estimates. Figure 7(c) provides the results of the significant places process based only on Homer's data. These are visitations to cluster instances which are classified by the system as being significant. The process begins with a default classifier based on training data sourced from the user population. The classifier is then refined by input from Homer. This figure shows these cluster visits having semantic labels. In this case these labels are solely provided by Homer himself. In Figure 7(d), a new significant location appears. This location represents the home of another user, Patty. Homer visited Patty's home and a raw cluster is created by this visitation (as seen in Figure 7(b)). However, since Homer had never visited this location before, although it is recognized as being significant a label can not be determined. Instead, to label this cluster the system executes Homer's buddy Patty's models on the data collected by Homer. Given a sufficiently good match, the labeling is performed and appears in visualizations such as Homer's log of his sensing presence accessible via the CenceMe portal. Importantly, the ability to recognize and label such visitations is then incorporated into Homer's significant place model.

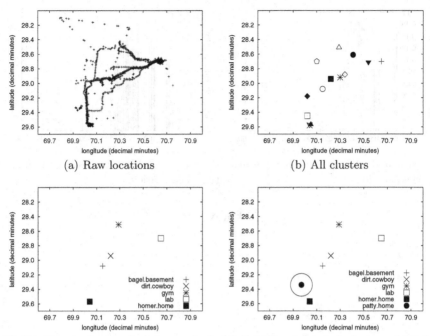

(a) Raw locations

(b) All clusters

(c) Significant clusters based on single user's data

(d) Additional labeled cluster added by cross-referencing against buddy's labeled significant clusters.

Fig. 7.

6.3 Health

By combining the output of the indoor/outdoor classifier and the mobility classifier with an external weather feed that provides a trace of the UV dose rate (i.e., erythemal irradiance) [24], we estimate a person's exposure to harmful UV radiation. This combination of classifiers tells us when a person is outside a building but not inside a motor vehicle. Figure 8(a) shows the estimated UV exposure of the person if indoor/outdoor classification is perfect, while 8(b) shows the estimated UV exposure given the actual classifier performance. The figures are in excellent agreement, underscoring the good performance of the classifier combination. Further, in presenting this result we show how simple classifiers can contribute to monitoring important human health features. As future work we will compare the estimated results against real measured exposure to further improve the classifier combination we use.

Similarly, we use the output of the mobility classifier to estimate how many calories a person is burning. In Figure 9, we show a comparison between the estimated cumulative calories burned using the classifier output and the actual activity, respectively. The plots reflect a real seven hour human activity trace. In both cases, the time spent walking, either inferred from the classifier or taken

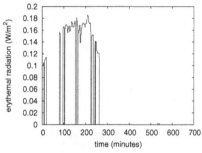 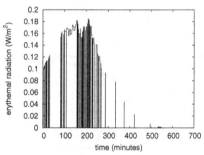

(a) Estimated UV exposure based on perfect classification.

(b) Estimated UV exposure based on actual classification.

Fig. 8.

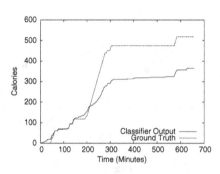

Fig. 9. A comparison of the estimated calories burned based on the ideal and actual performance, respectively, the mobility classifier

from the ground truth log, is multiplied by a standard average factor for calories burned while walking [20] to calculate the estimate.

6.4   BlueCel Applications

The flexibility of the BlueCel accessory sensor allows people to run many different applications that are of interest to them, with a single multi-purpose device. This external, Bluetooth-connected accelerometer offers advantages even over accelerometers that are integrated into mobile phones (e.g., Nokia 5500 Sport) since for many applications it is required or at least convenient that the accelerometer be in a different place than mobile phones are normally carried (e.g., pocket, hand bag). Further, the form factor of a mobile phone is too large to facilitate useful data collection for some applications.

To demonstrate the flexibility of the BlueCel approach, we implement and collect data from three simple and diverse applications. For each of these applications, the accelerometer signature in terms of combined three channel

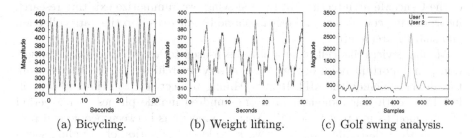

(a) Bicycling.   (b) Weight lifting.   (c) Golf swing analysis.

Fig. 10. Use of the BlueCel sensor to add a 3d accelerometer to any BlueTooth-equipped mobile device supports a number of applications. The plots in (a), (b) and (c) show BlueCel signatures, the combined magnitude of the three channel output of the accelerometer. A moving average with a window size of 25 samples is applied to the raw accelerometer samples for all three plots.

magnitude (i.e., $\sqrt{x^2 + y^2 + z^2}$) is plotted; sampling is at 37 Hz. Figure 10(a) shows a 30 second excerpt from a bicycle pedaling analysis experiment. The Blue-Cel is slipped inside a rider's sock. A user can easily do the same to determine his pedaling cadence, for training purposes or just for fun. Figure 10(b) shows the accelerometer signature for several repetitions of a weight lifting exercise. Though also periodic, the signature is quite distinct from that of the pedaling. We can easily log workout statistics by analyzing this signature [6]. As future work, with further processing of the signature we hope to provide an indication of whether the exercise is being performed properly in terms of range of motion. Figure 10(c) shows the golf swing signatures of two different users, one a novice and the other more experienced. The signatures are quite distinct, that of the experienced golfer (User 1) being more smooth and compact. A novice user might use such comparisons as a guide to iteratively analyze and modify his swing for improvement. These three are just a sample of possible applications, and users themselves have the freedom (due to the flexibility of the BlueCel model) to come up with their own applications.

7   Related Work

Much attention has been paid in the research community to the intersection of social networking and communication technology. In particular, cell phones have long been recognized as an ideal platform for pervasive applications (e.g., [58]). They are increasingly seen as a platform of choice for urban and people-centric sensing systems [32] [29] [31] [15] [35] [30] [4]. They are well suited for this domain due to their ubiquity, expanding suite of sensors and ability to interact with additional external sensors via short range radio. Further, given the increasing market penetration of cellular phones and the parallel trend of sole reliance on cell phones for telephonic service, they are likely to be carried at all times. Data collection and sharing via cell phones and similar mobile devices are key enablers of the CenceMe architecture. In the following, we summarize exisiting research incorporating cell phones and other commodity devices into the data collection and sharing architecture.

Mobile devices like cell phones have been used to learn about social connections, an important point of the CenseMe service architecture. Contributions in this domain include the MIT Reality Mining project. In [26], the authors collect BlueTooth neighbor data from one hundred mobile phones over a period of nine months in order to identify social connections between people and patterns in their daily life. Follow on work explores the notion of sampling social connections with data such as proximity and sounds [18], and applies principal component analysis to location/time series data and BlueTooth neighborhoods to learn social groups [25].

One use of data collected by cell phones is to facilitate context-aware communications, wherein availability data is collected from personal devices and shared with friends. In [8] the authors describe iCAM, a web-based address book for cellular phones. System users opt-in to share communication contexts via a web interface to expose the preferred method of communication at a given time (e.g., in person, email, home/work phone). Contexts are generated by cellular tower-based localization and manually configured schedule information. Registered rules govern the type of information exposed. The authors of [13] propose and build a system similarly aimed at choosing the best communication modality to interact with close friends and family. The system uses GPS location, accelerometer data to determine between walking and driving, and a microphone to determine between talking and silent. CenceMe goes beyond both [8] and [13] by taking as inputs a broader set of sensor feeds, learning patterns in each user's life automatically rather than relying on manually input schedule information, and outputting status information much richer than just current location and communication preference. Additionally, CenceMe is not limited to running on customized hardware (e.g., the PHS in [8], the WatchMe watch in [13]), but is integrated into popular social applications (using supported APIs) already running on commodity hardware (cellular phone, laptop computer, PDA, etc.).

The CenceMe notion of sensing presence sharing and logging differs from the idea of presence sharing and exchange presented in [22] and [23]. The latter idea refers to the beaconing of personal information using short range radio. The CenceMe notion incorporates in situ exchanges (e.g., via the buddy beacon service), but extends beyond this simple interaction. CenceMe is focused on the process of distilling the sensed presence of the individual from COTS devices and sharing the sensed presence irrespective of the actual proximity between users, largely based on social groups (e.g., buddies) via existing applications. Twitter [28] is more closely aligned with CenceMe in terms of sharing personal status information, but is typically limited to manually generated text-based status sharing. The primary benefit of [28] is the ability to aggregate and distribute these status messages from and to multiple points (i.e., cell phones, IM and the Web). CenceMe extends beyond sharing text messages and focuses on the automated distillation of users' personal sensed presence. The work presented in [17] provides a limited exploration of the sharing sensing presence between socially connected parties. This work shares simple moving/not moving status while investigating the utility and privacy concerns of sharing such status. As an indication of the demand for a more capable system, participants of the study requested that richer forms of presence sharing be offered [17].

Alternative proposed architectures that enable people-centric sensing often rely on specialize hardware, rather than commodity consumer devices like cell phones. The architecture proposed in [27] relies on users to attach numerous cheap radio tags to everyday objects with which they come in contact. Through proximity detection between user devices and object tags, activity recognition is possible. The Mithril project [64] is representative of those assuming more capable devices and is built around a linux PDA with multiple body-attached sensors. Similarly, the SATIRE project [16] builds a system with greater levels of sensing capability using "smart clothing". With SATIRE, networked mote-class devices are embedded in clothing such as a jacket. CenceMe uses heterogeneous COTS devices (e.g., cell phones) already in widespread use. CenceMe collects data from the sensors available on these devices (e.g., BlueTooth, GPS, accelerometer, camera, microphone), and in support of services and applications not possible with proximity data alone. The CenceMe architecture leverages the idea of integrating simple external sensors devices (e.g., the BlueCel device, c.f. Section 5.1) as application-specific add-ons to the cell phone. This model is similar to [30] and [14], both of which assume additional expansion boards are attached to a standard cell phone.

8 Conclusion

We have presented a detailed description of the CenceMe architecture. Through our prototype implementation we have demonstrated successful integration with a number of popular off-the-shelf consumer computer communication devices and social networking applications.

Acknowledgment

This work is supported in part by Intel Corp., Nokia, NSF NCS- 0631289, and the Institute for Security Technology Studies (ISTS) at Dartmouth College. The authors thank Hong Lu for his help with data collection, Ronald Peterson for building the BlueCel, and Peter Boda and Chieh-Yih Wan for their support of this project. ISTS support is provided by the U.S. Department of Homeland Security under Grant Award Number 2006-CS-001-000001. The views and conclusions contained in this document are those of the authors and should not be interpreted as necessarily representing the official policies, either expressed or implied, of the U.S. Department of Homeland Security.

References

1. Lester, J., et al.: Sensing and Modeling Activities to Support Physical Fitness. In: Proc. of Ubicomp Workshop: Monitoring, Measuring, and Motivating Excerise: Ubiquitous Computing to Support Fitness, Tokoyo (September 2005)
2. Kids say e-mail is, like, soooo dead. CNET (July 19, 2007), http://news.com.com/Kids+say+e-mail+is,+like,+soooo+dead/2009-1032_3-6197242.html
3. Lester, J., et al.: A Hybrid Discriminative/generative Approach for Modeling Human Activities. In: Proc. of the 19th IntlJoint Conf. on Artificial Intelligence, Edinburgh, pp. 766–722 (2005)
4. MetroSense Project Web Page, http://metrosense.cs.dartmouth.edu
5. Nike+, http://www.nikeplus.com
6. Chaudhri, R., et al.: Mobile Device-Centric Exercise Monitoring with an External Sensor Population (Poster abstract). In: Proc of 3rd IEEE Int'l. Conf. on Distributed Computing in Sensor Systems, Santa Fe (June 2007)
7. Nokia, http://www.nokia.com
8. Nakanishi, Y., Takahashi, K., Tsuji, T., Hakozaki, K.: iCAMS: A Mobile Communication Tool Using Location and Schedule Information. IEEE Pervasive Computing 3(1), 82–88 (2004)
9. Skype, http://www.skype.com
10. Moteiv Tmote Mini, http://www.moteiv.com/
11. Johnson, P., Kapadia, A., Kotz, D., Triandopoulos, N.: People-Centric Urban Sensing: Security Challenges for the New Paradigm. Dartmouth Technical Report TR2007-586 (February 2007)
12. Gaim/Pidgin, http://sourceforge.net/projects/pidgin
13. Marmasse, N., Schmandt, C., Spectre, D.: WatchMe: Communication and Awareness Between Members of a Closely-knit Group. In: Proc. of 6th Int'l. Conf. on Ubiq. Comp., Nottingham, pp. 214–231 (September 2004)
14. Lester, J., Choudhury, T., Borriello, G.: A Practical Approach to Recognizing Physical Activities. In: Proc. of 4th Int'l. Conf. on Perv. Comp., Dublin, pp. 1–16 (May 2006)
15. Abdelzaher, T., et al.: Mobiscopes for Human Spaces. IEEE Perv. Comp. 6(2), 20–29 (2007)
16. Ganti, R., Jayachandran, P., Abdelzaher, T., Stankovic, J.: SATIRE: A Software Architecture for Smart AtTIRE. In: Proc. of 4th. Int'l. Conf. on Mobile Systems, Applications, and Services, Uppsala (June 2006)
17. Bentley, F., Metcalf, C.: Sharing Motion Information with Close Family and Friends. In: Proc of SIGCHI Conf. on Human Factors in Computing Systems, San Jose, pp. 1361–1370 (2007)
18. Eagle, N., Pentland, A., Lazer, D.: Inferring Social Network Structure using Mobile Phone Data (in submission, 2007)
19. Matlab Library for Speaker Identification using Cepstrul Coeff., http://www.mathworks.com/matlabcentral/fileexchange/loadFile.do?objectId=8802
20. Food, Nutrition, Saftey and Cooking Site at the University of Nebraska - Lincoln, http://lancaster.unl.edu/food/
21. Borriello, G., Brunette, W., Lester, J., Powledge, P., Rea, A.: An Ecosystem of Platforms to Support Sensors for Personal Fitness. In: Proc of 4th Int'l. Workshop on Wearable and Implantable Body Sensor Networks, Aachen (March 2006)
22. Cox, L., Dalton, A., Marupadi, V.: Presence-Exchanges: Toward Sustainable Presence-sharing. In: Proc. of 7th IEEE Workshop on Mobile Computing Systems and Applications, Durham, pp. 55–60 (April 2006)

23. Cox, L., Dalton, A., Marupadi, V.: SmokeScreen: Flexible Privacy Controls for Presence-sharing. In: Proc. of 5th Int'l. Conf. on Mobile Systems, Applications and Services, San Juan, pp. 233–245 (June 2007)
24. USDA UV-B Monitoring and Research Program, http://nadp.nrel.colostate.edu
25. Eagle, N., Pentland, A.: Eigenbehaviors: Identifying Structure in Routine (2006)
26. Eagle, N., Pentland, A.: Reality Mining: Sensing Complex Social Systems. In: Personal Ubiq. Comp., pp. 255–268 (May 2006)
27. Lamming, M., Bohm, D.: SPECs: Another Approach to Human Context and Activity Sensing Research, Using Tiny Peer-to-Peer Wireless Computers. In: Proc. of 5th Int'l. Conf. on Ubiq. Comp. Seattle (October 2003)
28. Twitter, http://twitter.com
29. Burke, J., et al.: Participatory sensing. In: ACM Sensys World Sensor Web Workshop, Boulder (October 2006)
30. Pering, T., et al.: The PSI Board: Realizing a Phone-Centric Body Sensor Network. In: Proc of 4th Int'l. Workshop on Wearable and Implantable Body Sensor Networks, Aachen (March 2006)
31. Sensorplanet, http://www.sensorplanet.org/
32. Eisenman, S., et al.: Metrosense project: People-centric Sensing at Scale. In: Proc. of World Sensor Web Workshop, Boulder (October 2006)
33. Reddy, S., et al.: Image Browsing, Processing, and Clustering for Participatory Sensing: Lessons From a DietSense Prototype. In: Proc. of 4th Workshop on Embedded Networked Sensors, Cork (June 2007)
34. UbiFit Project, http://dub.washington.edu/projects/ubifit
35. Kansal, A., Goraczko, M., Zhao, F.: Building a Sensor Network of Mobile Phones. In: Proc of IEEE 6th Int'l. IPSN Conf., Cambridge (April 2007)
36. Facebook, http://www.facebook.com
37. MySpace, http://www.myspace.com
38. Apple iPhone, http://www.apple.com/iphone/
39. Witten, I.H., Frank, E.: Data Mining: Practical machine learning tools and techniques, 2nd edn. Morgan Kaufmann, San Francisco (2005)
40. Garmin Edge, http://www.garmin.com/products/edge305
41. Campbell, A.T., et al.: People-Centric Urban Sensing (Invited Paper). In: Proc. of 2nd ACM/IEEE Annual Int'l. Wireless Internet Conf., Boston (August 2006)
42. Symbian, http://www.symbian.com
43. Eisenman, S.B., et al.: The BikeNet Mobile Sensing System for Cyclist Experience Mapping. In: Proc. of 5th ACM Conf. on Embedded Networked Sensor Systems, Sydney, November 6-9, 2007, pp. 6–9. ACM Press, New York (2007)
44. CitySense, http://www.citysense.net/
45. Place Lab., http://www.placelab.org/
46. Gemmell, J., Bell, G., Lueder, R.: MyLifeBits: a Personal Database for Everything. Communications of the ACM 49(1), 88–95 (2006)
47. Zhou, C., et al.: Discovering Personally Meaningful Places: An Interactive Clustering Approach. In: ACM Trans. on Information Systems vol. 25(3) (2007)
48. Skyhook Wireless, http://www.skyhookwireless.com/
49. Liao, L., Fox, D., Kautz, H.: Extracting Places and Activities from GPS Traces using Hierarchical Conditional Random Field. Int'l. Journal of Robotics Research 26(1) (2007)
50. Krumm, J., Horvitz, E.: LOCADIO: Inferring Motion and Location from Wi-Fi Signal Strengths. In: Proc. of 1st Int'l. Conf. on Mobile and Ubiq. Systems: Networking and Services (August 2004)

51. Eisenman, S.B., Campbell, A.T.: SkiScape Sensing (poster abstract). In: Proc. of 4th ACM Conf. on Embedded Networked Sensor Systems, Boulder (November 2006)
52. Motorola L2, http://www.motorola.com/
53. Lane, N.D., Lu, H., Campbell, A.T.: Ambient Beacon Localization: Using Sensed Characteristics of the Physical World to Localize Mobile Sensors. In: Proc. of 4th Workshop on Embedded Networked Sensors, Cork (June 2007)
54. Sohn, T., et al.: Mobility Detection Using Everyday GSM Traces. In: Proc. of the 6th Int'l. Conf. on Ubiq. Comp. Orange County, pp. 212–224 (September 2006)
55. Anderson, I., Muller, H.: Practical Activity Recognition using GSM Dat. In: Technical Report CSTR-06-016, Dept. of Comp. Sci., Univ. of Bristol (July 2006)
56. iGoogle, http://www.google.com/ig
57. Kapadia, A., Henderson, T., Fielding, J.J., Kotz, D.: Virtual Walls: Protecting Digital Privacy in Pervasive Environments. In: Proc. of 5th Int'l. Conf. on Perv. Comp. Toronto, pp. 162–179 (May 2007)
58. Masoodian, M., Lane, N.: MATI: A System for Accessing Travel Itineary Information using Mobile Phones. In: Proc. of 16th British HCI Group Annual Conf., London (September 2002)
59. Google Gadgets, http://code.google.com/apis/gadgets
60. Choudhury, T., Basu, S.: Modeling Conversational Dynamics as a Mixed Memory Markov Process. In: Advances in Neural Information Processing Systems 17, pp. 218–288. MIT Press, Cambridge (2005)
61. Hightower, J., et al.: Learning and Recognizing the Places We Go. In: Proc. of the 7th Int'l. Conf. on Ubiq. Comp. Toyko, pp. 159–176 (September 2005)
62. Welbourne, E., Lester, J., LaMarca, A., Borriello, G.: Mobile Context Inference Using Low-Cost Sensors. In: Strang, T., Linnhoff-Popien, C. (eds.) LoCA 2005. LNCS, vol. 3479, Springer, Heidelberg (2005)
63. Kang, J., Welbourne, W., Stewart, B., Borriello, G.: Extracting Places from Traces of Locations. In: Proc. of ACM Int'l. Workshop on Wireless Mobile Applications and Services on WLAN Hotspots, Philadelphia (October 2004)
64. DeVaul, R., Sung, M., Gips, J., Pentland, A.: MIThril 2003: Applications and Architecture. In: Proc. of 7th Int'l. Symp. on Wearable Computers, White Plains (October 2003)
65. Ashbrook, D., Starner, T.: Using GPS to Learn Significant Locations and Predict Movement across Multiple Users. In: Personal Ubiq. Comp., pp. 275–286 (2003)
66. Liao, L., Fox, D., Kautz, H.: Location-Based Activity Recognition using Relational Markov Networks. In: Proc of IJCAI-05, Edinburgh (August 2005)

Mapping by Seeing – Wearable Vision-Based Dead-Reckoning, and Closing the Loop

Daniel Roggen, Reto Jenny, Patrick de la Hamette, and Gerhard Tröster

Wearable Computing Laboratory, ETH Zürich, Switzerland
droggen@ife.ee.ethz.ch
http://www.wearable.ethz.ch/

Abstract. We introduce, characterize and test a vision-based dead-reckoning system for wearable computing that allows to track the user's trajectory in an unknown and non-instrumented environment by integrating the optical flow. Only a single inexpensive camera worn on the body is required, which may be reused for other purposes such as HCI. Result show that distance estimates are accurate (6-12%) while rotation tends to be underestimated. The accumulation of errors is compensated by identifying previously visited locations and "closing the loop"; it results in greatly enhanced accuracy. Opportunistic use of wireless signatures is used to identify similar locations. No a-priori knowledge of the environment such as map is needed, therefore the system is well-suited for wearable computing. We identify the limitations of this approach and suggest future improvements.

1 Introduction

Wearable computers aims to empower users by providing them with information or support anytime and anywhere, proactively and in unobtrusive ways. By being "body-worn" wearable computers can sense the user's state, such as his gestures, activities or his location. This *contextual awareness* is the key mechanism that enables wearable computers to support users proactively and in an unobtrusive way [1,2]. The physical location of the user is an important contextual information that allows *location aware computing*, such as providing location-specific information in a wearable tourist guide [3], or learning a person's daily activity patterns [4,5].

Satellite positioning systems allow outdoor global localization, but they do not operate in difficult environments (e.g. indoor, urban canyons). Wi-Fi or GSM signals allow localization indoor as well as outdoor [6,7,8,9], but they generally need a map of the radio beacon location. Accurate localization is possible in environments instrumented with localization beacons (e.g. radio, ultrasound)[1]. Yet these approaches are not well suited for wearable computing where the user can freely move in an open-ended environment: a-priori knowledge (map) or instrumentation should be avoided in this case.

Absolute localization (e.g. satellite) combined with *dead-reckoning* using body-worn sensors is a way to address this issue. Dead-reckoning in wearable computing has been investigated with combination of compass to estimate the heading direction and velocity estimation methods with e.g. step counting [11,12,13].

---

[1] For a review of localization methods, especially indoors, see [10].

G. Kortuem et al. (Eds.): EuroSSC 2007, LNCS 4793, pp. 29–45, 2007.

In this paper we investigate for the first time in wearable computing a dead-reckoning method based on the integration of the *optical flow* from a camera placed on the front of the user to determine his *egomotion* (*vision-based path integration*). Vision-based dead-reckoning has the potential of being an inexpensive, low-power and highly integrated solution to dead-reckoning since a single camera is needed such as those used in cellular phones. This approach allows sensor reuse, e.g by using the same camera for vision-based HCI or even for taking pictures.

This approach is biologically inspired from insect flight and it has been used in mobile robotics for odometry, where it often benefits from controlled environments such as flat surfaces or good lighting conditions. In contrast, in a wearable setting the constraints on the environment are relaxed and noise in the optical flow induced by terrain geometry or camera angle variations caused by the movement of the user have to be considered. In this paper we translate known dead-reckoning methods using optical flow integration in a wearable computing application domain. Our objective is to investigate the challenges as well as the benefits and limitations of such an approach in this new application domain.

In order to cope with path integration errors, a path correction mechanism is implemented that "closes the loop" when identical locations are visited. Locations are identified from the signal strength of radio beacons opportunistically found in the environment. Yet the whole system does not need instrumented environments and operates without any a-priori knowledge, such as a map of the radio-beacons. It therefore fulfills the requirement for a scalable wearable mapping system.

This paper is organized as follows. In section 2 we explain how optical flow is computed and leads to egomotion. The setup is described in section 3. In section 4 we characterize the system in a variety of conditions and in section 5 we show path integration results for typical trajectories. In section 6 we discuss how path integration errors may be compensated by closing the loop when identical locations are visited. Finally we discuss the results in section 7 and conclude this paper in section 8.

2   Vision-Based Dead-Reckoning

Vision-based dead-reckoning operates by integrating the optical flow (or image velocity) registered on a camera. The optical flow is a 2-D motion direction vector field which is the projection of the 3-D velocities of the environment surface points onto the image plane. The trajectory of a camera moving in the 3-D space (*egomotion*) can be reconstructed from the optical flow knowing the environment structure (e.g. geometry of obstacles or open environment). Two steps must be carried out: 1) find the optical flow in a sequence of images, 2) find the egomotion corrsponding to this optical flow (i.e. dead-reckoning). Figure 1 illustrates the optical flow of typical motions.

Vision-based dead-reckoning has been studied in biology: insects such as blowflies or bees rely on optical flow to control flight and navigate [14]. The human brain also uses vision-based path integration to estimate travel distance and direction [15,16,17]. Optical flow has been used in mobile robots for navigation [18,19], direction estimation [20] or odometry [21], and in cars for road-navigation [22].

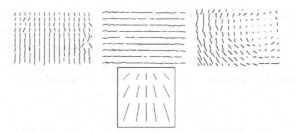

Fig. 1. Top: typical optical flows with a camera pointing vertically to the floor while moving forward (left), laterally (center), and rotating (right). The lines are the 2-D motion direction vector field on the camera. Bottom: schematic view of the optical flow with the camera moving forward and pointing with a 45° angle downwards.

2.1 Computing Optic Flow

A summary and comparison of the various techniques to estimate optical flow can be found in [23]. Here we use the Lucas-Kanade differential technique [24] because of its low complexity, with the modification proposed by [23][2] to estimate accuracy of the optical flow estimation.

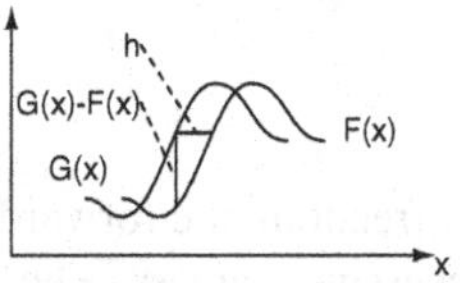

Fig. 2. The optic flow is the displacement h between $F(x)$ and $G(x)$ representing the light intensity in two successive frames

Differential technique in one dimension. In the one dimensional case the Lucas-Kanade method consists in finding the displacement h between two curves $F(x)$ and $G(x)$ representing the light intensity along the same line in two successive frames, where $G(x) = F(x + h)$ (fig. 2). The problem of finding h from $F(x)$ and $G(x)$ can be solved by a linear approximation of F in the neighborhood of x. For small h:

$$F'(x) \approx \frac{F(x+h) - F(x)}{h} = \frac{G(x) - F(x)}{h}$$

$$h \approx \frac{G(x) - F(x)}{F'(x)} \tag{1}$$

$$F'(x) = \frac{F(x + dx) - F(x - dx)}{2 \cdot dx}$$

This algorithm assumes h to be "small enough" in comparison to the frequency of $F(x)$ (see [24] for details). With fast displacements or low frame rate this condition may not be fulfilled.

---

[2] Sec. 2.1, paragraph referring to Lucas-Kanade's method in that paper.

Differential technique in two dimensions. In this work we implement exactly the Lucas-Kanade method as described in [23]. It is the 2D extension of the 1D approach described above. $I(x, t)$ is the intensity of a pixel at a position x at time t. We assume an image translation: $I(x, t) = I(x - v \cdot t, 0)$ with $v = (v_x, v_y)^T$ the displacement velocity. Equation (1) leads to the following *gradient constraint equation*[3]:

$$\nabla I(x, t) \cdot v + I_t(x, t) = 0 \qquad (2)$$

$I_t(x, t)$ is the temporal derivative of $I(x, t)$ and $\nabla I(x, t) = (I_x(x, t), I_y(x, t))^T$ is the spatial derivative. A least-square fit under this constraint is performed in a small neighborhood Ω centered on the desired pixel. This yields a solution to v that consists in simple 2D matrix multiplications of the form $v = (A^T W^2 A)^{-1} A^T W^2 b$ with A, W and b being respectively the gradient of I, a weighting factor giving more importance to the constraints in the center of the neighborhood, and the temporal derivative of I. The mathematical steps are left in [23].

In this work Ω is a 5x5 pixel window, the derivatives are computed with three-point estimation, and the window function $W(x)$ is separable and isotropic with the effective 1-D weights $(0.0625, 0.25, 0.375, 0.25, 0.0625)$. The eigenvalues $\lambda_1 > \lambda_2$ of $A^T W^2 A$ can be used to identify unreliable measurements. Here measurements with $\lambda_2 < 1$ are discarded. All other details are as in [23][4].

2.2 Computing Egomotion

We consider 2 relevant degrees of freedom: the forward displacement (*translation*) and the angular rotation between successive frames[5]. They are accumulated to obtain the travelled path. Figure 3 top illustrates the system parameters. The camera with the horizontal and vertical aperture angle α_h and α_v is placed at a fixed height h and is looking downwards with an angle φ. A trapezoidal area of the floor with angle ψ is projected onto the camera image which has a resolution of res_x by res_y pixels.

During rotation, the optical flow is horizontal for pixels corresponding to the horizon; pixels mapping to the ground show a slight curvature. This curvature is negligible because the camera height and inclination map an area of the ground far away from the rotation center and we assume the optical flow vectors to have identical magnitude v_x along the horizontal direction. The rotation angle is computed as follows:

$$\Omega = \frac{res_x \cdot v_x}{\psi} \qquad (3)$$

ψ is function of the camera angle φ: $\psi = 2 \cdot arctan(sin(\varphi)tan(\frac{\alpha_h}{2}))$.

---

[3] The link with the one-dimensional case is the following: $F(x) = I(x, t)$, $G(x) = I(x, t + \Delta t)$, $h = v_x \cdot \Delta t$, $F'(x) = I_x(x, t)$ the partial derivative along x.

[4] Sec. 2.1, paragraph referring to Lucas-Kanade's method in that paper.

[5] In our scenario the camera is fixed to the body and therefore more than 2 degrees of freedom (DoF) apply. While in principle the six DoF of the camera motion can be reconstructed from the optical flow [25], we consider here the 2 major DoF as a way to assess the performance of the system with the simplest motion model.

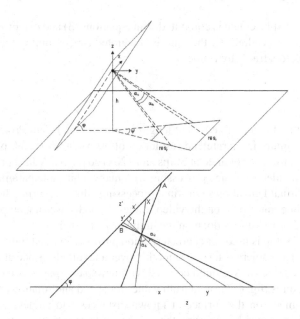

Fig. 3. Estimation of egomotion from optical flow. Bottom: tilted camera looking at the floor and mapping of ground points x and y onto the image x' and y'.

During translation with the camera moving forward and pointing downwards the x-component of the velocity vectors can be neglected without loss of information. The y-component of the velocity vectors have different magnitude at the top and bottom of the frame (see fig. 1 bottom) and must be taken into consideration. The mapping between pixels and the ground is found according to figure 3 bottom. The mapping between the optical flow and the ground motion is given in equation (4). α is the half aperture angle of the camera (in the figure α_1 and α_2 are the aperture angle measured from the point where the normal axis trough the camera crosses the image plane; in practice $\alpha_1 = \alpha_2 = \alpha$). z is the displacement on the ground in meters, h the camera height, z^* is the optic flow measured in pixel at position p_s (pixels) in the image, and p_m is the vertical image size in pixels. p_f is a factor to convert between pixel and meters: $p_f = (2tan(\alpha))/p_m$. The overall translation is the average of z computed for each motion vector with equation (4).

$$z = \frac{h}{cos(\varphi)} \left[\frac{tan(\alpha) - p_f p_s}{cos(\varphi) + sin(tan(\alpha) - p_f p_s)} + \right.$$

$$\left. \frac{tan(\alpha) + p_f p_s - p_f p_m + p_f z^*}{cos(\varphi) - sin(tan(\alpha) + p_f p_s - p_f p_m + p_f z^*)} \right] \tag{4}$$

In order to improve accuracy the optic flow is computed and averaged on 10 frames; only optic flow vectors which satisfy the reliably conditions are considered ($\lambda_2 > 1$).

The rotation and displacement is computed with equations (3) and (4). From the starting position $(0, 0)$ with heading $0°$, the path is integrated by combining the rotation and displacement vectors frame by frame.

3   Setup

Images are acquired from a fixed focal length Logitech QuickCam Pro 5000 webcam which has a maximum framerate and resolution of 30fps and 640x480 pixels. The results presented below are obtained at 30fps and 120x160 pixels. A Java program allows offline and batch video sequence processing with many adjustable parameters. A C filter plugin for Virtual Dub allows real-time processing while capturing the video and it superposes the integrated path on the video frames. In both cases uncompressed videos are acquired and processing is done on the luminance channel.

A "synthetic" setup is used for characterization purposes in good lighting conditions (high contrast). The camera is moved smoothly over a textured surface at a fixed height and with a fixed angle α. A ground covered with newspaper pages is used, as well as other floor structures (fig. 4) Ground truth values for speed and rotations were obtained from markings made on the surface at known distances and angles. The speed and rotation speed is obtained by counting the number of frames between the appearances of the markings. In the real-world setup the camera is fixed on the chest of the user, pointing forward and downwards, so that only the ground is visible in the frame (height 1m45, approximate angle $\alpha = 45°$).

Fig. 4. Floors: concrete, carpet, linoleum, parquet, and newspaper on the floor

4   Characterization

The system is characterized in the synthetic environment by moving the camera straight for a fixed distance of 30cm. Figure 5 left illustrates the distance travelled by the camera as estimated from the optical flow for various heights and effective speeds ($\alpha = 45°$). Figure 5 right illustrates the same with various floor types ($h = 54cm$, $\alpha = 45°$). The distance tends to be underestimated at higher speeds and a higher camera tends to allow faster speeds. The underestimation is caused at higher speeds by the image that tends

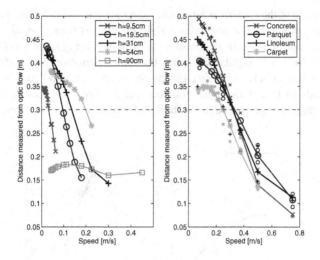

Fig. 5. Camera travel distance estimated from the optical flow for various effective speeds and camera heights (left) and floor types (right)

to blur and the "small displacement" condition that is not respected. With $h = 90cm$ the distance is constantly underestimated due to the blurred images acquired from the camera which has a focal length optimized for short distances (i.e. the fine textures disappear). The ground type has only limited impact on the results and similar results were achieved with the dark carpet and the lighter parquet textures.

Rotation estimation is characterized in the same way by rotating the camera smoothly at various rotation speeds until a 90° rotation is achieved. The camera angle is $\alpha = 45°$.

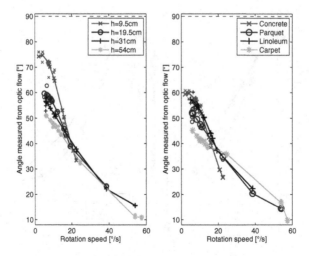

Fig. 6. Camera rotation angle estimated from the optical flow reported effective rotation speeds and various camera heights (left) and floor types (right)

The rotation angle of the camera as estimated from the optical flow is illustrated in fig. 6 left for various camera height and in fig. 6 right for various floor types ($h = 9.5cm$ in this case). The angle tends to be underestimated with increasing angular velocities. Furthermore even for small angular velocities the underestimation is already high. The reasons are similar as before: the small displacement assumption is not respected and the image tends to blur with higher rotation speeds. Even low rotation speeds correspond to a large image translation in terms of pixels. As a consequence the small

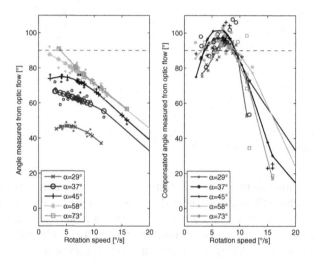

Fig. 7. Original (left) and mathematically compensated (right) angular measurement at various speeds and camera angles

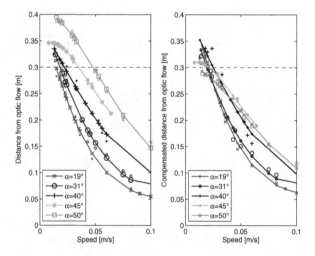

Fig. 8. Original (left) and mathematically compensated (right) distance measurement at various speeds and camera angles

displacement assumption is not respected even for low rotation speed, which explains the underestimation.

The results may be improved by a mathematical compensation based on the system characterization. The incremental angle and displacement estimated from the optic flow is corrected at each time step by a factor linearly proportional to the angular speed or the velocity. The correction funtion is determined from analysis of the system's behavior at various rotation speed and camera angles. If some parameters are fixed (e.g. fixed camera angle) this correction function can be further optimized for this case. Figures 7 and 8 illustrate the system characteristics without (left) and with (right) mathematical compensation. The real rotation and translation is 90° and 0.3m, the camera height 53cm.

5   Path Integration Results

Figure 9 illustrate the path integration result in the synthetic environment (floor textured with newspaper) when moving along a square and a circle at a slow (<0.02m/s, 8-10°/s) and higher speed. Camera parameters are h=9.5cm, $\alpha = 45°$. Dashed lines indicate the reference trajectory. The distance is well approximated from the optic flow, but inaccuracies in the estimation of the angle, especially with the higher speed, evidence the "closing of the loop" problem: the trajectory estimated from the optical flow ends away from the starting position, although the start end end point match in the reference trajectory. Figure 10 illustrates a more complex trajectory: a 1/80th scale of our laboratory building[6].

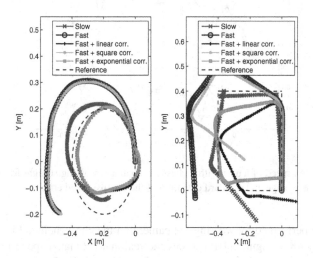

Fig. 9. Path integration while moving along a circle (left) and a square (right). Slow and fast refers to the motion speed. Corrected paths use the closing of the loop.

---

[6] Tests in the real building were not successful: no reliable motion vectors could be extracted due to the highly uniform texture of the carpet. A scaled synthetic environment was used insted with floor textured with newspaper.

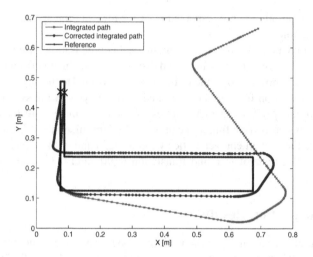

Fig. 10. Path integration while following the corridors of a laboratory building

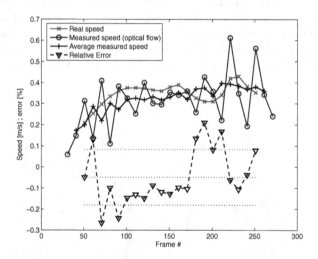

Fig. 11. Estimated speed of a user with a chest-worn camera walking straight for 5.5m. Dashed lines indicate the average and standard deviation of the relative speed error.

Tests were performed by wearing the camera on the chest (height 145cm, angle of approximately 48°). Figure 11 illustrates the real and estimated speed profile with a user walking straight on 5.5m on a concrete floor (fig. 4 left). The estimated travelled distance is 5.17 meters (relative error of 6%). The instantaneous speed measured by the optical flow reveals the periodicity of the footsteps. Speed under- or over-estimation depends mostly on the camera angle, which may be caused by the user bending. Figure 12 illustrates the trajectory of a user walking along a square of 10m sides (concrete floor). The estimated travelled distance was 43.4 meters whereas the real distance was 38.8

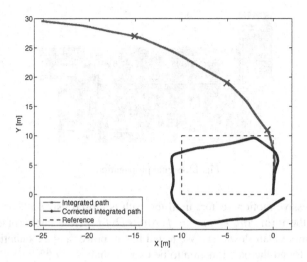

Fig. 12. Path integration of a user with a chest-worn camera walking along a square with 10m sides. The crosses on the integrated path indicate when the user rotates in the corners of the square.

meters (slightly below 40m because of the rounding of the corners), corresponding to a relative error of 12%. This error is higher than in the straight walk, and may be caused by the inaccuracies introduced by the rotation. The rotations are however strongly underestimated. This is explained by the high speed at which humans rotate while walking (30 °/s to 90 °/s) which far higher than what the current system can handle as characterized before.

6   Closing the Loop

When a user's start and end point of a trajectory coincide, the errors accumulated by dead-reckoning lead to the "loop closing" problem: the end point mapped from dead-reckoning is likely not to correspond with the start point. This problem may be addressed by recognizing locations previously visited and correcting the previously mapped trajectory in order to "close the loop" by matching the two visited locations. In other words, we use the knowledge that two visited locations are identical to correct a-posteriory the drift resulting from the integration of the optical flow.

Path integration errors may stem from errors in the estimation of the motion vector magnitude or orientation. Here we consider only the orientation that is the main source of errors[7]. We close the loop by iteratively increasing the angle difference $AngleDiff$

---

[7] We assume that the magnitude and orientation of motion vectors is under- or over-estimated in a systematic way. Previous results show that the errors in the estimation of the angular rotation dominate over the errors in the estimation of the the translation distance. For this reason we correct here only the orientation of the motion vectors with a forumla that attempts to compensate for the under- or over-estimation of the angular rotation. This approach benefits any type of trajectory and not only circular or rectangular path.

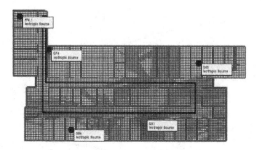

Fig. 13. Signal propagation

between successive motion vectors until the path is closed using a "corrected" angle difference of the form $AngleCorr = f(AngleDiff)$. Here faster rotations are underestimated more than slow ones. We tested linear, power and exponential correction functions and found the best function to be exponential: $A \cdot \left(e^{AngleDiff \cdot B} - 1\right)$ with $0.05 \leq A \leq 0.5$ and $0.5 \leq B \leq 1.5$ found by testing. A search algorithm finds the correction parameters that minimize the distance between two identified points of the trajectory through exhaustive search in the parameter space.

Figure 9 illustrates the correction functions when the camera moves along a circle and a square (same conditions as in section 5). Figures 12 and 10 use the exponential correction and illustrate the improved dead-reckoning with the closing of the loop algorithm.

In order to recognize previously visited locations, characteristic landmarks must be identified and compared against previously memorized ones (e.g. visual landmarks [26]). We show preliminary results in simulation were the "wireless signature" (signal strength of nearby WiFi stations) is used for this purpose[8]. We simulate the signal propagation of the 5 WiFi stations of the building where the trajectory of fig. 10 would take place using the Radio Propagation Simulator of Radioplan[9]. We record the signal intensities on the whole floor with a grid spacing of 0.25m and on the 617 points of the trajectory where at least two WiFi station were in range. The signature matching achieves a $< 5m$ average localization accuracy in this simulation (nearest match). This accuracy indicates how close the start and end point of a loop must be for them to be matched. In the trajectory of fig. 10, we compared the 20 first meters of the trajectory

---

[8] Visual landmarks would make the best use of the available camera. Recognizing visual landmarks is challenging and computationally intensive. Instead we use simpler wireless signatures. Our objective is to assess the implications of "closing the loop". The exact way in which previously visited locations is determined does not affect the algorithm correcting the trajectory nor the results. This approach to the use of wireless signatures is similar to [8,9]. In these work the wireless signature is used for accurate localization, by relying on a previously learned "wireless map" of the environment that indicates typical wireless signatures at each position. Here we avoid this wireless map measured a-priori. We simply reference current wireless signatures to past ones, without mapping them to absolute positions. Therefore no a-priori knowledge of the environment is needed.

[9] http://www.radioplan.com. We used the free version which simulates a limited number of reflections and penetrations.

with the 20 last meters of the trajectory, and found that the two points marked by the cross are the closest (0.65m) and should trigger the closing of the loop as illustrated.

7 Discussion

Alternative dead-reckoning methods mostly rely on step count integration, using accelerometers and compass [11,12,13]. In the experiments carried out here the system measured the travelled distance with an accuracy between 6% to 12% depending on the path. This compares well to the accuracies reported in [12], which vary between 1.3% and 29% depending on the body location of the system. Step count integration requires to know the length of the step (a-priori definition or inferred from the acceleration signal). Here we do not require this knowledge since the optical flow is considered; as such gait dynamics has in principle no influence on the performance of the system, which would work even if the user were on a bike or scooter.

A benefit of this vision-based approach is that a single camera sensor is needed, which may also be used for other wearable computing activities such as taking pictures or vision-based HCI. In particular in [27] a chest-mounted "FingerMouse" camera system was investigated which captures hand motion as an input for wearable computers. Further work may combine HCI and localization in a single chest-mounted camera system.

In this method we evidenced several trade-offs between parameters (camera placement, image parameters, motion) and system performance. In theory the system is not sensitive to floor type or color as long as contrast is perceived. In practice the spatial frequency of the floor is linked to the resolving power of the camera. Performance degrades at higher rotation/translation speeds. This is caused by the assumption of small displacement of the differential technique that is not verified at higher speeds. This assumption is also linked to the spatial frequency of the textures [24]. This can be compensated by increasing the framerate (results not presented here confirm this) or placing the camera higher. A larger ground area is projected in the image frame with a higher camera, which allows for faster motion (reduction in the spatial frequency of the textures and thus the small displacement condition is valid for higher speed). A too high height leads to difficulties in obtaining the motion vectors (e.g. blurred images due to the fixed focal length, or uniform texture such as carpet which mostly contains high spatial frequency textures). This may be compensated by a higher frame resolution.

Alternatively another algorithm to obtain the optical flow may be used, such as looking for features in successive frames. Block-matching as used in MPEG video encoding may be used for this purpose. By adjusting the block search neighborhood, the small displacement constraint may be alleviated and the system may operate at faster speeds. The computational requirements of block-matching are higher than those of the differential method, but optimized instruction sets (e.g. MMX) allow to compute these operations in real-time on any 1+GHz computer.

Other sources of errors stem from the hypotheses of our approach: the camera parameters (angle, height) are known, and the camera motion is only a combination of forward motion and rotation. Errors are introduced when these are not verified. Changes in camera height and angle lead to an over- or under-estimation of the speed, with the camera angle being the most important parameter. Uneven floor or tilted floor as well as body motion (leaning forward/backward) are similar to a change in camera angle. Torso rotation (e.g. looking sideways while walking) or bending sideways affect the system since they introduce optical flows that are not expected. The footsteps are easily distinguished from the instantaneous speed estimate (fig. 11) but are not a source of error since they are averaged over time. Interestingly this may lead to a visual approach to count the footsteps. Further improvement may be obtained by optimizing the mathematical compensation method for a particular camera angle or surface structure, at the loss generality. Alternatively image processing may detect typical surfaces to optimize the compensation parameters.

The system may be improved with additional sensors. A sensor giving the camera angle (e.g. accelerometer) may allow to take into consideration the instantaneous camera angle while integrating the path. A compass may allow to improve the rotation estimation accuracy from the optical flow. The floor geometry is important, yet recent findings suggest this approach may work even in extreme terrains [28]. Finally, a more complex model may allow to reconstruct the full 6D egomotion from optical flow [29,18,25]. In principle optical flow may thus be used in a wearable setting without constraint on the user motion, however it remains to be investigated what would be the tradeoffs of such algorithms.

Dead-reckoning runs in real-time at 160x120, 30fps on a 1.6GHz Intel Centrino CPU without any implementation optimization: all the compuation is done in floating point without use of vectorial instructions. Fixed-point optimization and use of vectorial instructions is liekly to allow a speedup of more than one order of magnitude and therefore allow higher resolution. A hardware optical flow implementation may also be investigated. Low-power analog VLSI chips directly give the optical flow at a high frame rate [30,31]. Digital hardware implementations of block matching may also allow low-power and high-speed processing (e.g. in [27] the authors present a block-matching approach to stereovision for HCI; the same block matching structure may be used to compute the optical flow).

We presented a method of opportunistic use of wireless signature to compensate for the errors accumulated by dead-reckoning in simulation. Future work will have to test this in reality. There are however a number of results suggesting that this approach may be applicable [8,9].

In summary, the system introduced here requires sufficient floor texture and contrast to operate. This precludes the use of such approach on uniform surfaces (depending on camera resolution), or in low lighting conditions (e.g. night). With the current egomotion model, the system performs best when the user is walking straight and performing rotations without significant sideways bending. Closing of the loop needs sufficiently distinct wireless signatures in the environment and is expected to work best in environments with dense mesh of wifi stations. With a low density of wifi stations the wireless signatures become less accurate (e.g. with a single isotropic wifi station the same signature is sensed at identical distance from the station regardless of the bearing).

8 Conclusion

The objective of this paper was to investigate how vision-based dead-reckoning methods may be applied in wearable computing. We were appealed by the fact that a single miniature and low-power camera may be used for this purpose. We introduced, characterized and tested a method for dead-reckoning based on the integration of the optical flow obtained from an inexpensive chest-mounted camera. Results show that the distances estimated from the optical flow are accurate for a walking subject (6-12% relative error) and comparable to methods integrating steps. The rotation tends to be underestimated due to the high angular speed at which users turn in relation to the framerate (30 fps). This calls for higher framerates in order to improve rotation estimation accuracy. We proposed a method to improve the path integration accuracy by opportunistically using wireless signatures to detect already visited location. The integrated path is corrected at this moment by "closing the loop" so that the visited locations match. We showed that the accuracy of the trajectory mapping was greatly enhanced in this way. A localization method for wearable computing needs to be "mapless" so that it can be used in any environment. This vision-based dead-reckoning method combined with the opportunistic use of wireless signatures does not require any a-priori map or knowledge about the environment. It thus satisfies this requirement and can be used anywhere.

The application in wearable computing is complex due to the free motion of the user and the unconstrained nature of the environment in which he may walk in. Our results are preliminary: further tests in realistic scenarios are needed, and we do not claim to a final vision-based dead-reckoning solution. Still, our results allow us to evidence a number of challenges, limitations and trade-offs which may shape future work. We highlighted the links between system accuracy, camera and environment parameters, and the user motions. Higher frame rates and higher camera resolutions improve the system accuracy, but improved characterization, improved optical flow models, as well as additional sensors may also improve the performance. We believe that there is more to this approach than what was presented here, as is also suggested by the numerous litterature describing the use of optical flow in mobile robotics. Our results evidence topics for further research. **Technical improvements** such as the use of higher resolution camera, higher framerate, and autofocus, may improve the system accuracy. A **motion model** with additional degrees of freedom may allow to better cope with motions typical of wearable setting. Comparative analysis may evidence more accurate or computationally-efficient methods to derive **optical flow**. Finally, we believe **sensor fusion** may best address the shortcomings of the current approach by using additional modalities. For instance the use of magnetometers to improve estimation of rotations may be investigated, as well as the use of accelerometers or inclinometers to take into account the camera angle in the optical flow model.

References

1. Dey, A.K., Abowd, G.D.: Towards a better understanding of context and context awareness. Technical Report GITGVU-99-22, Georgia Tech. (1999)
2. Lukowicz, P., Junker, H., Staeger, M., von Bueren, T., Troester, G.: WearNET: A distributed multi-sensor system for context aware wearables. In: Borriello, G., Holmquist, L.E. (eds.) UbiComp 2002. LNCS, vol. 2498, pp. 361–370. Springer, Heidelberg (2002)

3. Simcock, T., Hillenbrand, P., Thomas, B.H.: Developing a location based tourist guide application. In: Johnson, C., Montague, P., Steketee, C. (eds.) Conferences in Research and Practice in Information Technology, Australian Computer Society, pp. 177–183 (2003)
4. Patterson, D.J., Liao, L., Fox, D., Kautz, H.: Inferring high-level behavior from low-level sensors. In: Goos, G., Hartmanis, J., van Leeuwen, J. (eds.) Proc. of the 5th Int. Conf. on Ubiquitous Computing, pp. 73–89. Springer, Heidelberg (2003)
5. Ashbrook, D., Starner, T.: Using gps to learn significant locations and predict movement across multiple users. Personal and Ubiquitous Computing 7(5), 275–286 (2004)
6. Haeberlen, A., Flannery, E., Ladd, A.M., Rudys, A., Wallach, D.S., Kavraki, L.E.: Practical robust localization over large-scale 802.11 wireless networks. In: Proc. of the 10th annual international conference on Mobile computing and networking, pp. 70–84. ACM Press, New York (2004)
7. Cavalieri, S.: A novel approach for localisation based on wi-fi. In: Proc. of the 3rd Int. Conf. on Industrial Informatics (INDIN), pp. 234–239 (2005)
8. Otsason, V., Varshavsky, A., La Marca, A., de Lara, E.: Accurate gsm indoor localization. In: Beigl, M., Intille, S.S., Rekimoto, J., Tokuda, H. (eds.) UbiComp 2005. LNCS, vol. 3660, pp. 141–158. Springer, Heidelberg (2005)
9. La Marca, A., Chawathe, Y., Consolvo, S., Hightower, J., Smith, I., Scott, J., Sohn, T., Howard, J., Hughes, J., Potter, F., Tabert, J., Powledge, P., Borriello, G., Schilit, B.: Place lab: Device positioning using radio beacons in the wild. In: Proc. of Pervasive Computing, pp. 116–133. Springer, Heidelberg (2005)
10. Hightower, J., Borriello, G.: Location systems for ubiquitous computing. IEEE Computer, 57–66 (2001)
11. Lee, S.W., Mase, K.: Activity and location recognition using wearable sensors. Pervasive Computing 1(3), 24–32 (2002)
12. Randell, C., Djiallis, C., Muller, H.: Personal position measurement using dead reckoning. In: ISWC 2003, pp. 166–173. IEEE Computer Society, Los Alamitos (2003)
13. Schindler, G., Metzger, C., Starner, T.: A wearable interface for topological mapping and localization in indoor environments. In: Hazas, M., Krumm, J., Strang, T. (eds.) Proc. of the 2nd Int. Workshop on Location- and Context- Awareness, pp. 64–73. Springer, Heidelberg (2006)
14. Schilstra, C., Van Hateren, J.H.: Blowfly flight and optic flow I: thorax kinematics and flight dynamics. The Journal of Experimental Biology 202(11), 1481–1490 (1999)
15. Ellmore, T.M.: Human path integration by optic flow. Master's thesis, Department of Psychology, The University of Arizona (2004)
16. Frenz, H., Bremmer, F., Lappe, M.: Discrimination of travel distance from 'situated' optic flow. Vision Research 43(20), 2173–2183 (2003)
17. Frenz, H., Lappe, M.: Absolute travel distance from optic flow. Vision Research 45(13), 1679–1692 (2005)
18. Srinivasan, M.V., Chahl, J.S., Zhang, S.W.: Robot navigation by visual dead-reckoning: inspiration from insects. International Journal of Pattern Recognition and Artificial Intelligence 11(1), 35–47 (1997)
19. De Souza, G.N., Kak, A.C.: Vision for mobile robot navigation: A survey. IEEE Trans. on Pattern Analysis and Machine Intelligence 24(2), 237–267 (2002)
20. Kröse, B., Dev, A., Groen, F.: Heading direction of a mobile robot from the optical flow. Image and Vision Computing 18(5), 415–424 (2000)
21. Nagatani, K., Tachibana, S., Sofue, M., Tanaka, Y.: Improvement of odometry for omnidirectional vehicle using optical flow information. In: IEEE/RSJ Int. Conference on Intelligent Robots and Systems, pp. 468–473 (2000)
22. Giachetti, A., Campani, M., Torre, V.: The use of optical flow for road navigation. IEEE Transactions on Robotics and Automation 14(1), 34–48 (1998)

23. Barron, J., Fleet, D., Beauchemin, S.: Performance of optical flow techniques. International Journal of Computer Vision 12(1), 43–77 (1994)
24. Lucas, B., Kanade, T.: An iterative image registration technique with an application to stereo vision. In: Proc. of the 7th Int. Joint Conf. on Artificial Intelligence (IJCAI), pp. 674–679 (1981)
25. Nagle, M.G., Srinivasan, M.V., Wilson, D.L.: Image interpolation technique for measurement of egomotion in 6 degrees of freedom. Journal of the Optical Society of America A: Optics, Image Science, and Vision 14(12), 3233–3241 (1997)
26. Newman, P., Ho, K.: Slam- loop closing with visually salient features. In: IEEE Int. Conf. on Robotics and Automation (ICRA), pp. 635–642. IEEE Computer Society Press, Los Alamitos (2005)
27. de la Hamette, P., Tröster, G.: Architecture and applications of the fingermouse: a smart stereo camera for wearable computing HCI. Personal and Ubiquitous Computing (online first) (2007)
28. Campbell, J., Sukthankar, R., Nourbakhsh, I.: Techniques for evaluating optical flow for visual odometry in extreme terrain. In: Proc. of Intelligent Robots and Systems (IROS), pp. 3704–3711 (2004)
29. Srinivasan, M.V.: An image-interpolation technique for the computation of optic flow and egomotion. Biological Cybernetics 71(5), 401–415 (1994)
30. Liu, S.C.: A neuromorphic avlsi model of global motion processing in the fly. IEEE Trans. on Circuits and Systems 47(12), 1458–1467 (2000)
31. Stocker, A., Douglas, R.: Analog integrated 2D optical flow sensor with programmable pixels. In: Int. Symposium on Circuits and Systems ISCAS, pp. 121–138 (2004)

The Design of a Pressure Sensing Floor for Movement-Based Human Computer Interaction

Sankar Rangarajan, Assegid Kidane, Gang Qian, Stjepan Rajko, and David Birchfield

Arizona State University, USA
{Sankaranaraya.Rangarajan, Assegid.Kidane, Gang.Qian,
Srajko, Dbirchfield}@asu.edu

Abstract. This paper addresses the design of a large area, high resolution, networked pressure sensing floor with primary application in movement-based human-computer interaction (M-HCI). To meet the sensing needs of an M-HCI system, several design challenges need to be overcome. Firstly, high frame rate and low latency are required to ensure real-time human computer interaction, even in the presence of large sensing area (for unconstrained movement in the capture space) and high resolution (to support detailed analysis of pressure patterns). The optimization of floor system frame rate and latency is a challenge. Secondly, in many cases of M-HCI there are only a small number of subjects on the floor and a large portion of the floor is not active. Proper data compression for efficient data transmission is also a challenge. Thirdly, locations of disjoint active floor regions are useful features in many M-HCI applications. Reliable clustering and tracking of active disjoint floor regions poses as a challenge. Finally, to allow M-HCI using multiple communication channels, such as gesture, pose and pressure distributions, the pressure sensing floor needs to be integrable with other sensing modalities to create a smart multimodal environment. Fast and accurate alignment of floor sensing data in space and time with other sensing modalities is another challenge. In our research, we fully addressed the above challenges. The pressure sensing floor we developed has a sensing area of about 180 square feet, with a sensor resolution of 6.25 sensels/in^2. The system frame rate is up to 43 Hz with average latency of 25 ms. A simple but efficient data compression scheme is in place. We have also developed a robust clustering and tracking procedure for disjoint active floor regions using the mean-shift algorithm. The pressure sensing floor can be seamlessly integrated with a marker based motion capture system with accurate temporal and spatial alignment. Furthermore, the modular and scalable structure of the sensor floor allows for easy installation to real rooms of irregular shape. The pressure sensing floor system described in this paper forms an important stepping stone towards the creation of a smart environment with context aware data processing algorithms which finds extensive applications beyond M-HCI, e.g. diagnosing gait pathologies and evaluation of treatment.

1 Introduction

Movement-based human-computer interaction (M-HCI) systems are receiving increasing attention recently due to their immediate applications in a number of areas

G. Kortuem et al. (Eds.): EuroSSC 2007, LNCS 4793, pp. 46–61, 2007.

with significant impact in our daily lives, e.g., biomedical (real-time monitoring of patient rehabilitation and providing guidance), culture and arts (interactive dance performances), education (encouraging collaborative and embodied learning through real-time visual and audio feedback based on the movement of the users). M-HCI systems read and respond to the movement of the user. In order to enable M-HCI system to understand the user's movement robustly and accurately, it is important to augment the user's environment with novel sensors, paving way for enhanced human interaction with computers on the basis of position, static poses, dynamics gestures and movement qualities. Pressure sensing plays a vital role in M-HCI systems. Every movement has certain motivation driven physical effort attached. Pressure sensing systems aid to understand and comprehend the nuances of such a physical effort thereby exploring the inherent nature of the human body as a powerful communication medium. Pressure sensing systems aiming at such M-HCI applications require a large sensing area (for unconstrained human movement), high sensor densities (for detailed and accurate representation of interacting objects), high frame rate and small sensing latency (for real time application), modular, scalable and portable design (for easy reconfiguration to suit external environments) and lastly integrability with other sensing modalities (for multimodal HCI).

In related prior work, various pressure sensing systems had been developed to capture and view pressure information associated with human movement across a floor. A detailed performance comparison study of those existing pressure sensing systems in terms of the desired features are listed in Table 1.

Table 1. Performance comparison table of existing pressure sensing floor systems

Sensor System	University	Year	Sensing method	Sensing area (Sq.feet)	Frame rate (Hertz)	Sensor density (sensor/sq.inch)	Data resolution (Number of bits)	Integrability	Modular	Portable
MIT Magic Carpet [1]	MIT Media Labs	1997	Piezoelectric wires	60	60	0.06	8	Yes	No	Yes
LiteFoot [2]	University of Limerick Ireland	1997	Optical proximity sensors	42.25	100	0.3	NA	No	No	No
ORL Active floor [3]	Oracle Research Lab	1997	Load cells	10.76	500	0.01	16	No	No	No
High resolution pressure sensor [4] distributed floor	University of Tokyo, Japan	2002	Binary switch	43	15	10.57	1	No	Yes	No
Z-Tiles [5]	University of Limerick Ireland & MIT Media Lab	2004	FSR	NA	100	0.5	12	No	Yes	Yes
Floor Sensor system [6]	University of Southampton, UK	2005	Binary switch	15.68	22	1.3	1	No	No	No
AME Floor I [7]	Arizona State University 2004-05	2004-05	FSR	9	10	0.44	8	Yes	No	No
AME Floor II [8][9]	Arizona State University 2005-06	2005-06	FSR	60	33	6.25	8	Yes	Yes	No
PROPOSED PRESSURE FLOOR			FSR	180	43	6.25	8	Yes	Yes	No

Color coding

1

2

3

NA - NOT AVAILABLE

The ranking in each dimension (column) is color-coded such that the best system is in dark green, the second best in lighter green, and the third in very light green. MIT

Magic Carpet [1] and LiteFoot [2] had fairly large sensing area and frame rate but were limited by poor sensor densities. ORL active floor [3] used load cells which lack the capability of detailed pressure measurement and cannot be used for applications requiring high sensor densities. High resolution pressure sensor distributed floor [4] has the best sensor density so far but was a binary floor (poor data resolution) that just detects presence or absence of pressure and does not give any measurement of pressure values on an analog scale. Z-tiles floor space [5] utilized a modular design, had high frame rate and data resolution but again suffers from low sensor density. Floor sensor system [6] is a low cost design but again a binary floor with poor data resolution. Also most of the sensing systems except [1] were stand alone systems and lacked the capability to be integrated in a multimodal environment which is vital requirement for our application. In-shoe sensors [10] have also been considered for force and pressure measurements but they have a limited scope of foot pressure measurement only. Also the in-shoe systems tend to alter the subject's actual pressure application due to their tendency to alter foot orientations by close contact.

It is quite obvious that all the sensing systems listed above have at least one serious limitation rendering it unsuitable to meet our application goals. It is worth mentioning that two generations of pressure sensing floor systems were developed with very similar goals as ours at the Arts, Media and Engineering (AME) Program at Arizona State University, namely, AME Floor I [7] and AME Floor II [8, 9] listed at the bottom of the table. It is apparent from the comparison table that the second generation did see pronounced feature improvements over the first generation. AME floor I [7] was a smaller prototype floor with 256 force sensing resistors arranged in less dense sensor matrix. During tests [7] it was found that there were large zones of no pressure detection during several activities. Also the scan rate was low deeming it unsuitable for real time M-HCI applications. These shortcomings were addressed by AME floor II [8, 9] with high sensor densities and high frame rate. Although AME Floor II [8, 9] showed significant advances and extended capabilities over AME floor I [7], it covered only a fraction of the sensing area required for our application, showed high sensing latency and lacked user friendliness. Also it showed preliminary multimodal integrable capabilities in temporal domain only and not spatial domain.

To fully address these issues, we have developed our own ingenious and in-house pressure sensing floor system PROPOSED in this paper and listed in the last row of table 1. Our proposed floor system is characterized by large sensing area, higher frame rate, smaller latency, enhanced user friendliness, spatial and temporal integrability with other sensing modalities in a multimodal environment, modular/scalable design etc. and thereby matching our ideal pressure sensing demands for real time M-HCI application. Comparison with other systems reveals that our proposed system in this paper ranks among the top 3 in most of the dimensions of the performance criteria. Although there are four systems with frame rates higher than that of our system, the sensing area and sensor resolutions of these systems are much lower than our system.

In this paper, we present system level description of our proposed pressure sensing floor followed by a discussion on hardware and software developments. Then we discuss the design methodologies for integration of the floor system with the marker based motion capture system as a first step towards the creation of a smart multimodal environment. The paper finally concludes with two interesting applications that we are currently exploring upon using such a powerful multimodal sensing set-up.

2 Pressure Sensing Floor Overview

The pressure sensing floor system consists of 96 networked pressure sensing units arranged in a rectangular matrix of 12 rows x 8 columns (shown in Fig. 1) spanning a total sensing area of 180 square feet. Each unit consists of a pressure sensing mat and associated supporting floor hardware. Each mat is in the size of 19"x17" embedded with a sensel array of 48x42 = 2016 sensels, resulting in a sensel resolution of 6.25 sensels/in^2. Force Sensing Resistors (FSR's) are used as sensel elements on the mat having an active area of 6mm x 6mm and made using pressure sensitive polymer between conductive traces on sheets of Mylar. The resistance of FSR is of the order of Mega ohms in the absence of pressure and drops to few kilo ohms when pressure is applied. Such a modular, scalable and networked architecture makes the floor readily reconfigurable and suitable for installations such as hallways, walk paths, and even spaces of irregular shape. We have estimated the cost of our 'one mat' system to be $600 and we can easily build smaller floor with fewer mats (as required by the application) at very low costs.

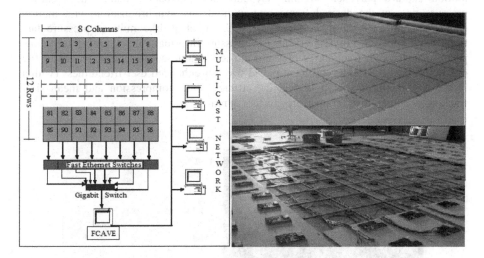

Fig. 1. Floor System overview and related network architecture (Left). Top view of the floor (Top right). Skeletal view of the floor (bottom right).

The floor hardware [9] comprises of microcontroller, multiplexers, A/D converter and Ethernet enabled rabbit controller which are all wired together on a processing and control board. The block diagram of the floor hardware is shown in Fig. 5. The microcontroller scan routine generates the timing and control signals for all hardware components to coordinate and sequence their operation. By optimizing microcontroller firmware, our pressure sensing system is made to run up to 43 Hz with average latency of 25 milliseconds. The pressure sensors of single mat are multiplexed to read the analog pressure values and then fed to A/D converter to produce 8 bit digital pressure data. The rabbit controller collects the digital pressure data of all sensors on one mat to produce a single mat packet and then transmits the packet over the network. All the pressure sensing units are assigned static IP addresses and they form a local private network. The data output of the rabbit controllers travels through two layers of network switches to the host computer. The power and sync clock are daisy chained and distributed to all the mats on floor. The pressure sensing floor is synchronized with the motion capture system by an external sync clock.

Floor Control And Visualization Engine (FCAVE) software has been developed at the host computer for floor control and visualization. FCAVE has an interactive graphical user interface (GUI) with various control buttons and indicators (shown in Fig. 2) and it is programmed to respond dynamically to user input. This software receives the raw pressure data packet for each mat separately, assembles the data of all 96 mats, assigns an incremental frame number and creates floor data frame which is then ready for further processing. FCAVE software has two operating modes namely 'live mode' and 'playback mode'. As the name implies, real time data collection and processing is done in the 'live mode' whereas offline data processing from a recorded pressure data file is usually done in the 'playback mode'. Furthermore playback can be done in synchronous and asynchronous ways. Synchronous playback streams the recorded pressure data synchronous with the motion capture playback stream. Asynchronous playback streams the recorded pressure data at the desired frame rate without any synchronization with motion capture system. FCAVE also offers various other controls like multicast pressure data to users on network, grayscale display of pressure information, set noise filter value, perform mean shift tracking of pressure clusters ,frame counter reset, record to file etc. FCAVE Software development paved way for enhanced user-friendliness (with a lot of GUI features shown in Fig. 2), efficient data compression and mean shift tracking of active, disjoint pressure clusters in real time.

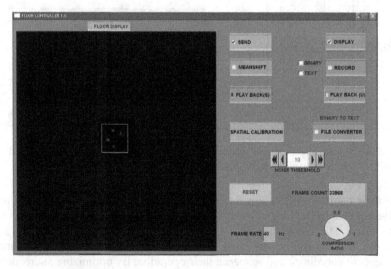

Fig. 2. Interactive Graphical User Interface (GUI) of the FCAVE software. The square box on the left highlights the pressure footprints.

3 Optimization of System Latency and Frame Rate

Small latency is critical for real time sensing systems used in M-HCI applications. Latency is defined as the time lag between the time instant of the true event and the time instant the pressure data pertaining to the true event arrives at end users on a multicast network. Labview software application and National Instruments (NI-DAQ 6020E) data acquisition hardware are used to measure and quantify the overall latency and the latency at each and every point along the data path.

The overall system latency is the sum of two components namely intrinsic latency and extrinsic latency. Intrinsic latency is defined as the latency induced by the sensor scanning process. Each sensing unit has a pressure mat with 2016 sensors and an associated hardware control board for pressure data collection and signal conditioning. All sensors are scanned sequentially from sensor 1 to sensor 2016 to read the pressure values. There is an inherent delay for the scanning process to complete and pressure packet to be produced. This delay is called as the intrinsic latency which is present due to lag in various hardware components on the hardware control board. The microcontroller generates the sensor scan signals and the scan routine incorporates all the hardware component delays. Thus total execution time of the microcontroller scan routine T_{scan} determines the frame rate F ($F = 1/ T_{scan}$) of the system. After a complete mat scan of 2016 sensors, the pressure data packet for that mat is produced. Extrinsic latency is defined as the time taken for such a pressure data packet to reach the end users on the multicast network and it accounts for the network transmission delay and FCAVE software delay. Our main focus was intrinsic latency reduction since it constitutes a major portion of the system latency and directly impacts the frame rate of system. Due to sequential scanning process, the intrinsic latency is direct function of the active sensor location given by a sensor address (An active sensor would be one that has pressure applied on it and sensors are addressed sequentially from 1 to 2016). A mathematical relationship is first established which gives an expected range of the intrinsic latency values based on the system scan rate and active sensor location. From this theoretical model, we know what latency distribution to expect when pressure is applied on a particular sensor location and later we did latency experiments to verify the same. The following section presents the mathematical relationship between intrinsic latency, frame rate and active sensor location.

3.1 Intrinsic Latency, Frame Rate, and Active Sensor Location

Let's assume that the system is running at a frame rate F and the time taken for one complete scan cycle of N sensors ($N = 2016$ in our case) is T_{scan}. We define pressure sensors applied with active load as *active sensors*. Let L be the address of such an active sensor. We are interested in finding the intrinsic latency related to this sensor at L. Let U be the address of the sensor currently being scanned at the time instant when the pressure application occurs on sensor L. Let X_L and X_U be time elapsed since the start of the scan until the sensor L and sensor U are reached respectively by the sequential scan routine, i.e.

$$X_L = \frac{1}{N}L \times T_{scan} \,, \ X_U = \frac{1}{N}U \times T_{scan} \tag{1}$$

According to the relationship between X_U and X_L, there are two different cases to be considered which are pictorially represented in Fig. 3.

- Case 1: $X_U \leq X_L$, pressure applied on sensel L is registered in the current scanning cycle.
- Case 2: $X_U > X_L$, pressure applied on sensel L is registered in the next scanning cycle.

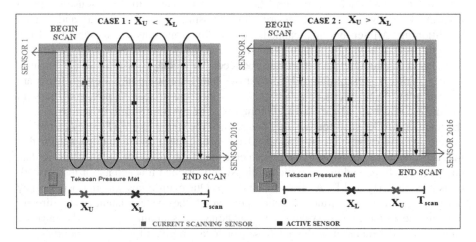

Fig. 3. Sequential mat scan process and depiction of Case 1 and Case 2

Hence, given L, the intrinsic latency τ caused by system scan is a function of X_U,

$$\tau(X_U) = \begin{cases} T_{scan} - X_U \,, \text{ when } 0 \leq X_U \leq X_L \\ 2T_{scan} - X_U \,, \text{ when } X_L < X_U < T_{scan} \end{cases} \tag{2}$$

Since X_U assumes a uniform distribution in $[0, T_{scan}]$ it can be easily shown that τ is uniformed distributed in the range given below:

$$T_{scan} - X_L < \tau \leq 2T_{scan} - X_L \tag{3}$$

Therefore, the mean intrinsic latency for the sensel at L is given by

$$\tau_m = 1.5T_{scan} - X_L \tag{4}$$

Thus the mean intrinsic latency is a direct function of T_{scan} and active sensor location X_L. Furthermore, since L can also be treated as a uniform random variable between 1 and N, the mean average intrinsic latency of all sensels on a mat is given by

$$E\{\tau_m\} = 1.5\,T_{scan} - E\,\{X_L\} = T_{scan} \tag{5}$$

Latency experiments have been conducted to verify the theoretical model derived above. Pressure is applied on a set of fixed sensor locations on the mat and the mean system latency is computed for 100 trials. Fig. 4 indicates the correlation between the theoretical and practical data sets when the system is running at 40Hz. The systematic offset between the theoretical and practical data sets is due to limited accuracy of experimental measurement of latency in milliseconds.

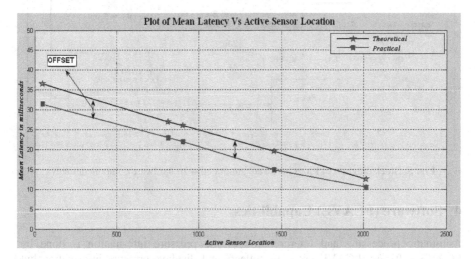

Fig. 4. Plot of mean latency vs. active sensor location

It is apparent from equation (5) that we can minimize intrinsic latency by minimizing T_{scan}, or equivalently maximizing the frame rate ($F = 1/T_{scan}$). Hence efforts were invested to increase the frame rate and reduce intrinsic latency which is described in the following section.

3.2 Maximization of Frame Rate

Frame rate of floor system is determined by the speed of hardware components on the hardware control board. Every hardware component has certain delay or lag associated with it. The microcontroller scan routine incorporates all the hardware component delays and accordingly generates the control signals. The sum of all hardware component delays gives minimum T_{scan} required whose reciprocal gives the maximum achievable frame rate. Fig. 5 shows the block diagram of floor hardware annotated with delay values for each hardware component explaining how we had achieved a maximum frame rate of 43 Hz in our proposed floor system from an old value of 33 Hz in AME Floor II (our precursor work). It is important to note that suffix (II) on Fig. 5 refers to AME Floor II whereas suffix (P) refers to floor system proposed in the paper. The block diagram quantifies the time savings obtained on each hardware component in the proposed system relative to AME Floor II. These time savings and hence increase in frame rate are obtained by choosing high performance hardware components and doing a more refined timing analysis on each component to determine their operational delay.

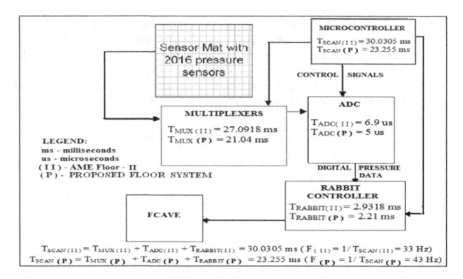

Fig. 5. Block diagram of Floor hardware annotated with hardware component delays

4   Software (FCAVE) Capabilities

Automatic processing and interpretation of the pressure information is done by FCAVE software. FCAVE not only collects and displays pressure information but also possess the capability of doing data compression and person location tracking by mean shift algorithm.

4.1 Data Compression

Each pressure mat has 2016 sensors and each sensor in turn sends one byte of pressure data at 43 Hz. Thus each mat data packet size adds to 2017 bytes which includes 2016 bytes of pressure data and one byte of frame number. The data volume from the entire floor comprising of 96 mats is a whopping 8.4 MB/sec. Usually, except a small area where the subject is in contact with the pressure sensing floor, most of the sensors do not have any load acting on them. Consequently a large proportion of the sensor data are null values of pressure or noise serving no interest to applications. Also there has been slight random noise observed in few sensors because of the nature of the sensing material which reports small values of pressure. Hence a simple but elegant compression algorithm is implemented by FCAVE to filter out all pressure values below the chosen noise threshold and pack only "active" sensor values and their addresses (location on floor system matrix) to be sent out to the end users on the network. Compression ratio as high as 0.9 is observed under normal case with five subjects which proves significant data volume reduction on the network.

It is known that compressed data packet comprise of only active sensor values and their address whereas the uncompressed data packet comprise of all sensor values (arranged in a sequence) and no address information since its address is implied by its location in the data packet. Thus the compression algorithm adds an additional overhead of sensor address which works well for low user activity with less active sensors. However as the user activity on the floor increase or when large numbers of sensors are active, the packet size also grows and a point is reached when compressed data volume exceeds uncompressed data volume. It is determined that this breakeven point is generally high and beyond bounds for normal usage. However, we are currently working on a dynamic compression scheme whereby the system is context aware and detects the extent of user activity and makes a decision whether to do compression or not.

4.2 Mean Shift Tracking of Pressure Clusters

Context awareness is the vital part of any smart environment. Perceiving context means sensing the state of the environment and users and it can be done with regard to a person or an activity. This may involve a variety of tasks such as person recognition, person location tracking, activity detection, activity recognition, activity learning etc. The primary step to accomplish the above tasks is to develop an efficient tracking procedure that shall ascertain the person location on the floor and also shift in the pressure gradient. The latter may lead to the study of various pressure patterns tied to each and every user activity. A mean shift algorithm is used to achieve the above mentioned goal. Mean shift is a simple iterative procedure that shifts each pressure data point to the average of the pressure data points in the neighborhood.

4.2.1 Mean Shift: An Introduction

Mean shift is the process of repetitively shifting the center t to the sample mean. The sample mean of samples S under a kernel $K(x)$ centered at t, with sample weights $w(s)$, can be found using this equation:

$$m(t) = \frac{\sum_{s \in S} K(s-t)w(s)s}{\sum_{s \in S} K(s-t)w(s)} \tag{6}$$

where m(t) is the new sample mean [11]. In [12] it's proven that if the kernel $K(x)$ has a convex and monotonically decreasing profile $k(\|x\|^2)$, then the center t will converge onto a single point. The kernel used in our tracking algorithm is the truncated Gaussian kernel which is the combination of the flat kernel and Gaussian kernel. The truncated Gaussian kernel is given by

$$(F_\lambda G_\beta)(x) = \begin{cases} e^{-\beta\|x\|^2} & if \ \|x\| \le \lambda \\ 0 & if \ \|x\| > \lambda \end{cases} \tag{7}$$

where λ is the radius of the Gaussian kernel and β is the Gaussian kernel coefficient.

4.2.2 The Clustering/Tracking Algorithm

The algorithm is iterated for every frame of pressure data. Each and every frame of pressure data contains information about the location of pressure and value of pressure at that location. The pressure values constitute the weights and pressure location constitutes the data points that need to be iterated using the mean shift algorithm. The full algorithm for finding and tracking the pressure clusters is given below.

1) For the first frame of pressure data or new cluster formation, cluster centers and the data points are one and the same i.e. the center set T is the same as the data set S, and both evolve with each iteration using the mean shift formula in equation (6) and truncated Gaussian in equation (7). Data points are clustered through the blurring process [11] using the observed pressure data as the weight used in (6). Once the process has converged, the data set will be tightly packed into clusters, with all of the data points located closely to the center of that cluster. (The process is said to the converged either after the maximum number of iterations defined by the algorithm or earlier when the mean shift of centers becomes less than the convergence threshold) After convergence, each cluster has a 'center' and 'label' associated with it. All data points not associated with any cluster center are classified as or orphan pressure points.

2) For every subsequent pressure data frame, centers from the previous frame are updated through the mean shift algorithm (6) using current observed pressure values as weights and checked for convergence. In practice, entirely new data points resulting in new cluster centers (new labels) can occur which is computed in step (3).

3) Calculate the number of orphan pressure points. If the number of orphan pressure points exceeds a chosen threshold then repeat step (1) to find new cluster centers. Orphan pressure points fewer than the chosen threshold are discarded.

4) Perform mean shift using the new set of cluster centers (repeat steps 2 & 3).

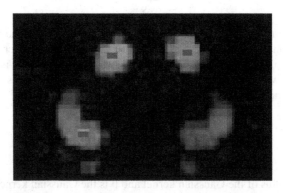

Fig. 6. Snapshot showing clustering and tracking by mean shift on left foot and right foot. Two pressure clusters are formed for each foot (one for heel and one for toe) and cluster centers are depicted by red dots.

5 System Integration for Multimodal Sensing

Multimodal systems have always proved to be robust and effective than independent uni-modal systems because it provides wide varieties of information for better realization and assimilation of the subject movement in capture space and also allows the users to interact multimodally with the system. They provide high redundancy of content information which leads to high reliability. After the completion of the pressure sensing floor, efforts have been put in to integrate the floor with the motion capture system to create a smart environment.

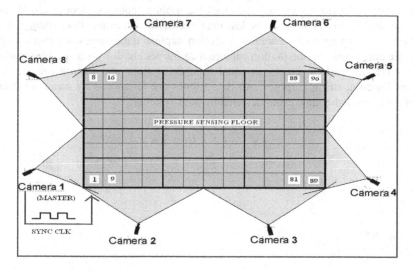

Fig. 7. Multimodal sensing set up of floor and motion capture system

A common capture volume (12' by 15') is first created within the sensing capabilities of the floor and motion capture system. The motion capture cameras are arranged around a capture volume and the floor forms a part of the capture volume as shown in Fig. 7. The location of the floor with respect to the coverage area of the cameras is important when pressure data about some movement needs to be interpreted with the marker. The pressure floor and motion capture system are integrated with respect to time and spatial domains. A subject moving in the capture space is sensed by both systems and they give information about the location and activity of the subject. Motion capture data contains the 3D location coordinates of the markers in physical space whereas the pressure data contains the pressure values and 2D location. Both sensing systems have independent coordinate set and hence spatial alignment by means of coordinate transformation becomes essential to ascertain the location of the subject in common capture space. Also any activity done by the subject is being detected by both systems simultaneously and hence both sensing modalities must operate synchronously. Thus time synchronization and spatial alignment are critical for two data sets to be highly correlated to ensure holistic inference.

5.1 Time Synchronization

Time synchronization of the floor and motion capture system is achieved by means of a common sync clock. This sync clock is generated by the master camera of the motion capture system and is used to trigger the scan of the floor. This sync clock is also used by the motion capture system to control the camera shutters. In this way, the scan of the floor and the camera image capture can be synchronized in time domain. Also the motion capture system is always set to run at multiples of the floor frequency. The common sync clock runs at the frequency of the motion capture system and that clock is down sampled by a factor to generate the scan frequency (or frame rate) of the floor. Currently we clock the motion capture system at 120 Hz and the floor at 40 Hz (which should be less than maximum achievable floor frame rate of 43 Hz). This arrangement generates 3 motion capture data frames for every single pressure date frame. So the motion capture data frames are down-sampled (redundant frames are dropped) to create an equal number of floor and motion capture data frames for comparison purposes. All data frames are referenced by means of frame numbers to track the same event detected by both systems.

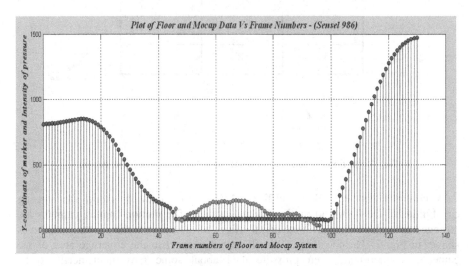

Fig. 8. Plot of Pressure and marker data v.s. frame numbers

A simple experiment is conducted to test the time synchronization of the floor and motion capture system. A mallet with a single marker on its head is banged on a single pressure sensor of the floor from a fixed height. The vertical coordinate (Y- coordinate) of the marker and pressure value on that sensor are monitored over time. Ideally the pressure sensor value should peak when the marker coordinate is at the lowest position (ground level). Fig. 8 gives the time-sampled plot of the sensor pressure value (green dots) and marker height (pink dots). The results obtained agree with our expectation thereby demonstrating a perfect frame alignment between the floor and motion capture system.

5.2 Spatial Alignment

The floor coordinate system is a two dimensional system in sensor units whereas the motion capture coordinate system is a three dimensional system in mm units. Hence it is essential to implement coordinate transformation between the floor and motion capture system so that we can view the events in one coordinate space for ease of inference and visualization. A spatial calibration procedure is in order to align the floor and motion capture system in physical space. Firstly the motion capture system is calibrated and stabilized. Three reflective markers are placed on the edge of the floor in order to get the boundary co-ordinates of the floor in motion capture coordinate space. Using this information, we compute three co-ordinate transfor-mation parameters namely rotation, translation and scaling. These parameters constitute the coordinate transformation matrix which is then applied to each and every floor coordinate to get the respective coordinate in the motion capture system. The converse also can be computed to view the data in the floor coordinate space alone. Spatial alignment computations are done by FCAVE software in real time.

6 Applications in Multimodal Movement Sensing and Analysis

6.1 Balance Analysis

Falling is one of the major health concerns for elderly people and incidence of falls is high for persons aged over 75. Hence an efficient fall detection system is necessary to detect potential situations of fall and signal the user of an impending fall or alert for assistance after the person is immobilized by fall. The state of body balance is the feature of interest in fall detection systems. We have collected data of different on-balance and off-balance body postures and currently evaluating on a fall detection algorithm. The state of body balance is characterized by center of gravity (COG) and center of pressure (COP). COG is computed from the motion capture data by assigning weight to each marker and computing the weighted mean. If the weight of each marker represents the weight of the body mass around that marker, the weighted mean is a good approximation of the center of gravity. Similarly the COP is the weighted mean of all the pressure data points. The subject's overall state of balance is determined by the relative positions of the COG and COP. If the COG is directly above the COP, the subject is in a state of balance. In other words, the subject is in a state of perfect balance when the projection of the COP and COG on a horizontal plane superimposes each other. As COP and COG moves away from each other, the subject slowly transitions into a state of off-balance. Thus it is obvious that time synchronization and spatial alignment of both sensing systems are critical for such an exercise. Since feelings of balance are visceral in human beings, such a quantitative approach paves way to tie the behavior of the system to a sensation/feeling that is very internal and apparent to the user and thereby complementing human computer interaction.

6.2 Gesture Recognition

This multimodal sensing system has also been used to drive a gesture recognition system that uses both kinematics and floor pressure distribution to recognize gestures.

Such a gesture recognition system can distinguish gestures that have similar body shapes but have different body weight distributions as shown in Fig. 9. These two gestures are recognized as one and the same by marker based motion capture system due to similar body shape. Hence pressure sensing becomes vital to distinguish between such gestures. The ability of the gesture recognition system to read and analyze both body kinematics and pressure distributions encourages users to communicate with computers in expressive ways.

Fig. 9. Snapshots of two gestures with similar body shape but different weight distribution

7 Conclusions and Future Work

We have successfully designed, developed and deployed a pressure sensing floor system with a higher frame rate, less latency, high sensor resolution, large sensing area that can provide us with real time data about the location and amount of pressure exerted on the floor. The floor has been integrated and synchronized with the marker based motion capture system to create a smart environment for M-HCI application. Our present direction is towards extending the context aware capabilities of the floor system. We are currently working on improving the fall detection algorithm by collecting data for various on-balance and off-balance body postures and analyzing them. We are also working on an algorithm to distinguish between the left foot/right foot and heel/toe on the basis of shape. Shape descriptors such Fourier, Hu moments come in handy for such an analysis. Such intelligence to the floor to recognize and distinguish the left /right foot and heel/toe paves way for recognizing gestures with varying foot contact. Also most of the gait pathologies are reflected by abnormal pressure patterns localized to either the toe or heel. Hence results of the above work could find extensive usage in gesture recognition, video gaming, rehabilitation work etc.

References

[1] Paradiso, J., Abler, C., Hsiao, K., Reynolds, M.: The Magic Carpet: Physical Sensing for Immersive Environments, Ext. Abstracts CHI 1997, pp. 277–278. ACM Press, New York (1997)
[2] Griffith, N., Fernström, M.: LiteFoot: A floor space for recording dance and controlling media. In: Proceedings of the 1998 International Computer Music Conference, International Computer Music Association, San Francisco, U.S.A., pp. 475–481 (1998)

[3] Addlesee, M., Jones, A., Livesey, F., Samaria, F.: The ORL Active Floor. IEEE Personal Communications 4(5), 35–41 (1997)

[4] Morishita, H., Fukui, R., Sata, T.: High Resolution Pressure Sensor Distributed Floor for Future Human-Robot Symbiosis Environments. In: Intl. Conference on Intelligent Robots and Systems, Switzerland, IEEE, Los Alamitos (2002) (Ref: 0-7803-7398-7/02 @ 2002)

[5] Richardson, B., Leydon, K., Fernström, M., Paradiso, J.: Z-Tiles: building blocks for modular, pressure-sensing floorspaces. In: Extended Abstracts of the 2004 conference on Human factors and computing systems, Vienna, Austria, pp. 1529–1532 (2004)

[6] Middleton, L., Bus, A.A., Bazin, A.I., Nixon, M.S.: Floor Sensor system for gait recognition, University of Southampton. In: UK Fourth IEEE workshop on Automatic Identification Advanced Technologies (AutoID 2005), pp. 171–176 (2005)

[7] Kidané, A., Rodriguez, A., Cifdaloz, O., Harikrishnan, V.: ISAfloor: A high resolution floor sensor with 3D visualization and multimedia interface capability. AME Program, AME-TR-2003-11p (2003)

[8] Srinivasan, P., Birchfield, D., Qian, G., Kidané, A.: A Pressure Sensing Floor for Interactive Media Applications. In: Proc. of ACM SIGCHI International Conference on Advances in Computer Entertainment Technology (ACE), Valencia, Spain, pp. 278–281 (June 2005)

[9] Srinivasan, P.: Design of a Large area pressure sensing floor. A thesis presented for the requirements Master of Science Degree, Arizona State University (May 2006)

[10] Paradiso, J., Hsiao, K., Benbasat, A., Teegarden, Z.: Design and Implementation of Expressive Footwear. IBM Systems Journal 39(3 & 4), 511–552 (2000)

[11] Cheng, Y.: Mean Shift, Mode Seeking and Clustering. IEEE Transactions on Pattern Analysis and Machine Intelligence 17(8), 790–799 (1995)

[12] Comaniciu, D., Ramesh, V., Meer, P.: Real-Time Tracking of Non-Rigid Objects using Mean Shift. In: IEEE Conf. Computer Vision and Pattern Recognition (CVPR 2000), Hilton Head Island, South Carolina, vol. 2, pp. 142–149 (2000)

Sensing Motion Using Spectral and Spatial Analysis of WLAN RSSI

Kavitha Muthukrishnan, Maria Lijding, Nirvana Meratnia, and Paul Havinga

University of Twente, Faculty of Computer Science
Computer Architecture Design and Test for Embedded Systems group
P.O. Box 217, 7500 AE, Enschede, The Netherlands
{k.muthukrishnan,m.e.m.lijding,n.meratnia,
p.j.m.havinga}@ewi.utwente.nl

Abstract. In this paper we present how motion sensing can be obtained just by observing the WLAN radio signal strength and its fluctuations. The temporal, spectral and spatial characteristics of WLAN signal are analyzed. Our analysis confirms our claim that 'signal strength from access points appear to jump around more vigorously when the device is moving compared to when it is still and the number of detectable access points vary considerably while the user is on the move'. Using this observation, we present a novel motion detection algorithm, *Spectrally Spread Motion Detection (SpecSMD)* based on the spectral analysis of WLAN signal's RSSI. To benchmark the proposed algorithm, we used *Spatially Spread Motion Detection (SpatSMD)*, which is inspired by the recent work of Sohn et al. Both algorithms were evaluated by carrying out extensive measurements in a diverse set of conditions (indoors in different buildings and outdoors - city center, parking lot, university campus etc.,) and tested against the same data sets. The 94% average classification accuracy of the proposed *SpecSMD* is outperforming the accuracy of *SpatSMD* (accuracy 87%). The motion detection algorithms presented in this paper provide ubiquitous methods for deriving the state of the user. The algorithms can be implemented and run on a commodity device with WLAN capability without the need of any additional hardware support.

1   Introduction

Ubiquitous computing is emerging as an exciting new paradigm with a goal to provide services anytime anywhere. Context is a critical parameter of ubiquitous computing. The ubiquitous computing applications make use of several technologies to infer different types of user context. The context cue that we are interested in is users motion being either *'moving'* or *'still'*. Knowledge about whether a user or device is still or in motion can serve a number of purposes. The mobility status is an important part of a users context and is thus of interest to context-aware systems, e.g. [13]. Knowing the users motion is very useful for activity recognition and localization [11].

It is possible to use specialized sensors and beacons, such as accelerometer, pedometer or motion sensor [2] to measure certain aspects of users motion. Although they offer precise motion inference, they are too expensive and obtrusive for wide deployments. Specialised motion tracking and motion detection systems make use of a range

G. Kortuem et al. (Eds.): EuroSSC 2007, LNCS 4793, pp. 62–76, 2007.

of sensing technologies [14]. Although such dedicated systems typically provide highly accurate information concerning position and even orientation they require additional hardware, which is often unwieldy, impractical or simply not available. Hence, solutions which do not require a dedicated hardware for determining the state of the user are gaining popularity [3][1][8][9]. Particularly, there is a growing interest in using the existing infrastructure and technology to derive motion status.

In the last few years, reading emails and surfing the internet whilst on the move at public places, has become an accepted part of daily life with WLAN standard being integrated into laptop, handhelds and cellular phones. Location based services, using WiFi tags in settings like hospitals to track equipments and patients are currently the trend [13], [12] and localization using WiFi [12] [5] [7] is considered as a value added application to WiFi networks. In this paper, we provide -*Motion inference* as a useful mechanism which adds incremental value to WiFi networks. The main advantages of the proposed motion detection are: *(i)* deducing motion of the user using the existing infrastructure (*WiFi*) without a need of additional hardware, as it offers a pure software based solution *(ii)* preserving user privacy, as motion inference is performed locally at the client device.

1.1   Motivation

The motion detection algorithms that we present in this paper provide ubiquitous methods for deriving the state of the user. The algorithms can be implemented and run on a commodity device with WLAN capability without the need of any additional hardware support. The recognition of device/users motion status will lead to a number of applications, including personalized assistants, smart monitoring and surveillance systems, as well as motion analysis in sports and medical domain. In the section below we envision three use cases of the motion detection algorithm presented in this paper.

A. Enhancing WLAN localization accuracy. Applying the motion information to WLAN localization algorithm may result in accuracy improvement in the existing Wireless LAN positioning systems, such as RADAR [5], Flavour [7] and Ekahau [12]. These systems essentially use received signal strength typically measured at the mobile device to estimate location. In general, signal samples fluctuate even in environments with few moving objects. Thus resulting in an unstable location estimation even while the user or the device is still. In order to overcome errors due to instability in RSSI used for location estimation, most positioning algorithms use some kind of time averaging to smoothen the RSSI values. Inspite of time averaging the instability still exists, resulting in lesser accuracy. The performance and hence accuracy of signal strength-based localization can be improved if knowledge about the user state, i.e. moving or still is incorporated as a part of localization algorithm itself. As an example, when the user state is deduced as still, all the unwanted jumps caused by the location estimation can be ignored. When the user state is deduced as moving, an appropriate motion model can be added as a part of the location algorithm. Therefore after each location update the new location estimate can move by a maximum distance depending on user's speed. In this way, many false location estimations can be curtailed, leading to better accuracy.

B. Healthy living at Office. Office workers tend to lead a very sedentary life, thereby making them very prone to various diseases like Repetitive Strain Injury (RSI). Recently health centers such as, Rehabilitation Center Het Roessingh (RRD) [15] is promoting healthy living in the office environment with a help of providing personalized assistance, which determines the user's activity level and use it to give timely feedback, warnings or advice to the users. For instance, if the user is continuously sitting for a long time, the system can give feedback or encourage the user to move, if the user is already moving, it can keep a count of how much the person has moved and so forth. With just a mobile handheld, an employee would be able to gather information about his daily activity level and get appropriate feedbacks. Motion recognition forms vital part of such applications, by improving the condition of the employee at work.

C. Activity recognition. An impressive amount of research falls under the umbrella of activity recognition. Be it for elderly care support or for social coordination [8], motion detection forms an important part. The WLAN radio by itself can sense motion, and it can potentially be also part of the sensor ensemble to improve recognition performance. Adding machine learning algorithms might result in identifying more states, thus making it very useful for fine-grained activity recognition.

The applications that are described above do not necessarily benefit from accurate and complete information about the mobility status. For the purposes described above it is sufficient to know whether the user is moving or not. However, all the above mentioned applications requires *reliability*- low false alarm rate, *low latency*- inferring motion status without much delay, instantaneously at the best and *light computation*-to efficiently run on mobile hand held devices.

1.2 Contributions and Outline

The key contributions of our work are:

1. For the first time, the spectral characteristics of WLAN signals RSSI are analyzed. Our analysis results in a conclusion that "when a device is moving, signal strengths of all heard access points vary much greater and more obvious compared to when a device is still and the number of detectable access points vary considerably when the device is moving". This analysis is presented in Section 3.2.
2. We propose a novel motion detection algorithm called *Spectrally Spread Motion Detection (SpecSMD)*. In this algorithm, the WLAN signal in time series is transformed to frequency domain by employing Fast Fourier Transform. By analyzing the spectral width of the fourier transformed signal, a two-state classification scheme is used to deduce if a user is moving or still. We discuss the algorithm in Section 4.1. The experimental results are outperforming the existing motion detection algorithms in use today by achieving an overall classification accuracy of approximately 94.1%. The evaluation of *SpecSMD* is addressed in Section 5.2.
3. Additionally, we present *Spatially Spread Motion Detection (SpatSMD)* which is inspired by [8]. We discuss it in Section 4.2. *SpatSMD* achieves an average classification accuracy of 86.7% in deducing the correct motion status.

Section 2 presents the related work. We outline the data collection process used in gathering the WLAN signal traces in Section 5.1. Section 6 concludes the paper and presents ongoing work.

2   Related Work

It is possible to infer the state of the user by the usage of external hardware such as an accelerometer. Randell et al. [2] demonstrated the possibility of distinguishing various states of the movement such as walking, climbing and running using a 2D accelerometer. Patterson et al. [9] take the velocity readings from GPS measurements and infer the transportation mode of the user, for instance walking, driving, or taking a bus using a learning model. The model learns the traveler's current mode of transportation as well as his most likely route, in an unsupervised manner. It is implemented using particle filters and is learned using Expectation-Maximization. The learned model can predict mode transitions, such as boarding a bus at one location and disembarking at another. The research now is focusing on finding a methodology to obtain the same information using the existing infrastructure or capability of the device.

A promising alternative to usage of specialized hardware is to investigate what can be obtained by measuring signal received from existing infrastructures (either WLAN or GSM). Since WiFi access points and WiFi clients are ubiquitous, this technique is very attractive. In this line, Krumm et al. [3] classified a user as either moving or still based on the variance of a temporally short history of signal strength from currently the strongest access point. This classification had many transitions, hence it was smoothened over the time with a two-state hidden markov model (HMM) resulting in an overall accuracy of 87%. Privacy is enhanced compared to systems that compute context on a central server, since the context inferences rely only on client-side data and computations.

Anderson et al. [1] use GSM cellular signal strength levels and neighboring cell information to distinguish movement status. The classification of the signal patterns is performed using a neural network model resulted in an average classification accuracy of 80%. The authors trained the neural network initially and demonstrated a proof of concept by implementing it at run time on a cell phone. However the initial training did not work in all the environments as signal strength fluctuations were different in different environments.

Recently Sohn et al. [8] published a similar technique for detecting users motion using signal traces from GSM network. Their motion detection system yields an overall accuracy of 85%. They extracted a set of 7 features to classify the user state as either still, walking, or driving.

3   802.11 Radio Channel Characteristics

In this section, we begin with a description of our scanning process used to log WLAN signal readings for studying the radio channel characteristics. We then discuss the RF signal propagation and the noisy wireless channel characteristics. Subsequently we analyze the temporal, spectral and spatial characteristics of RSSI of the WLAN signal which forms the basis of the motion detection algorithm presented in this paper. To study the signal propagation and perform measurements, we used a HP IPAQ pocketPC with built-in WLAN card. We used a Spotter [6] [7] to capture the RSSI from each access point.

3.1 Scanning Process

The IEEE 802.11 standard defines a mechanism by which RF energy is to be measured by the circuitry on a wireless NIC. In 802.11b/g/a, this numeric value is an integer with an allowable range of $0 - |255|$ and called as the Received Signal Strength Indicator (RSSI). The IEEE 802.11 does not require that a chipset vendor use all 255 values, so each vendor will have a specific maximum value. For example, Cisco chooses RSSI-max as 100 while the atheros chipset use 60 as the maximum value. Thus, the *Spotter* [6] [7] measures a signal strength value between 0 and $|255|$ for each of the access points in the vicinity. Figure 1(B) shows an output example of *Spotter*. The spotter outputs a list of the MAC addresses of access points accompanied with the time-stamped signal strength observed in the scan (probe response frames). Any WLAN device using the *Spotter* can scan for nearby access points. Additionally the user can log which activity he was performing at the time of measurement and specify where the measurement was made (see Figure 1(A)). This phase is typically known as *Scanning phase*.

Tradeoff between Scanning rate vs battery-life. Continuous scanning and logging WLAN access points can expend a devices battery in few hours. While continuous scanning provides rich depiction of a users dynamic environment, most individuals expect their personal devices to have more standby times. Hence, for a longer battery life, the scanning interval should be as large as possible. However, with a very large scanning period, the motion updates are also much less frequent. Therefore we have used spotter to scan the environment once every 4 seconds.

3.2 Analyzing Temporal, Spectral and Spatial Characteristics of WLAN Signal

The IEEE 802.11b standard operates in the 2.4 GHz, Industrial-Scientific-Medical (ISM) band. Since the ISM band is open to anyone, radio systems operating in this band must cope with several unpredictable sources of interference, such as baby monitors, garage door openers, cordless phones, other radio such as bluetooth and microwave ovens (the strongest source of interference). There are two types of variations associated with the wireless channel [10], -*Small scale variations* and *Large scale variations*. Small scale variation are equivalent to the correlated fading experiences in the wireless channel for close sampling distances. This happens when the receiver moves over a very small distance. Handling small scale variations is very challenging. Large scale variation on the other hand describes the characteristics of the channels structure when the distance between the transmitter and the receiver increases considerably. The large scale variation is significant as over long distances the signals get attenuated. The indoor radio signal propagation is often subjected to reflection, refraction, diffraction and absorption by structures and even human bodies, thus signal propagation suffers from severe multi-path fading effects in an indoor environment [10]. As a result, a transmitted signal can reach the device through different paths, each having its own amplitude

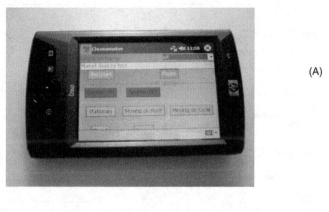

(A)

Measurment Fri May 25 12:14:44 CEST 2007   (B)

25-05-200712:14:45   149   Start Sequence:Building Waaier
25-05-200712:14:45   149   Motion:Moving
000136079de0,-90
000cf6164f6c,-90
00116b267fd8,-73
0001e3d43a8d,-53
0001e3da0a55,-90
00147f54a4ff,-74

Fig. 1. (A). Screen shot of WiFi Spotter Application, illustrating the data collection diary, (B). An example of Spotter output.

and phase. These different components combine and reproduce a distorted version of the original signal. Moreover, changes in the environmental conditions, such as temperature or humidity affect the signals to a large extent.

Temporal Characteristics. This subsection describes how the wireless channel changes over time when the user is standing still and moving. Figure 2(A) presents signal strength and its variations over time from all the heard access points. We measured (numerous measurements were performed at different settings (indoors/outdoors) and representative data is used for plotting) the signal strength a little over 2 minute time interval. From the figures it is obvious that even at a fixed position the signal strength varies, however the variation is not more than 10 - 15 dBm. This happens mainly because motion of people or equipments that are present in indoor environments make the radio channel nonstationary in time, that is the channel's statistics change even when the transmitter and receiver are fixed. But when the user is moving (refer Figure 2(B)) the signal variations can be as high as 45 - 60 dBm. Hence the variation in signal due to the motion of device has greater influence than the variation due to the dynamics of the changing environment, i.e. motion of people and objects. One can see from the figure that, when the user is still, the fading of the signal occurs as bursts lasting few seconds. However, for moving case the fading of signal is more and persistent. Furthermore, it is clear from Figure 3 that at a fixed location the access points that are detectable do not

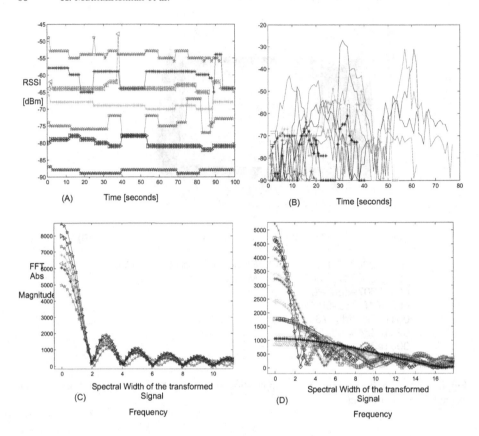

Fig. 2. (A). Temporal characteristics of signal when user is 'still', (B). Temporal characteristics of signal when user is 'moving', (C). Spectral characteristics of signal when user is 'still', (D). Spectral characteristics of signal when user is 'moving'.

vary too much, meaning the number of signal strength samples received from the same access point does not vary much. In the case of moving, the number of signal strength samples received from the access point varies as the number of access point detectable at a place varies as the user moves.

Spectral characteristics. Analysing the signals spectral characteristics might give a different representation. Hence for converting the WLAN signal in time series to frequency domain we used Fast Fourier Transform (FFT) which enables us to view the frequency representation of the signal. The FFT of N points x_n is defined as follows:

$$X_k = \sum_{n=0}^{N-1} x_n e^{-2\pi i k n/N} \quad k = 0, ... N - 1 \qquad (1)$$

where X_k is the kth coefficient of the FFT and x_n denotes the nth sample of the time series which consists of N samples and $i = \sqrt{-1}$. In our case, x_n is the RSSI time series obtained from the Spotter.

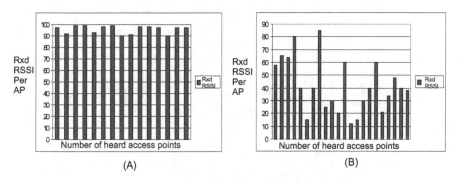

Fig. 3. Variation in the number of samples received when (A). Still (B). Moving.

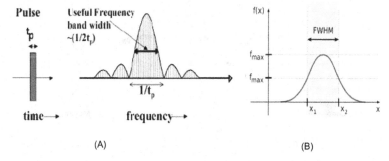

(A)   (B)

Fig. 4. (A). Schematic representation of a rectangular pulse in time and frequency domain, short duration pulses produces a large bandwidth , (B). Full Width Half Maximum, corresponding to peak width at 50% peak height.

The schematic representation of how a signal looks in time and frequency domain is illustrated in Figure 4(A). As a rule of thumb, the more concentrated the time domain, the more spread out the frequency domain. In particular, if we "squeeze" a function in time, it spreads out in frequency and vice-versa. Therefore, a function which equals its Fourier transform strikes a precise balance between being concentrated and being spread out. Also Figure 4(B) illustrates Full Width Half Maximum (FWHM) that corresponds to peak width of the FFT signal at 50% peak height.

Figure 2(C) & (D) represent the variation of signal strength from all the heard access points in the frequency domain when the user is still and when the user is moving. It is evident that although signal strength vary even while the user is still, this variation is reflected in all the heard access points uniformly as there is a well defined peak with a narrow spectral width in the frequency domain from all the access points, despite the fact that there is difference in the fourier amplitude from each of the heard access point. But when the user is moving, there is no well defined peak from all the access point in the frequency domain indicating that the variation in the signal strength happens more often and not in all the heard access points in the same manner. Furthermore, one can see the effect of spectral broadening from a significant number of access points when the user is moving, resulting in a higher full width half maximum. This phenomenon

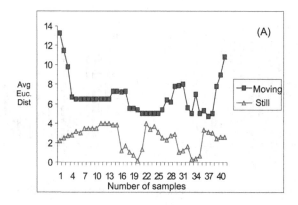

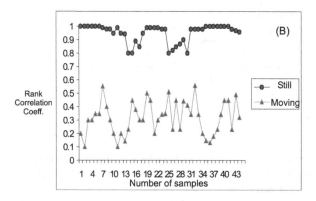

Fig. 5. (A). Spatial characteristics of signal, when user is 'still', and 'moving' (B). Rank correlation Coefficient when the user is still and moving.

happens mainly due to two reasons: *(i)* the variation in the signal strength is large in case of moving user and *(ii)* number of access points detectable vary with distance resulting in too few received samples from the access points. This confirms that both the temporal and spectral analysis lead to the similar conclusions but give a different view of representation. However the advantage of spectral analysis is that even when the number of samples are few, meaning that the received signals are captured only for an instance of time, it still exhibits the same behavior.

Spatial characteristics. Here we present our observation of the signal strength fluctuations reflected in the euclidean space. In particular we want to see: *(i)* whether any signal correlation for spatially seperated signals exists for the case when user is moving and *(ii)* how the signal correlates over time for the case when the user is still.

The observation is based on the same principle as fingerprinting-based location systems [5], which state that the signals observed from the access points are consistent in time but variable in euclidean space. Conceptually the Euclidean distance should give a cue on whether a user is moving or still. We conducted experiments to analyze the signal fluctuation and to see how it varies in space. We used Euclidean distance between WLAN measurements as a measure to find out if there is any difference between a moving and a still user. Figure 5(A) illustrates the average Euclidean distance between WLAN measurements and shows that the average Euclidean distance between WLAN measurements are proportional to the state of the movement. When the user is still, the Euclidean distance is relatively small (< 4), when the user is moving the Euclidean distance is higher (typically > 4). In a generalized form, the euclidean distance between any two measurements can be written as:

$$\sqrt{\sum_{i=1}^{n}(S_{i(X)} - S_{i(X+C)})^2} \tag{2}$$

where i is the index of the access point used, X is the index of measurement and C is a constant which can vary from 1 till 10 (refer to Section 4.2 for more details on choice of C).

We estimate rank correlation coefficient using the Spearman Rank Correlation Coefficient [4]. The rank correlation coefficient between any two measurements represent how closely the signals are ranked. It takes the value between 0 and 1. Ranking closer to 1 indicates that the measurements are similar and hence the user is still and when the user is moving the ranking is close to 0. Figure 5(B) presents how the rank correlation coefficient behaves when the user is still and moving.

With these observations presented in Section3.2 as the basis, we now present our motion detection algorithms in the next section.

4   Motion Detection Algorithms

4.1   Spectrally-Spread Mobility Detection (SpecSMD)

In this subsection we present a novel motion detection algorithm which is based on our observations explained in section 3.2. The algorithm tracks how many access points have spectral width that is exceeding a certain threshold within the window of readings. Intuitively a larger spread indicates that the device is moving.

Algorithm 1 presents the pseudocode of the Spectrally Spread Motion Detection algorithm *(SpecSMD)*. The input to the algorithm is the spotter output comprising of the WLAN readings. The WLAN signal reading in time series is converted into frequency domain using Fast Fourier Transformation (FFT). As a next step, the full width half maximum i.e. peak width of the FFT signal at 50% peak height is calculated for each of the MAC entries in the spotter output. We are interested in a two-state classifying scheme in which state 1 and state 2 are dealing with deducing still and moving states, respectively. To make a decision based on the full width half maximum, a threshold is set. Whenever a MAC entry exceeds the FWHM threshold, the algorithm treats this as an outlier and increments a counter. If the number of the counter is more than a certain threshold the algorithm returns the user state as Moving, otherwise it returns Still. The number of the counter indicates that many of the access points that are detected have a spectral width exceeding certain threshold. This algorithm is implemented in MATLAB

Algorithm 1. Pseudocode Spectrally Spread Motion Detection

```
Initialize Counter =0;
Initialize WindowSize;
for ENTRY= 1;ENTRY < TOTAL MAC ENTRIES;ENTRY++ do
   Compute FAST FOURIER TRANSFORM (USING EQUATION1)
   Calculate Full Width Half Maximum FWHM
   if FWHM> FWHMThreshold then
      Increment Counter;
   end if
end for
if COUNTER > COUNTERTHRESHOLD then
   return User is MOVING;
else
   return User is STILL;
end if
```

Algorithm 2. Pseudocode Spatially Spread Motion Detection

```
Initialize WindowSize;
for every pair of WLAN readings defined within the WindowSize do
   for MACENTRY=1;MACENTRY < END OF MAC ENTRIES do
      Compute the Euclidean Distance;
   end for
   Compute Mean Euclidean Distance (MeanED)
end for
if MEANED > MEANTHRES then
   return User is moving;
else
   return User is still;
end if
```

and currently performs offline analysis to classify the state of the user's motion. The fast fourier transformation results in a complexity of NLog(N), where N refers to the number of samples. We use smaller values of N (we used N=6) hence the computation load is negligible.

4.2 Spatially-Spread Motion Detection Algorithm (SpatSMD)

The algorithm presented in this subsection is based on our observation explained in section 3.2. We used the spotter to record the connected and the neighboring access points. The algorithm tracks the euclidean distance between the WLAN readings using a sliding window of WLAN readings, as defined by the window size. Intuitively larger euclidean distance indicates that the device is moving. *SpatSMD* uses Mean euclidean distance as a metric to distinguish between still and moving states. We found that euclidean distance between consecutive measurement does not provide any useful information to infer the state of the user (inspite of lower sampling frequency 0.25 Hz i.e. compared to 1Hz sampling frequency in [8]). In our experiments we found the optimal C value as 5 (refer Section 5 for details). Since we are interested in using only a two state classification (to distinguish between still and moving) it suffices that we just use Mean Euclidean distance over a window of measurements, where the values are calculated between measurements and then averaged together. The pseudocode of spatially-spread motion detection algorithm is presented in Algorithm 2.

5   Experimental Evaluation of SpecSMD and SpatSMD

5.1   Data Collection Phase

Two members of our research team collected WLAN network traces while doing their daily activities for about a week. Each data collector carried a HP IPAQ pocket PC running spotter for recording readings from nearby access points and logging them. Data collectors recorded their mobility activities using a custom diary application running on the PDA that allowed them to indicate whether they were walking, driving, cycling or in one place. Each time a measurement of spotter was logged the associated activity performed at that instance was recorded as ground truth manually (refer Figure 1(A)) from the pull down menu of the Spotter application to make the diary logging more accurate. Data collection was performed at common places such as city center, parking lot, university campus and indoor at the office, canteen and home. In all, the spotter logs contained WLAN traces of different periods ranging from 1 minute to over 15 minutes.

5.2   Results and Discussion

In this subsection, we evaluate how accurately the *SpecSMD* and *SpatSMD* can differentiate between moving and still states.

SpatSMD Algorithm. Out of the logs that were collected during the data collection phase, a representative data, collected from diverse range of settings is plotted in the Figure 6(A). The logs included, data with various access points densities, to see the impact of the number of access point on the results. The least number of access points in the data collected was 4, including the traces collected at home in several experiments. Figure 6(A) shows the accuracy results of the *SpatSMD* for C=1, C=5 and C=10. The higher the value of C, the higher is the lag associated in making the decision. Since we sample once in every 4 seconds, for C=10, we need to obtain data for atleast 40 seconds to make the decision. In order to reduce the computation latency, we should keep C small. Hence we chose C=5 as an optimal value as it performs better than C =1. The data collected from the "Outdoor By Foot" category (refer Figure 6(A)), was captured in dense vegetation area near farms and heard access point was 4. A reason for the low accuracy could be because of the dense vegetation or could be the number of heard access points. The average classification accuracy obtained with *SpatSMD* algorithm (86.7%) is slightly better than the one reported by Sohn et al. [8](85%). A reason for the better performance of our algorithm could be the kind of radio technology we are using (i.e. WLAN instead of GSM) and a different sampling frequency (0.25Hz as opposed to 1Hz). However, Sohn et al. achieved this accuracy for a three state classification scheme. Perhaps employing other features as reported in Sohn et al. might result in a significant improvement in the accuracy. This is yet to be investigated. We are yet to investigate the performance of the algorithm under low access point density (< 4).

SpecSMD Algorithm. Figure 6(B) shows the accuracy results of the *SpecSMD* for various values of the *FWHM Thresholds* and *Counter Thresholds*. We set the threshold based on a trial-and-error method, by testing the various threshold values on different data sets that were gathered from a wide range of settings. As explained before FWHM

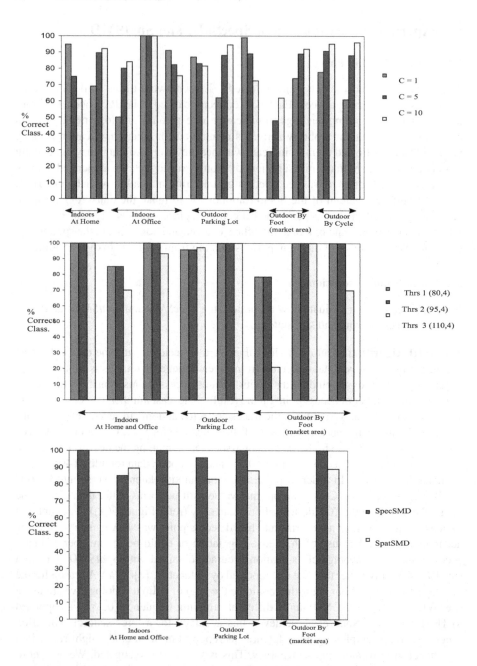

Fig. 6. (A). Accuracy results across different experimental set up for SpatSMDA Algorithm for various C values, with an average correct classification accuracy of 78.6%(C=1), 86.7%(C=5) and 87%(C=10), (B). Accuracy results across different experimental set up for SpecSMDA Algorithm for various FWHM and counter threshold, Thrs 1: FWHM=80, Ctr Threshold =4, Thrs 2:FWHM= 95, Ctr Threshold = 4 and Thrs 3:FWHM =110, Ctr Threshold =4, (C). Accuracy comparison between SpecSMD and SpatSMD, against same data sets.

refers to full width half maximum i.e. peak width of the FFT signal at 50% peak height. We collected at least 6 samples, before FFT is computed. Hence the spotter should be ON at least for 24 seconds to be able to run the motion detection algorithm explained above. This is comparable with *SpatSMD* (with C=5, it takes approximately the same time).

SpecSMD performs very well for all experimental settings by achieving an overall classification accuracy of 94.1%, clearly outperforming all the existing motion inference algorithms. It is thus reasonable to conclude that the schemes detection accuracy and performance is significant after being tested across various settings. Fine tuning the thresholds might even further increase the accuracy. Figure 6(C) shows the one-to-one comparison in the results obtained from both algorithms tested against the same datasets. The performance of *SpecSMD* (tested on same WLAN traces) is good even in the experiments in which *SpatSMD* was achieving only 50% (refer Figure 6(C)). The data collected from the "Outdoor By Foot" category, was taken in a dense vegetation area near farms and the access point density was 4, this is perhaps why the accuracy is low. Nevertheless, the performance of the SpecSMD is better than SpatSMD (78% against 50%). The accuracy obtained is much higher than the accuracy reported by Krumm et al. [3] using an algorithm based on the temporal variation of RSSI.

6   Conclusion and Outlook

The motion detection algorithms presented in this paper provide ubiquitous methods for deriving the state of the user. The algorithms can be implemented and run on a commodity device with WLAN capability without the need of any additional hardware support. Compared to accelerometer, this is a software solution so it could be easily applied to laptops, PDA and cellphones with WiFi capability. Compared to GSM based motion detection, WiFi based motion detection has a different application setting, like for instance improving WLAN positioning accuracy. In this paper for the first time, spectral characteristics of WLAN signals RSSI is analyzed. Based on the analysis we present a novel motion detection algorithm, *Spectrally Spread Motion Detection (SpecSMD)*. We benchmarked the proposed algorithm, against *Spatially Spread Motion Detection (SpatSMD)*. Both algorithms were evaluated by carrying out extensive measurements in a diverse set of settings (both indoors and outdoors). *SpecSMD* achieved a classification accuracy of 94.1%, outperforming the accuracy of *SpatSMD* (86.7%) and all the existing motion detection in the literature. Although context like moving or still can be deduced from external sensors like accelerometer, the techniques presented in this paper are promising as both these algorithms use only the radio signal that the wireless device should have to perform its normal operation. Both algorithms are reliable, have low latency and are computationally efficient to be able to run on mobile hand helds.

Ongoing work includes implementing both algorithms at run time on a PDA. We are interested in combining the techniques with machine learning algorithms to generate more states of classification which is useful for fine grained activity recognition. We are also currently working on incorporating the presented schemes as a part of WLAN localization algorithm.

Acknowledgements

We thank Andre Kokkler from Computer Architecture for Embedded Systems (CAES) group at the University of Twente for supporting us during temporal and spectral analysis of WLAN signal. This work is part of the *Smart Surroundings* project, funded by the Ministry of Economic Affairs of the Netherlands under the contract no. 03060.

References

1. Anderson, I., Muller, H.: Context Awareness via GSM Signal Strength Fluctuation. In: the 4th International Conference on Pervasive Computing, Late breaking results. Oesterreichische Computer Gesellschaft, pp. 27–31 (May 2006)
2. Randell, C., Muller, H.: Context awareness by analysing accelerometer data. In: MacIntyre, B., Iannucci, B. (eds.) The Fourth International Symposium on Wearable Computers, pp. 175–176. IEEE Computer Society, Los Alamitos (2000)
3. Krumm, J., Horvitz, E.: LOCADIO: Inferring Motion and Location from Wi-Fi Signal Strengths. In: First Annual International Conference on Mobile and Ubiquitous Systems: Networking and Services (Mobiquitous 2004), pp. 4–13 (August 2004)
4. Weisstein, E.W.: Spearman Rank Correlation Coefficient. From MathWorld–A Wolfram Web Resource, http://mathworld.wolfram.com/
5. Bahl, P., Padmanabhan, V.N.: Radar: An in-building rf-based user location and tracking system. In: Proceedings of the IEEE Infocom 2000, Tel-Aviv, Israel, vol. 2, pp. 775–784 (March 2000), http://citeseer.ist.psu.edu/bahl00radar.html
6. LaMarca, A., Chawathe, Y., Consolvo, S., Hightower, J., Smith, I., Scott, J., Sohn, T., Howard, J., Hughes, J., Potter, F., Tabert, J., Powledge, P., Borriello, G., Schilit, B.: Place Lab: Device Positioning Using Radio Beacons in the Wild. In: Gellersen, H.-W., Want, R., Schmidt, A. (eds.) PERVASIVE 2005. LNCS, vol. 3468, pp. 225–242. Springer, Heidelberg (2005)
7. Muthukrishnan, K., Meratnia, N., Lijding, M., Koprinkov, G., Havinga, P.: WLAN location sharing through a privacy observant architecture. In: 1st International Conference on Communication System Software and Middleware (COMSWARE), IEEE Communication Society Press, Los Alamitos, California (2006) (paper no 43)
8. Sohn, T., Varshavsky, A., LaMarca, A., Chen, M.Y., Choudhury, T., Smith, I., Consolvo, S., Hightower, J., Griswold, W.G., de Lara, E.: Mobility Detection Using Everyday GSM Traces. In: Proceedings of the Eighth International Conference on Ubiquitous Computing. Irvine, California, pp. 212–224 (September 2006)
9. Liao, L., Patterson, D.J., Fox, D., Kautz, H.: Inferring High-Level Behavior from Low-Level Sensors. In: Dey, A.K., Schmidt, A., McCarthy, J.F. (eds.) UbiComp 2003. LNCS, vol. 2864, pp. 73–89. Springer, Heidelberg (2003)
10. Hashemi, H.: The Indoor Radio Propagation Channel. In: Proceedings of IEEE, vol. 81(7) (July 1993)
11. Hightower, J.: The Location Stack, Phd Thesis, University of Washington (2004)
12. Exahau Positioning System, http://www.ekahau.com
13. Air Location II,
    http://www.engadget.com/2006/10/02/hitachis-employee-tracking -airlocation-ii-tag-w-wifi-enabled-rf/
14. Welch, G., Foxlin, E.: Motion tracking: No silver bullet, but a respectable arsenal. IEEE Comput. Graph. Appl. 22(6), 24–38 (2002)
15. Smart Surroundings, http://wwwes.cs.utwente.nl/smartsurroundings/

Inferring and Distributing Spatial Context

Clemens Holzmann

Johannes Kepler University Linz
Department of Pervasive Computing
Altenberger Straße 69, 4040 Linz, Austria
clemens.holzmann@jku.at

Abstract. An increasing number of computationally enhanced objects is distributed around us in physical space, which are equipped – or at least can be provided – with sensors for measuring spatial contexts like position, direction and acceleration. We consider spatial relationships between them, which can basically be acquired by a pairwise comparison of their spatial contexts, as crucial information for a variety of applications. If such objects do have wireless communication capabilities, they will be able to build up an ad-hoc network and exchange their spatial contexts among each other. However, processing detailed sensor information and routing it through the network lowers their battery lifetime or even may exceed the capabilities of embedded systems with limited resources. Thus, we present a novel and efficient approach for inferring and distributing spatial contexts in multi-hop networks, which builds upon qualitative spatial representation and reasoning techniques. Simulation results show its behavior with respect to common network topologies.

1   Introduction

People are nowadays interacting with an increasing number of real-world objects with embedded computing capabilities like vehicles, household appliances, notebook computers, mobile phones and portable music players. As they are by nature distributed throughout physical space, their *inherent spatial properties* as well as *spatial relationships* between them are valuable context information for a variety of applications. We refer to technology-enriched physical objects as *artifacts* in the following, and use the term *spatially-aware* if they are able to acquire and use spatial context information. They usually contain an embedded processing platform, wireless communication capabilities, a power management unit and possibly sensors and actuators.

A simple vehicular application scenario for the computational use of spatial relations can be seen in Figure 1. It shows four vehicles approaching a crossroads, whereas vehicle *b* sends information about its current position and moving direction as well as its relations to others in vicinity – namely that *c* and *d* are close behind and moving in the same direction – to vehicle *a*. Upon receiving this information, *a recognizes* that *b* is in front and for example alerts the driver if it is still moving too fast. Moreover, from the information that *b* is in front of

G. Kortuem et al. (Eds.): EuroSSC 2007, LNCS 4793, pp. 77–92, 2007.

Fig. 1. Application scenario vehicle-to-vehicle communication

a, and c and d are behind b, a can *infer* that two more vehicles with priority will soon be approaching from the right hand side.

In this paper we study how such autonomous and spatially-aware artifacts recognize spatial relationships to others in their vicinity, and how this knowledge can be distributed among artifacts out of communication range (i.e. packets cannot be delivered directly, but only via other artifacts). In [1], we have presented an approach that is based on the idea that each artifact exchanges its own spatial context with others in proximity, recognizes spatial relationships by comparing its own with received context information, and infers relations to artifacts out of range by exploiting the transitivity-property of spatial relations.

We present an extension of this approach by using *qualitative spatial representation and reasoning* techniques in the following, which comprises several aspects that are surveyed in Section 2. In this regard, we point out which ones we consider particularly useful with respect to *resource efficiency* in order to cope with limited battery lifetime and processing constraints of embedded systems. A comparison of related approaches is discussed in Section 3, and a new spatial calculus for *composing* qualitative positional and directional relations is presented afterwards. In Section 4, we finally propose an algorithm that builds upon this calculus for *distributing* spatial relationship information throughout a multi-hop network of artifacts, and discuss simulation results showing its behavior by means of different network topologies with varying numbers of nodes.

2   Spatial Representation and Reasoning

2.1   Static Spatial Contexts

We observe that an increasing number of real-world objects with integrated computing capabilities is distributed around us in physical space. For this reason, they basically do have a certain *position*, *direction* and *extension*, which can be changed through *translation*, *rotation* and *scaling*, correspondingly [2]. We refer to the first three properties as *static spatial context*, as they describe an artifact's spatial situation at a particular point in time, while the latter three are referred to as *dynamic spatial context*, as they describe how its static situation is changing at that point in time.

Similarly to [3], we classify the spatial context of an artifact both in terms of its inherent characteristics and with respect to other objects, which describe its *spatial properties* independently of other artifacts and its *spatial relations* to others, respectively. In Table 1, the static characteristics of an artifact, namely its *position* (i.e. where it is located) and *direction* (i.e. how it is positioned) as well as its *topology* (i.e. a description of parts of which it consists of) and *extension* (i.e. its shape and size), are classified along these two categories. The scope of our work is on *positional and directional relations* among artifacts, their spatial extension, inherent topology (i.e. holes and separations) and thus also topological relations like containment and overlapping as well as extensional relations between them are not considered. The main reason is that taking into account the spatial extension of artifacts requires much more computational resources [4], which often exceeds the capabilities of *embedded systems* with limited resources.

Table 1. Static characteristics of an object's spatial situation

	Inherent spatial properties	Spatial relations to artifacts
Position	geographic position	*orientation and distance relations*
Direction	intrinsic direction axis	*relations between direction axes*
Topology	holes and separations	spatial arrangement
Extension	shape and size	relation between extensions

2.2 Qualitative Spatial Representation

The computational processing of spatial relations requires a formal representation, wherefore the mathematics of Euclidean space probably comes to mind first [5]. However, such precise *quantitative* approaches have numerous disadvantages compared to *qualitative* representations, especially with regard to resource-constrained embedded systems. First, qualitative models allow to deal with coarse and imprecise spatial information, which is an important property as exact sensor information is often not available or precise answers are not required [5] [6]. Second, processing quantitative knowledge is more complex and thus computationally more expensive [3] [7]; moreover, quantitative models are often intractable or even unavailable [4]. A huge field of research is *qualitative spatial representation and reasoning* [4] [8], which is concerned with *abstracting* continuous spatial properties and relations of the physical world, and *inferring* knowledge from the respective qualitative representations.

In order to represent spatial relations in a qualitative way, it is necessary to decide on a certain kind of spatial primitive first. We decided to use *points* as abstractions of physical artifacts and define relations between these basic spatial entities in a two-dimensional plane. For both orientation and directional relations, a common approach is to partition the 360° range into intervals, where each one of the respective regions is associated with a certain relation. In the case of *orientation relations*, which describe where a certain object (i.e. the primary object p) is placed relative to another object (i.e. the reference object r) [4] [8], the space around the reference object is partitioned and the relation is denoted by the region in which the primary object is located.

Directional relations on the other hand relate the direction of the primary object, as given by its intrinsic direction axis, with that of the reference object; therefore, the space around the primary object is partitioned according to the direction axis of the reference object, and the region in which the direction axis of the primary object points denotes the directional relation. The most common representation systems used are cone- or projection-based [9], where we consider the cone-based system as the most suitable with regard to embedded systems, as it easily allows to change the *granularity* of relations just by adding or removing axes and thus allows to cope with senors of different accuracy.

For representing *distance relations*, we use Euclidean distances and assume an *isotropic space*, where points at the same distance are connected with concentric circles. Each of the qualitative distances conforms to an interval of quantitative ones [3], defining the qualitative relation between the reference object and the primary object; the number of intervals again determines the granularity of the relations. Figure 2 shows common qualitative representations of orientation and distance relations, which partition the space in the eight cardinal directions *north, east, south, west, north-east, south-east, south-west,* and *north-west,* and in the five distance levels *very close, close, commensurate, far* and *very far,* respectively [3] [9].

Fig. 2. Cone- and projection-based directions, and qualitative distances [3]

Another issue related to qualitative spatial representations are *frames of reference,* which influence the semantics of spatial relationships. For orientation relations, the frame of reference fixes the front-side of the reference object and thus defines its reference direction. A distance frame of reference is presented in [10], which is however not important in the following. According to [3], our scope is on *intrinsic* reference frames, where the relation is given by inherent properties of the reference object like its intrinsic direction axis, and on *extrinsic* frames of reference, which are determined by external factors like the earth reference frame; in this regard, its scale defines distances between objects and the North Pole serves as a fixed reference point for orientation relations. In both cases, the reference frame is centered in the reference object (i.e. referred to as *egocentric* [11]), as we only consider artifacts that recognize spatial relations with respect to themselves and never between other artifacts. *Deictic* frames of reference, which represent relations from an external viewpoint, are thus out of scope. The resulting four types of spatial relations can be seen in Figure 3.

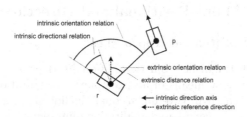

Fig. 3. Intrinsic/extrinsic positional and directional relations

2.3 Qualitative Spatial Reasoning

Many applications for the use of spatial relations can be found in literature, for example their visualization on a display [12], triggering of certain actions when entities get into spatial proximity [13], or reasoning [2] [4] about spatial configurations. Our scope is on the latter, namely to use *qualitative reasoning* techniques for inferring relationship information among artifacts. Qualitative spatial reasoning is commonly realized in form of calculi over sets of jointly exhaustive and pairwise disjoint spatial relations (i.e. non-overlapping relations covering the whole space), which are in turn defined over sets of spatial entities (cf. Section 3.1).

A relation R between two objects x and y (i.e. $(x, y) \in R$) is often denoted as $R(x, y)$, and it is read as "x is in relation R to y". A spatial calculus consists of a domain D containing the spatial entities, a finite set $\mathcal{BR}$ of n-ary base relations on the domain and the powerset $\mathcal{R}$ of these base relations, as well as a set of operations [14]. The result of an operation may be the union of multiple base relations, wherefore the operations of a calculus have to be defined for all possible unions of base relations. We use the following *operations on binary relations* for inferring and distributing relations in Section 4.1, where $R, S \in \mathcal{R}$ [14]:

- Union: $R \cup S = \{x | (x \in R) \vee (x \in S)\}$
- Intersection: $R \cap S = \{x | (x \in R) \wedge (x \in S)\}$
- Composition: $R \circ S = \{(x, z) | \exists y \in D : (x, y) \in R \wedge (y, z) \in S\}$

Of particular interest for this work is the *composition* of relations [4] [8]: given the relation between two objects x and y as well as between y and z, what is the relation between the objects x and z? It may result in a set of neighboring relations, which means that any of them can be the relation between x and z; such a set is referred to as compound relation [8]. The results are commonly stored in *composition tables*, which define the resulting relations of all possible compositions of base relations; compound relations of $\mathcal{R}$ can be computed as the union of the compositions of base relations. In contrast to the set-theoretic operations union and intersection, the composition has to be computed from the *meaning* of the respective relations [6]. A related concept in qualitative spatial reasoning is that of *conceptual neighborhood*; two relations are conceptual neighbors if and only if they can be *directly* transformed into each other (i.e. without passing other relations) [15]. In Section 3.2, an *iconic notation* of the neighborhood structure is used for defining the composition table.

3   Reasoning About Positional and Directional Relations

3.1   Comparison of Related Approaches

In Section 2.2, we distinguished four types of qualitative spatial relations which are shown in Figure 4 by means of an exemplary configuration, where two artifacts p and r are placed in two-dimensional Euclidean space. For orientation and direction relations, a cone-based qualitative representation with four equally sized sectors is used, and distance relations partition the space around the reference object r in circular ranges of the same size – except the outer range which is open. Solid arrows represent intrinsic direction axes, and the dotted one an extrinsic reference direction. The resulting relation of artifact p with respect to r is written boldface for the example in Figure 4. In addition to the relations shown in Figure 4, the identity relations *straight-front* (for orientation), *here* (for distance) and *same-dir* (for direction) can be defined. While the former two are however practically impossible due to sensor inaccuracies and objects that have a physical extension respectively, the latter corresponds to using an extrinsic frame of reference.

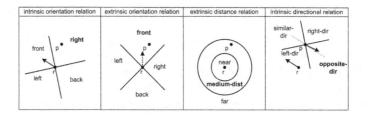

Fig. 4. Qualitative positional and directional relations

As stated in Section 2.3, qualitative spatial reasoning is commonly realized in form of calculi over sets of qualitative spatial relations. Many of such *qualitative spatial calculi* have been developed during the past decades, mainly for topological or positional reasoning; however, they are often not fully specified, and mostly no implementation is made available [14]. Table 2 shows a comparison of popular qualitative spatial calculi which are classified according to the four relation types presented above. Many of them incorporate the spatial context orientation only, for example the ternary *double-cross calculus* presented by Freksa [7] [14] and the *flip-flop calculus* of Ligozat et.al. [16], which describe the deictic orientation of a point in the plane with respect to a vector that is given by two further points. However, they can represent intrinsic orientation relations by viewing them as positional binary relation between a dipole and an isolated point. On the other hand, there are only few calculi dealing with spatial distance only, as for example Hernández et.al. [10] who particularly addressed the composition of distances depending on intrinsic orientation relations.

Early work that combines orientation and direction relations is that of Abdel-moty et.al. [2], which allows for representing extrinsic and intrinsic orientation relations by computing the intersections of orientation lines; directional relations are represented through the inverse orientation relation. The binary *dipole relation algebra* of Moratz et.al.[17] uses straight line segments, which are formed by a pair of points at a time, for representing orientation relations between objects with an intrinsic direction axis. A continuative calculus of Moratz et.al. is the *oriented point relation algebra* [18], where oriented points are used instead of dipoles, and the granularity is adjustable with a single parameter; the exact set of base relations thus depends on the chosen level of granularity. A similar approach with arbitrary granularity is presented by Renz et.al. in [9], which developed the *star calculus* for relating two points in a plane with respect to an extrinsic reference direction.

However, there is only few existing work about the *combination* of orientation and distance relations. Zimmermann et.al. [19] add distance to their ternary calculus for representing intrinsic orientation, and show how distance information restricts the possible orientation relations; this dependency is also shown by Sharman in [20]. Clementini et.al. [3] show the interplay between orientation and distance relations, but do not present a calculus for homogeneous reasoning about orientation and distance relations. To the best of our knowledge, there is no existing work which deals with *compositional reasoning* about combined orientation and distance relations as presented in Section 3.2, neither taking into account directional relations nor without considering them. Moreover, in Section 4 we apply this composition for distributing relationships among autonomous artifacts, which also seems to be new.

Table 2. Comparison of approaches for reasoning about static spatial relations

	extr. orient.	intr. orient.	extr. dist.	intr. direct.
Freksa [7] [14]		x		
Ligozat et.al. [16]		x		
Hernández et.al. [10]			x	
Abdelmoty et.al. [2]	x	x		x
Moratz et.al. [18] [17]		x		x
Renz et.al. [9]	x			x
Zimmermann et.al. [19]		x	x	
Clementini et.al. [3]		x	x	
Proposed approach	x	x	x	x[1]

3.2   Composition of Positional and Directional Relations

Motivated by characteristics of pervasive and ubiquitous computing applications, primarily the distribution of huge numbers of artifacts in the real world which do have limited processing, storage and communication resources, we are addressing a combined analysis of position and direction relations within a single framework

---

[1] For composing positional relations with intrinsic orientation (cf. Section 3.2).

in the following. We discuss the *composition of static positional and directional relations*, which is used in the subsequent section for inferring relations between artifacts out of range by repeatedly applying it to triples of artifacts.

We developed *composition tables* for the four orientation base relations *front*, *right*, *back* and *left* as well as the three distance base relations *near*, *medium-dist* and *far*. The result of a composition operation depends on the directional relation between the involved objects, which can be *similar-dir*, *right-dir*, *opposite-dir* and *back-dir*; the relation *same-dir* is additionally considered, meaning that x and y do have exactly the same direction in space (e.g. due to using the extrinsic earth reference frame). Thus, dedicated composition tables for distance and orientation relations are required, depending on the artifact's intrinsic direction. We thus get a total number of 12 base relations in the case of an extrinsic, and 48 in the case of an intrinsic reference direction.

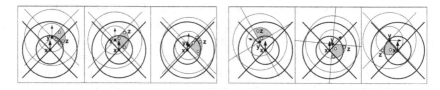

Fig. 5. Composition of the two positional relations *front(y,x)* ∧ *near(y,x)* and *right(z,y)* ∧ *near(z,y)*, with the directional relations *same-dir(y,x)* (left) and *similar-dir(y,x)* (right). Possible alternative positions of z are shown with white dots.

How the composition of positional relations among three artifacts x, y and z is acquired can be seen in Figure 5, exemplarily for the *extrinsic* earth reference frame (*same-dir(y,x)*) and with a *similar direction* between the objects x and y (*similar-dir(y,x)*). While the former composition results in just four possible relations due to the range of possible positions (i.e. *front(z,x)* ∧ *near(z,x)*, *front(z,x)* ∧ *medium-dist(z,x)*, *right(z,x)* ∧ *near(z,x)* or *right(z,x)* ∧ *medium-dist(z,x)*), the latter results in a set of even seven relations as a consequence of the additional range of possible alternative directions of object y with respect to x. Due to a lack of publication space, just extrinsic orientation and distance are dealt with in the following.

Figure 6 shows the *separate* composition tables for extrinsic orientation and distance, wherefore an *iconic notation* is used (cf. Section 2.3). The four orientation base relations are visualized with black dots indicating their orientation, and the three distance base relations are visualized with filled areas indicating their possible ranges; disjunctions of base relations, which represent possible alternative relations, are visualized by superimposing their icons. However, it can be seen that the composition operation often leads to *coarse results*; e.g., the compositions of *front* and *left* or of *near* and *medium-dist* result in the union of all orientation or distance base relations, respectively, which are referred to as *universal relations* and represent the complete lack of knowledge about the spatial relation between two artifacts. The composition tables for *combined* orientation and distance relations can be seen in Figure 7, partially for six base relations at a time; all others can be acquired by simply rotating the table. The combined consideration allows for more accurate conclusions [8]; for example, composing the extrinsic distance relations *medium-dist(z,y)* and *far(y,x)* only yields the universal relation, but taking into account the orientation relations *front(z,y)* and *front(y,x)* results in just one distance relation, namely *far(z,x)*.

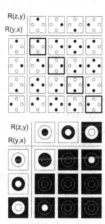

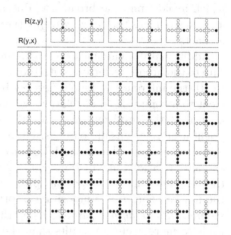

Fig. 6. Composition tables for extrinsic orientation (top) and distance relations (bottom), which are algorithmically managed separately

Fig. 7. Extraction of the composition table for an algorithmically combined extrinsic orientation and distance relations management

4   Distribution of Spatial Relationship Information

4.1   Distribution Algorithm

In this section, we address the question of how a whole *collective* of artifacts, namely *all* artifacts which are connected directly or via multiple hops, can be provided with an awareness about positional relations among each other. We distinguish two general approaches therefore: (i) exchanging *quantitative sensor data* among all artifacts, or (ii) exchanging sensor data between artifacts within communication range only, together with their knowledge about *qualitative spatial relations* to other artifacts. We refer to data packets which contain these spatial contexts as the artifacts' *self-descriptions*. While the former one can be done by simply *flooding* them throughout the network, we developed a new approach for the latter.

The basic idea is that a certain artifact starts to *broadcast* its self-description containing quantitative sensor data (e.g. its position from a GPS receiver) to others in vicinity, which *recognize* qualitative spatial relations to the broadcasting one, put them in their own self-descriptions and broadcast them, too. This initial broadcast may be for example due to significant *changes* of its sensor readings as a result of movement, or periodically after a certain time period has elapsed. An artifact in turn broadcasts its own self-description either upon receiving that of another artifact the first time, or if its qualitative spatial relations to others changed. This process of distributing relations terminates if no artifact recognized further changes in its qualitative spatial relations. The broadcasting step can be *delayed* by performing broadcasts in short fixed intervals only, which avoids multiple broadcasts due to successively received self-descriptions and thus reduces the *induced traffic* (i.e. the total number of packets received by artifacts).

Additionally, an artifact may *infer* further relations by *composing* its relationship to the broadcasting one with those contained in the received self-description. In [1], the inference is done by processing the *transitive closure*, which is equivalent to a composition of relations where all three – the two composed relations as well as the resulting one – are the same; the respective cases for extrinsic orientation relations are emphasized in Figure 6. Although this approach is universal in the sense that it can be applied to arbitrary relations, it is quite limited as many relations like distance and intrinsic orientation are not transitive. We thus extended it by *composing* spatial relations as described in Section 3.2. In the first version, the composition of orientation and distance relations is *algorithmically managed separately* using the composition tables of Figure 6, which leads to more accurate results than the previous approach as the composed relations need not be the same. The best results – i.e. those which constrain the resulting possible relations most, particularly with regard to distance relations – are acquired by an *algorithmically combined management* of orientation and distance relations using the composition table of Figure 7.

Algorithm 1 describes the operations an artifact performs upon receiving a self-description, whereas the *composition* step can be one of the three described above. If the composed relations R and S are compound ones, the composition result RS is the *union* of the compositions of base relations, whereas resulting universal relations are not stored. In order to retain the most accurate result, RS is eventually *intersected* with relations to the respective artifact that are possibly contained in the local self-description. An example therefore is given in Figure 8 by means of a simple network topology, where each node represents an artifact and edges between the nodes indicate that they are within communication range.

4.2   Simulation and Discussion

We have *implemented* the flooding algorithm as well as the proposed one with its three ways for composing spatial relations as described Section 4.1, and simulated them with the J-Sim[2] simulation environment using different network topologies with varying numbers of nodes. The simulation was done without taking into account certain wireless communication technologies or transmission protocols, just the protocol logics have been implemented. The aim was to compare our

---

[2] http://www.j-sim.org/

Algorithm 1. artifact x receives self-description of artifact y

---

1: **if** self-description of y received the first time **then**
2:     recognize qualitative relation of y to x and put it to self-description of x;
3: **end if**
4: $R \leftarrow$ get qualitative relation of y to x from self-description of x;
5: **for all** artifacts z which are in relation to y **do**
6:     **if** $z \neq x$ **then**
7:         $S \leftarrow$ get qualitative relation of z to y from self-description of y;
8:         $RS \leftarrow$ perform composition $R \circ S$;
9:         **if** self-description of x already contains relation of z to x **then**
10:             intersect known relation with composition-result RS;
11:         **else**
12:             put composition-result RS to self-description of x;
13:         **end if**
14:     **end if**
15: **end for**
16: **if** first self-description received **or** relations in self-description of x changed **then**
17:     broadcast self-description of x;
18: **end if**

---

algorithm and the flooding approach both with regard to the achieved spatial relation awareness of all artifacts after its termination, and the traffic induced therefore due to broadcasts of self-descriptions.

We first simulated the four algorithms with the topology shown in Figure 8, the resulting spatial awareness of all $n = 5$ artifacts can be seen in Figure 9. With the *flooding* approach, each artifact gets to know the self-descriptions of all others in the network, and it is possible to compute exactly one base relation from the sensor data of each artifact. Flooding is thus the most accurate algorithm, resulting in a total number of $n * (n - 1) = 20$ relations, and it causes a total number of 50 received self-descriptions over all n artifacts. The *transitive closure* algorithm on the other hand is the least accurate one, mainly for two reasons. First, all three relations have to be the same for processing the transitive closure, wherefore artifact a is not able to infer any relation to artifact c and vice versa, as the orientation on the path from a to c is changing (i.e. *right(b,a)* and *back(c,b)*). Second, the spatial relations have to be transitive, which is not the case e.g. for distances; for this reason, no distance relation can be inferred from the artifacts c, d and e to a, and from e to b and c. It results in 58 relations, as a missing distance or orientation relation corresponds to 3 or 4 base relations (i.e. the respective *universal relations*), and no positional relation at all (e.g. from c to a) corresponds to their product with 12 alternative positional relations. The simulation also showed that it induced a *traffic* of 36 received self-descriptions; with delayed broadcasts, it could be reduced to 25 which is 50% compared with flooding. In both cases, the simulation was started once from each artifact and the average of the resulting traffic was taken. Additionally, we investigated the *relative accuracy* achieved by a certain algorithm, which we define as the complement of the ratio between the difference of the

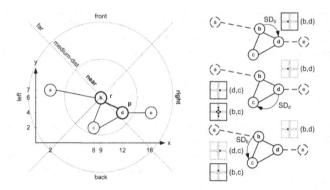

Fig. 8. Topology of the simulated network (left), and an example scenario (right) showing the composition of the relations $R(b,d)$ and $R(d,c)$ as well as the intersection of the resulting relation $R(b,c)$ with the more accurate one which is recognized due to a received self-description SD_b of artifact b

actual and the minimum number of relations, and the difference of the maximum and the minimum number of relations. For the transitive closure algorithm, it is $1 - (58 - 5 * 4 * 1)/(5 * 4 * 11) = 83\%$ compared with flooding.

Taking into account the *separate composition* of orientation and distance, more accurate conclusions can be drawn. For example, artifact a is now able to infer the two alternatively possible base relations $right(c,a)$ and $back(c,a)$ as defined in the composition table of Figure 6, and artifact c is able to narrow

Algorithm	Artifact a	Artifact b	Artifact c	Artifact d	ARTIFACT e
Flodding	R(b,a): medium-dist∧right	R(a,b): medium-dist∧left	R(a,c): medium-dist∧left	R(a,d): far∧left	R(a,e): far∧left
	R(c,a): medium-dist∧right	R(c,b): near∧back	R(b,c): near∧front	R(b,d): near∧left	R(b,e): medium-dist∧left
	R(d,a): far∧right	R(d,b): near∧right	R(d,c): near∧right	R(c,d): near∧left	R(c,e): medium-dist∧left
	R(e,a): far∧right	R(e,b): medium-dist∧right	R(e,c): medium-dist∧right	R(e,d): near∧right	R(d,e): near∧left
Transitive Closure	R(b,a): medium-dist∧right	R(a,b): medium-dist∧left	R(a,c): -	R(a,d): *left*	R(a,e): *left*
	R(c,a): -	R(c,b): near∧back	R(b,c): near∧front	R(b,d): near∧left	R(b,e): *left*
	R(d,a): *right*	R(d,b): near∧right	R(d,c): near∧right	R(c,d): near∧left	R(c,e): *left*
	R(e,a): *right*	R(e,b): *right*	R(e,c): *right*	R(e,d): near∧right	R(d,e): near∧left
Separate Composition	R(b,a): medium-dist∧right	R(a,b): medium-dist∧left	R(a,c): *left* ∨ *front*	R(a,d): left	R(a,e): left
	R(c,a): *right* ∨*back*	R(c,b): near∧back	R(b,c): near∧front	R(b,d): near∧left	R(b,e): *(near ∨medium-dist)* ∧ left
	R(d,a): right	R(d,b): near∧right	R(d,c): near∧right	R(c,d): near∧left	R(c,e): *(near ∨ medium-dist)* ∧ left
	R(e,a): right	R(e,b): *(near ∨medium-dist)* ∧ right	R(e,c): *(near ∨medium-dist)* ∧right	R(e,d): near∧right	R(d,e): near∧left
Combined Composition	R(b,a): medium-dist∧right	R(a,b): medium-dist∧left	R(a,c): (near∧left) ∨ (medium-dist∧left) ∨ (far∧left) ∨ (near∧front) ∨ (medium-dist∧front) ∨ (far∧front)	R(a,d): *(medium-dist ∧ left)* ∨ *(far ∧ left)*	R(a,e): *(medium-dist ∧ left)* ∨ *(far ∧ left)*
	R(c,a): (near∧right) ∨ (medium-dist∧right) ∨ (far∧right) ∨ (near∧back) ∨ (medium-dist∧back) ∨ (far∧back)	R(c,b): near∧back	R(b,c): near∧front	R(b,d): near∧left	R(b,e): (near∧left) ∨ (medium-dist∧left)
	R(d,a): *(medium-dist ∧ right)* ∨ *(far ∧ right)*	R(d,b): near∧right	R(d,c): near∧right	R(c,d): near∧left	R(c,e): (near∧left) ∨ (medium-dist∧left)
	R(e,a): *(medium-dist ∧ right)* ∨ *(far ∧ right)*	R(e,b): (near∧right) ∨ (medium-dist∧right)	R(e,c): (near∧right) ∨ (medium-dist∧right)	R(e,d): near∧right	R(d,e): near∧left

Fig. 9. Comparison of simulation results for the topology shown in Figure 8, where the columns show the acquired spatial relation awareness of the artifacts $a \ldots e$ after termination of the distribution process

down the possible distance relations to artifact *e* to *near(e,c)* and *medium-dist(e,c)*. It results in a total number of 42 relations, and induces a traffic of 38 self-descriptions without and 26 with using delayed broadcasts. The accuracy raises to 90% and the traffic for delayed broadcasts to 52%. Nevertheless, the composition of distances often results in the universal relation, which does not provide any information about the spatial distance between artifacts. With a *combined composition* of orientation and distance relations as shown in Figure 7 however, the accuracy of the relations between some artifacts can be increased. For example, while artifact *a* is only able to infer the relation *right(e,a)* due to the resulting universal relation by composing the distance relations *medium-dist(b,a)*, *near(d,b)* and *near(e,d)* without combining orientations and distances, their combined consideration allows to exclude the distance relation *near(e,a)*. The total number of relations can eventually be refined to 38, the induced traffic is the same with 38 and 26 self-descriptions, respectively. The accuracy thus raises to 92% and the traffic remains 52%, which means that a higher accuracy is achieved with the same traffic necessary.

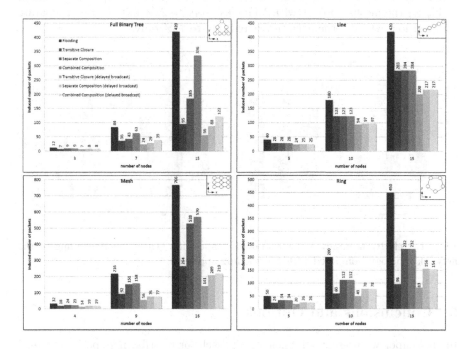

Fig. 10. Traffic induced by different algorithms depending on the network topology

Figure 10 shows the induced traffic for the common network topologies *full binary tree*, *line*, *mesh* and *ring*, with varying numbers of nodes at a time; the topologies are included as small images in the diagrams. First, it can be seen that it is in any case reduced by using qualitative composition in comparison to flooding. The transitive closure algorithm is always equal or below the traffic

induced by a separate or combined composition of positional relations, which is
due to the smaller number of inferred relations and thus the fewer broadcasts.
Second, the induced traffic for separate and combined composition is quite different for the binary tree topology, whereas it is virtually the same for line,
mesh and ring topologies. Third, using delayed broadcasts significantly reduces
the traffic with an increasing number of nodes, for example to less than 40% for
the mesh and binary tree topology with 15 nodes. We also experimented with
complete graphs and *star* topologies, leading to similar results; combined composition with delayed broadcast even allows to decrease the traffic to less than
30% for a complete graph topology.

The respective *relative accuracies* for the four topologies are finally shown
in Figure 11. First, it can be seen that the transitive closure approach leads to
the least accuracy, as it only supports a subset of the possible compositions.
Second, the accuracy decreases with an increasing number of nodes, which is
due to the coarser composition results coming along with the higher number of
hops between artifacts. Third, the percentage-wise reduction of traffic is in all
simulated scenarios higher than the loss of accuracy; for example, the induced
traffic for the mesh-topology with 16 nodes drops to 27% in the case of separate
composition, whereas the accuracy is reduced to 73% only.

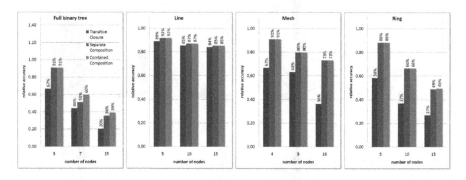

Fig. 11. Relative accuracy of the proposed composition algorithms depending on the
network topology, compared with the flooding approach

5   Conclusions and Outlook

In this paper we present an efficient approach for distributing spatial contexts
in multi-hop networks, which builds upon qualitative spatial representation and
reasoning techniques. We argue for four types of spatial relationships we consider particularly useful regarding autonomous artifacts with limited resources:
extrinsic orientation and distance relations, which we used for evaluating the proposed approach by simulation means, as well as intrinsic orientation relations
that rely on directional relations between artifacts. A new spatial calculus for
composing qualitative positional and directional relations is presented, which

allows to infer spatial relations over multiple hops without the need for exchanging quantitative sensor data. This is achieved by repeatedly applying the composition operation to triples of artifacts, wherefore both the algorithmically separate and combined management of orientation and distance relations have been investigated. We propose an algorithm for inferring and distributing qualitative relationship information, which has been implemented and evaluated using a Java-based simulation environment. The simulation results show the feasibility of the presented algorithm, the reduction of network traffic compared with exchanging sensor data among all artifacts as well as the achieved accuracy of relationship information depending on the network topology and the number of nodes.

With regard to future work, we plan to put our focus on dynamic spatial relations as well as their combined consideration with static ones. Another issue is to investigate the impacts of moving artifacts and changing sensor data in detail, which both lead to relation changes and build up the basis for spatial reasoning over time.

References

1. Holzmann, C., Ferscha, A.: Towards collective spatial awareness using binary relations. In: 3rd International Conference on Autonomic and Autonomous Systems, ICAS 2007, Athens, Greece, June 19-25, 2007, pp. 19–25. IEEE CS Press, Los Alamitos (2007)
2. Abdelmoty, A.I., El-Geresy, B.: An intersection-based formalism for representing orientation relations in a geographic database. In: 2nd ACM Workshop on Advances In Geographic Information Systems, Workshop at CIKM 1995, Gaitherburg, MD, USA, December 1-2, 1994, pp. 44–51. ACM Press, New York (1994)
3. Clementini, E., Felice, P.D., Hernández, D.: Qualitative representation of positional information. Artificial Intelligence 95(2), 317–356 (1997)
4. Cohn, A.G., Hazarika, S.M.: Qualitative spatial representation and reasoning: An overview. Fundamenta Informaticae 46(1-2), 1–29 (2001)
5. Hobbs, J.R., Narayanan, S.: Spatial representation and reasoning. In: Encyclopedia of Cognitive Science, MacMillan, London (2002)
6. Moratz, R., Dylla, F., Frommberger, L.: A relative orientation algebra with adjustable granularity. In: Workshop on Agents in Real-Time and Dynamic Environments at IJCAI 2005, Edinburgh, Scotland (July 30 - August 5, 2005)
7. Freksa, C.: Using orientation information for qualitative spatial reasoning. In: Frank, A.U., Formentini, U., Campari, I. (eds.) Theories and Methods of Spatio-Temporal Reasoning in Geographic Space. LNCS, vol. 639, pp. 162–178. Springer, Heidelberg (1992)
8. Hernández, D.: Qualitative Representation of Spatial Knowledge. LNCS, vol. 804. Springer, Heidelberg (1994)
9. Renz, J., Mitra, D.: Qualitative direction calculi with arbitrary granularity. In: Zhang, C., W. Guesgen, H., Yeap, W.-K. (eds.) PRICAI 2004. LNCS (LNAI), vol. 3157, pp. 65–74. Springer, Heidelberg (2004)
10. Hernández, D., Clementini, E., Felice, P.D.: Qualitative distances. In: Kuhn, W., Frank, A.U. (eds.) COSIT 1995. LNCS, vol. 988, pp. 45–57. Springer, Heidelberg (1995)

11. Klatzky, R.L.: Allocentric and egocentric spatial representations: Definitions, distinctions, and interconnections. In: Freksa, C., Habel, C., Wender, K.F. (eds.) Spatial Cognition. LNCS, vol. 1404, pp. 1–18. Springer, Heidelberg (1998)
12. Hazas, M., Kray, C., Gellersen, H.W., Agbota, H., Kortuem, G., Krohn, A.: A relative positioning system for co-located mobile devices. In: 3rd International Conference on Mobile Systems, Applications, and Services, MobiSys 2005, Seattle, Washington, USA, June 6-8, 2005, pp. 177–190. ACM, New York (2005)
13. Ferscha, A., Hechinger, M., Mayrhofer, R., dos Santos Rocha, M., Franz, M., Oberhauser, R.: Digital aura. In: Ferscha, A., Mattern, F. (eds.) PERVASIVE 2004. LNCS, vol. 3001, pp. 405–410. Springer, Heidelberg (2004)
14. Dylla, F., Frommberger, L., Wallgrün, J.O., Wolter, D.: SparQ: A toolbox for qualitative spatial representation and reasoning. In: Freksa, C., Kohlhase, M., Schill, K. (eds.) KI 2006. LNCS (LNAI), vol. 4314, pp. 79–90. Springer, Heidelberg (2007)
15. Freksa, C.: Conceptual neighborhood and its role in temporal and spatial reasoning. In: Workshop, I.M.A.C.S. (ed.) IMACS Workshop on Decision Support Systems and Qualitative Reasoning, Toulouse, France, March 13-15, 1991, pp. 181–187. Elsevier Science Publishers, Amsterdam (1991)
16. Scivos, A., Nebel, B.: The finest of its class: The natural point-based ternary calculus for qualitative spatial reasoning. In: Freksa, C., Knauff, M., Krieg-Brückner, B., Nebel, B., Barkowsky, T. (eds.) Spatial Cognition IV. LNCS (LNAI), vol. 3343, pp. 283–303. Springer, Heidelberg (2005)
17. Moratz, R., Renz, J., Wolter, D.: Qualitative spatial reasoning about line segments. In: 14th European Conference on Artificial Intelligence, ECAI 2000, Berlin, Germany, August 20-25, 2000, pp. 234–238. IOS Press, Amsterdam (2000)
18. Moratz, R.: Qualitative spatial reasoning about oriented points. Technical Report SFB/TR 8 Report No. 003-10/2004, University of Bremen, Bremen, Germany (October 2004)
19. Zimmermann, K., Freksa, C.: Qualitative spatial reasoning using orientation, distance, and path knowledge. Applied Intelligence 6(1), 49–58 (1996)
20. Sharman, J.: Integrated Spatial Reasoning in Geographic Information Systems: Combining Topology and Direction. PhD thesis, University of Maine (May 1996)

Context Sensitive Adaptive Authentication

R.J. Hulsebosch[1], M.S. Bargh[1], G. Lenzini[1], P.W.G. Ebben[1], and S.M. Iacob[2]

[1] Telematica Instituut, PO Box 589, 7500 AN, Enschede, The Netherlands
{Bob.Hulsebosch, Mortaza.Bargh, Gabriele.Lenzini,
Peter.Ebben}@telin.nl
[2] Decis Lab, PO Box 90, 2600 AB, Delft, The Netherlands
Sorin.iacob@icis.decis.nl

Abstract. We exploit the ability to sense and use context information to augment or replace the traditional static security measures by making them more adaptable to a given context and thereby less intrusive. We demonstrate that by fusing location information obtained from various sources that are associated to the user and are available over time, the confidence in the identity of the user can be increased considerably. In fact, the level of confidence in the identity of the user is related to the probability that the user is at a certain location. This probability is used as a measure to parameterize the authentication level of the user making it thereby much more adaptive to changing situational circumstances. In this paper we describe the theoretical background for a context-sensitive adaptation of authentication and the design and validation of the system that we have developed to adaptively authenticate a user on the basis of the location of his sensed identity tokens.

Keywords: Authentication; context awareness; adaptive; probability.

1 Introduction

In traditional security systems, security services are pre-configured to a static behavior and cannot be adapted dynamically to new constraints. This limitation is due to two main shortcomings: the adoption of non-adaptive behavior and the inability of considering context information.

To address the first shortcoming, i.e. non-adaptive behavior, security services must be flexible and able to cope with different situations. Adaptive security mechanisms (e.g., [1], [2]) are able to dynamically respond to environmental changes by re-configuring their security functions. Moreover, they support the idea that security can be more effective if variable levels of security are presented to users and to systems. For each security level a certain threshold must be fulfilled, which may be absolute or statistical; thresholds indicate degrees of security with respect to assurance, availability, execution efficiency, etc. Adaptive security solutions are known to ensure a high level of usability (e.g., they avoid absolute identity verification), realism (e.g., their access control mechanisms are fine-grained), sensibility to external constraints (e.g., power limitations may influence the choice of encryption algorithm), and ability to deal with exceptional situations (e.g., emergencies are treated differently).

G. Kortuem et al. (Eds.): EuroSSC 2007, LNCS 4793, pp. 93–109, 2007.

One means of ensuring security adaptation is parameterization [3], [4], [5], [6]. Parameterization of security implies the ability of identifying levels of security. For instance, for each security service such as authentication, authorization, confidentiality, and integrity, levels of security are expressed. Moreover, one should also be able to compute a value corresponding to a security level, i.e. the performance of the security function should be made measurable. Parameterization is one approach to making security adaptive. Another approach would be to adapt the security level by means of structural changes of the system. Being the simplest method, we focus on parameterized adaptation of security services and in particular authentication.

Regarding the second shortcoming, i.e., the inability of considering contextual information, security services require a context-aware infrastructure for the detection and interpretation of context information in order to allow for a controlled security adaptation when needed [7]. Context information can include any sort of data such as human factors (user habits, mental state, social environment, task-related activities), the physical environment (location, network connectivity, battery power), business data (goal-directed activities, trust), and time. What context information is relevant for a certain situation is not fully predictable and depends on the specific application. A security context can be defined as the information collected from the user and his application environment that is relevant to the security infrastructure of both the user and the application [8]. Context information thus forms, besides the traditional security services, an additional important element of the security context. An illustrative example is the use of location and velocity information to infer that a user is a train traveler and therefore is granted access to services offered in the train [9].

In this paper, we propose to combine parameterization and context-awareness to control security adaptation. We call this paradigm *context-sensitive adaptive security*. Its goal is to optimize the security functionality for a given situational context in a non-intrusive way. In fact, we can imagine a system that, by constantly monitoring and analyzing context information, is able to maintain the desired security level and to respond to new security constraints that may arise from changes in the situational context. We believe that systems can achieve a higher trustworthiness, security, usability, and flexibility by adding the ability to automatically adapt their security functionality depending on changes in the situational context.

To support and evaluate this idea, we have set-up an authentication experiment where different sources of location information contribute to evaluating the degree of authentication of a user. In fact, we have devised and prototypically implemented a location-aware component that combines user identity tokens with location information extracted from an arbitrary set of sensors. Different sensed identity tokens (e.g., RFID badge or Bluetooth-enabled mobile phone) are associated to location information and are fused to calculate the probability that the user is at a certain location. This probability is used by an application to determine the user's authentication level: the lower the probability, the lower the authentication level. We show that by fusing various sources of location information that are available over time, the confidence in the identity of the user can be effectively evaluated.

The structure of the paper is as follows. Section 2 introduces and discusses the security features of context-sensitive adaptive authentication. Section 3 discusses the sensor fusing probability algorithm that serves our goal and also provides several simulation results to illustrate and validate the behavior of the algorithm. Section 4 describes the design of our system to support context sensitive adaptive authentication. This is explained by means of an application scenario that we have implemented. Section 5 discusses several essential features of our system. Section 6 compares our approach with related work in the field. Finally, Section 7 presents the conclusions of our work and future outlook.

2 Authentication with Context

In computer security, authentication is the process of attempting to verify the digital identity of the user. In a ubiquitous context-aware computing environment, users can authenticate themselves using a variety of means with a variable degree of reliability. User authentication means can be classified into the following three classes:

1. *what the user is* (e.g., fingerprint or other unique biometric identifiers);
2. *what the user has* (e.g., ID card, security token, or cell phone);
3. *what the user knows* (e.g., a password, a pass phrase or a PIN).

Most of today's widely available authentication solutions can not guarantee very high quality user identification. For instance, if a user enters the right username/password combination, there is still a certain amount of uncertainty on whether we are really dealing with this user; the combination could have been eavesdropped. Even the use of biometric identification solutions is not 100% accurate; there is always a chance for a false positive or negative. Clearly, the assumption that a user's identity can be verified with absolute certainty is unrealistic in most of the scenarios, but the confidence on the user's identity can increase with the adoption of clever strategies. Generally, the combination of methods such as a bankcard and a PIN (called "two-factor authentication") or the username/password authentication solution with the biometric identification, results in a more reliable user identification. Potentially, the more solutions that can be used to authenticate the user, the stronger the system's confidence in that user's identity will be.

Formally, if A_1, A_2, ..., A_n, are the confidence values associated with different authentication methods (e.g., RFID, username/password, Bluetooth, biometrics) then, under the assumption that all authentication methods have yielded a positive outcome, the overall confidence OC associated with the composite authentication solution may, using e.g. probability theory, be calculated with the following formula [10]:

$$OC = 1 - (1 - A_1)(1 - A_2)....(1 - A_n)$$ (1)

Here, A_i's are authentication values in the real interval [0, 1], where 1 expresses the highest confidence, and 0 the lowest. Informally, Eq. (1) says that sources with low confidence have a weak impact, while sources with a high confidence bring to a higher OC. For example, if the authentication confidence of an RFID badge is A_{RFID} = 0.80 and that of a Bluetooth (BT) device is A_{BT} = 0.60, the resulting OC is 0.92.

Although Eq. (1) represents a significant improvement with respect to single source authentication, the use of combined identification sources is always reliable as well. For example, what if the RFID badge and the BT device of the same user are almost simultaneously used at two completely different locations? In addition, what if the time interval between two different authentication sources of the same user is long? Moreover, we also note that the determination of meaningful authentication confidence values for each authentication technique (i.e., the values $A_1,...,A_n$) proves difficult and is strongly application specific [11]. As a solution, we propose to look at the context of the authentication process, specifically, location and time. In our vision, location and time constitute the fourth authentication class, namely *"where the user is, and when"*. Thus, in addition to combined identification inputs, the use of sensor information allows the system to reason about the belief in the composite information to come to a higher authentication status.

User authentication information derived from sensors in the environment can result in a significant enhancement of the confidence strength of the identification service. For instance, if a RFID reader at the entrance of the building has identified a user via his RFID badge, and at the same time, a BT dongle at the third floor of the building has identified the BT-enabled Personal Device Assistant (PDA) of the same user, granting the user access to confidential files via his PDA should be restricted. This restriction arises from the contradictory location information of the two identification measures and it results into a lower accuracy of the user identification and therefore into a lower authentication level. On the contrary, if the locations match, the confidence in the identity of the user should be higher. In other words, the overall confidence in the identity of the user is also influenced by the location and time associated with the respective RFID and BT identifiers.

In this paper, we approximate the authentication confidence with the probability of the user being at a given location at a particular time. Thus:

$$OC \approx P_m \qquad\qquad (2)$$

Here, P_m is the probability of the user being at a certain location based upon the composite location information of different sensed identity tokens that are associated to him. In the case of our RFID and BT example, P_m is the probability that the user is at the location where he has requested access to resources; that probability is based upon the locations of the sensors that have sensed the user's RFID and BT tokens.

3 Location Sensor Fusing

In the following sections we describe and discuss the algorithm used to calculate the probability P_m according to the available location information of that user.

3.1 Theoretical Study

We start with some notation and with the formal statement of the problem.

Sensors and Cells. Our setting is a region T (e.g., a building or a city). The location of a user u who is somewhere in T, can be detected via his personal devices by different sensor sources $S^1,...,S^n$, where indexes $1,...,n$ stands for type of sensors. Each sensor type S^X is a set of sensors $S_1^X, S_2^X,..., S_{|X|}^X$ with *non-necessarily disjoint* coverage regions or cells of $C_1^X, C_2^X,..., C_{|X|}^X$, respectively. For example, a S^{RFID}

source can include n sensors $S_1^{RFID}, S_2^{RFID}, ..., S_n^{RFID}$ whose cells are $C_1^{RFID}, C_2^{RFID}, ..., C_n^{RFID}$. The area of any arbitrary region A is denoted by $\|A\|$.

Our first assumption regards the area covered by each sensor source.

Assumption 1. For each source S^X, cells $C_1^X, C_2^X, ..., C_{|X|}^X$ partition the whole T. With $S_i^X = 1$ we denote the event that user u is detected by sensor S_i^X. This event indicates that the sensor detects the user's corresponding device in cell C_i^X. A consequence of Assumption 1 is that if $S_i^X = 1$, then $S_j^X = 0, \forall j \neq i$. This can be regarded as a quantization of the user location to one cell of the sensor source.

Sensor Error Model. Generally errors are associated with such a quantization process. Three error causes can be identified. The first error depends on the reliability of sensors themselves. For instance, a BT device can be detected within five meters from a dongle 95% of the time. In addition, sensors also have a probability of misidentification, i.e. the sensor incorrectly says the device is in the area or misses the presence of the user. The second error depends on the probability of the user carrying the device corresponding to that sensor source (e.g., RFID reader). All location sensing technologies rely on the user carrying or using a certain device like a RFID badge, BT-enabled PDA or smart phone, WLAN enabled laptop or even a keyboard. So knowing the location of the device implies that the location of the user is known as well. Finally, the third error is introduced by the "freshness" of the sensor information; the older the information the less reliable it is.

Indeed, most product specifications of location sensing technologies give the conditional probability that the device is correctly detected if it is present in its cell. Let's denote this probability by $P(S_i^X = 1 | u \in C_i^X) = q_i^X$. The probability of the complement event, i.e., $p'_i^X = P(S_i^X = 0 | u \in C_i^X) = 1 - q_i^X$, is called "false negative" probability. In addition, location technologies have a probability of misidentification, that is $P(S_i^X = 1 | u \notin C_i^X) = p_i^X$. This probability is called "false positive" probability. The aforementioned three sources of errors are contributors to the false positive and false negative probabilities p'_i^X and p_i^X.

Observation. *Referring to Assumption 1, we can also release that* $C_1^X, C_2^X, ..., C_{|X|}^X$ are mutually disjoint.

In fact, this requirement is a way of obtaining a quantization of the user location to one cell of the sensor source. This quantization is needed to cope with two or more sensors (of the same type) detecting the same user at different locations. We note that quantization can be obtained at the sensor source level as well by selecting just one sensor in case of inter-type conflict. A detailed explanation about how to carry out this quantization is out of our scope here.

Depending on the outcomes of the sensors of a certain type and on the quantization method used, one can derive per sensor type p^X and q^X values from p_i^X and q_i^X values. For example, consider the case where only one sensor gets triggered and the pairs (p_i^X , q_i^X) are the same for $i : 1...|X|$. Then, for that sensor type, we have:

$$q^X = P(S_i^X = 1|u \in C_i^X) = q_i^X(1-p_i^X)^{|X|-1} \text{ and } p^X = P(S_i^X = 1|u \notin C_i^X) = p_i^X(1-q_i^X)(1-p_i^X)^{|X|-2}.$$

From this point on we will base our calculation on these p^X and q^X values that are independent of the individual sensors.

Proposition 1. For each $R \subseteq C_i^X$, $P(S_i^X = 1|u \in R) = q^X$. For each $R \subset \overline{C}_i^X$, where $\overline{C}_i^X$ denotes the complement set of C_i^X in T , $P(S_i^X = 1|u \in R) = p^X$.

The following assumption regards the typology of source errors that we admit in our setting.

Assumption 2. We consider only false positive errors in our setting. This assumption follows the quantization process imposed by each sensor type. That is our sensor fusion method considers only those sensors types that have detected a user's presence and each sensor type detects a user's presence in only one of their cells.

Our last assumption concerns the independence of sensors. Though strictly speaking sensors are not mutually independent, the following (weaker) conditional independency is reasonable true for most sensors setting.

Assumption 3. Sensors are conditionally independent, that is:

$$If P(S_i^X = 1|u \in C_i^X) = q^X \text{ then } \forall j,Y \ P(S_i^X = 1|u \in C_i^X, S_j^Y = 1) = q^X$$
$$If \ P(S_i^X = 1|u \notin C_i^X) = p^X \text{ then } \forall j,Y \ P(S_i^X = 1|u \notin C_i^X, S_j^Y = 1) = p^X$$

Informally, the position of the user inside or outside of a sensor's cell determines the behavior of the sensor. In other words, the behavior of the sensor is independent of whether or not other sensors of different types are triggered.

Fusing Sensor Sources. Fusing n sensor sources concerns the computation of the probability that user u is in a region of interest I , given that n sources $(S_{i_1}^1,...,S_{i_n}^n) \subseteq S^1 \times ... \times S^n$ have indicated that the user is in their cells $C_{i_1}^1,...,C_{i_n}^n$, respectively (i.e., $S_{i_1}^1 = 1,..., S_{i_n}^n = 1$). Thus we are interested in the probability

$$P_m = P(u \in I|S_{i_1}^1 = 1,...,S_{i_n}^n = 1) = \frac{P(u \in I, S_{i_1}^1 = 1,...,S_{i_n}^n = 1)}{P(S_{i_1}^1 = 1,...,S_{i_n}^n = 1)} \quad (3)$$

In the following we derive two relations for the numerator and denominator of Eq. (3). Note that the effect of region I appears only in the numerator relation.

Denominator. Cells $C_{i_1}^1,...,C_{i_n}^n$ are like an order-n Venn diagram that includes n simple closed curves in the T plane. These curves partition the T plane into maximum 2^n connected and disjoint regions $R_1,...,R_K$, where $K \le 2^n$, $\sum_{k=1}^{K}\|R_k\|=\|T\|$ or $\cup_{k=1}^{K}R_k = T$, $\forall k \ne k': R_k \cap R_{k'} = \phi$; and $R_k \subset C_{i_j}^j$ XOR $R_k \subset \overline{C}_{i_j}^j$ for $j = 1...n$. Term $P(S_{i_1}^1 = 1,..., S_{i_n}^n = 1)$ can be rewritten as follows:

$$\sum_{k=1}^{K} P(u \in R_k) \cdot P(S_{i_1}^1 = 1,..., S_{i_n}^n = 1 | u \in R_k) = \sum_{k=1}^{K} P(u \in R_k) \prod_{j=1}^{n} (q^j)^{\alpha_{kj}} (p^j)^{1-\alpha_{kj}} \quad (4)$$

Here $\alpha_{kj} = 1$ if $R_k \subset C_{i_j}^j$ and $\alpha_{kj} = 0$ if $R_k \subset \overline{C}_{i_j}^j$. In Eq. (4), we used the independency condition of sensors because either $R_k \subset C_{i_j}^j$ or $R_k \subset \overline{C}_{i_j}^j$ for $j = 1...n$. Assuming a uniform distribution for user location in all regions, we have:

$$P(S_{i_1}^1 = 1,..., S_{i_n}^n = 1) = \frac{1}{\|T\|}\sum_{k=1}^{K} \|R_k\| \cdot \prod_{j=1}^{n}(q^j)^{\alpha_{kj}} (p^j)^{1-\alpha_{kj}} \quad (5)$$

The time complexity of Eq. (5) grows exponentially in n; hereto we need to calculate Eq. (4) or (5) for all disjoint regions obtained from intersections of n cells $C_{i_1}^1,...,C_{i_n}^n$.

Numerator. The numerator can be written as:

$$P(u \in I, S_{i_1}^1 = 1,..., S_{i_n}^n = 1) = P(u \in I)P(S_{i_1}^1 = 1,..., S_{i_n}^n = 1 | u \in I) \quad (6)$$

In the second term of Eq. (6) it is given that user $u \in I$. The intersections of cells $C_{i_1}^1,...,C_{i_n}^n$ with region I are like an order-n Venn diagram with n simple closed curves in the I plane. These closed curves partition the plane into maximum 2^n connected and disjoint regions $R_1^I,...,R_{K'}^I$, where $K' \le 2^n$; $\sum_{k=1}^{K'}\|R_k^I\|=\|I\|$ or $\cup_{k=1}^{K'}R_k^I = I$, $\forall k \ne k': R_k^I \cap R_{k'}^I = \phi$; and $R_k^I \subset C_{i_j}^j \oplus R_k^I \subset \overline{C}_{i_j}^j$ for $j = 1...n$. Here $\oplus$ denote exclusive disjunction. The term $P(S_{i_1}^1 = 1,..., S_{i_n}^n = 1 | u \in I)$ in Eq. (6) can be rewritten as follows:

$$\sum_{k=1}^{K'}P(u \in R_k^I | u \in I)P(S_{i_1}^1 = 1,..., S_{i_n}^n = 1 | u \in I, u \in R_k^I) = \sum_{k=1}^{K'}P(u \in R_k^I | u \in I)\prod_{j=1}^{n}(q^j)^{\alpha_{kj}}(p^j)^{1-\alpha_{kj}} \quad (7)$$

In which: $\alpha_{kj} = 1$ if $R_k^I \subset C_{i_j}^j$ and $\alpha_{kj} = 0$ if $R_k^I \subset \overline{C}_{i_j}^j$. Eq. (7) uses the independency condition of sensors because either $R_k^I \subset C_{i_j}^j$ or $R_k^I \subset \overline{C}_{i_j}^j$, $j = 1...n$. In

case of uniform distribution for user location in all regions, we have:
$P(S_{i_1}^1 = 1,...,S_{i_n}^n = 1 | u \in I) = \frac{1}{\|I\|} \sum_{k=1}^{K} \|R_k^I\| \cdot \prod_{j=1}^{n} (q^j)^{\alpha_{kj}} (p^j)^{1-\alpha_{kj}}$. Thus from Eq. (6):

$$P(u \in I, S_{i_1}^1 = 1,...,S_{i_n}^n = 1) = \frac{1}{\|T\|} \sum_{k=1}^{K} \|R_k^I\| \cdot \prod_{j=1}^{n} (q^j)^{\alpha_{kj}} (p^j)^{1-\alpha_{kj}} \qquad (8)$$

Assuming that $(I \subset C_{i_j}^j) \oplus (I \subset \overline{C}_{i_j}^j)$, $j = 1...n$ (i.e., the area of interest does not overlap with both areas of each cell) then from Eq. (6) and the independency condition of sensors we can directly derive:

$$P(u \in I, S_{i_1}^1 = 1,..., S_{i_n}^n = 1) = P(u \in I) \prod_{j=1}^{n} P(S_{i_j}^j = 1 | u \in I, S_{i_1}^1 = 1,..., S_{i_{j-1}}^{j-1} = 1)$$

$$= P(u \in I) \prod_{j=1}^{n} P(S_{i_j}^j = 1 | (u \in C_{i_j}^j) \oplus (u \in \overline{C}_{i_j}^j)) = P(u \in I) \prod_{j=1}^{n} (q^j)^{\alpha_j} (p^j)^{1-\alpha_j} \qquad (9)$$

Here $\alpha_j = 1$ if $I \subset C_{i_j}^j$ and $\alpha_j = 0$ if $I \subset \overline{C}_{i_j}^j$. The time complexity of Eq. (8) grows exponential in n, but with the simplification we used in Eq. (9), it becomes linear in n.

3.2 Simulations

To illustrate the principle of location sensitive adaptive authentication we simulated two extreme situations: overlapping and non-overlapping location sensor information. The first situation assumes a BT device that is sensed in an area that is completely covering the area of interest I. Additionally, there is a second identity token, an RFID badge, that starts with zero coverage and slowly starts overlapping with I. The second situation deals with a BT device that is detected in an area completely outside I whereas the RFID badge slowly starts overlapping with I.

For the simulations the following input data was used: T = 5000 m^2, I = 16 m^2, BT's cell (C_I) is a circle with a fixed radius of 8 m, p = 0.01 and q = 0.95, and RFID's

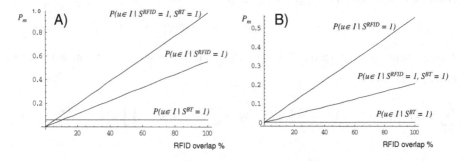

Fig. 1. Individual and fused RFID and BT identity probability as a function of the RFID cell area overlap with the area of interest I. For Fig. 1A the BT identity token has a cell area that is constantly overlapping with I; for Fig. 1B the BT identity token cell area has no overlap with I.

cell (C_2) is a circle with a fixed radius of 1 m, and the relative $p = 0.0005$ and $q = 0.99$. Fig. 1A shows the outcome of P_m for the overlapping situation. Clearly observable is the strong increase of P_m in the case that the RFID and BT location sensor information agree with each other. In case of maximal overlap, the individual BT or RFID identification probabilities of 7% or 55%, respectively, sum up to a 'fused' probability of 98%. In case of a conflicting BT identity token, the fused RFID and BT probability drops considerably to less than 25% (Fig. 1B).

4 Design

This section describes the design of a system that uses location information to determine and dynamically adapt the authentication level of a user. The goal of our implementation is to demonstrate and validate the context-sensitive adaptive authentication scheme. The location information from multiple different location sensors is used to calculate P_m, i.e., the probability of the user being in a certain location of interest I, which is supposed to be the location from where the user forwards his access request. The result is used to determine the authentication level of the user and to modify his authorization level accordingly.

For obtaining sensor location information we used the Context Management Framework (CMF) described in [12]. The CMF enables processing and exchange of heterogeneous context information collected from various sensors, is distributed over multiple administrative domains, and stems from different protocol layers. Examples of context information supplied by the CMF include location coordinates via GPS receivers, WLAN access points associations, RFID reader data, BT scan measurements, desktop keyboard typing, and Outlook Calendar meetings.

4.1 Message Flow

For calculating the probability values we implemented the User Location Probability Calculator (ULPC). The ULPC is a context aware component that collects and

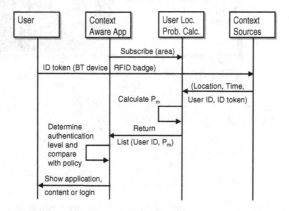

Fig. 2. Message flow for location probability based adaptive authentication

reasons about context information of users obtained from the CMF. Fig. 2 shows the message flow for application service access that relies on a location probability based authentication method.

Application service access is provided by the ULPC in collaboration with the Context Provider (i.e. CMF). The application subscribes itself to the ULPC for obtaining probability measures of users in a certain area of interest. When a user is detected by location sources of the Context Provider, this information is sent to the ULPC together with a timestamp indicating the time the user has been detected. The ULPC caches this information for all sensor types (e.g. RFID, BT, WLAN, keyboard, etc.) and uses equations (5) and (9) to determine the probability a user is in the area of interest for each new input it receives. The outcome of the calculation is communicated to the application that uses it to determine the actual level of authentication. If other persons are in the area of interest as well, their probability will be communicated as well to the application. If the level of authentication is not sufficiently high the application may ask the user to provide stronger identification information by e.g. presenting a username/password window or by asking for performing an iris scan. In case of multiple persons, the confidentiality of the information shown may be harmed and therefore, the information will be removed from the screen. This is an example of context-aware adaptive confidentiality.

4.2 Buddy Spotter Application

To demonstrate the concept of location sensitive adaptive authentication we build a 'Buddy Spotter' application that allows users to locate their buddies or colleagues. A screen dump of the application's user interface is shown in Fig. 3.

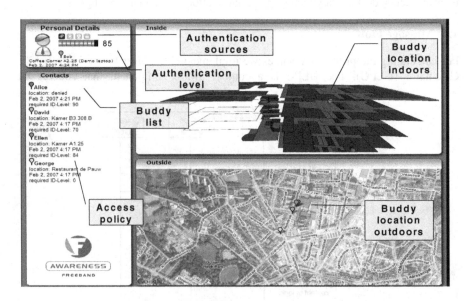

Fig. 3. Screen dump of the Buddy Spotter application

The right-hand panels provide information regarding the user's buddy locations in- and outside the office building. The upper left-hand panel shows the authentication level of the user by means of an "ID-level" display. The ID-level display informs the user about the confidence the application has in his identity and also what sensor information was used to come to this (e.g. by means of BT, RFID, etc.). The lower left-hand panel shows the buddy list of the user. The buddies can specify an ID-level that is required prior to getting access to their location information. If the user's ID-level is not sufficiently high, the location of the buddy shall not be shown. This functionality allows the buddies to preserve their privacy to a certain extend.

5 Discussion

Our location sensitive adaptive security solution may raise questions regarding time dependency, trustworthiness and usability. This section discusses them.

5.1 Dynamicity of Authentication Level

Due to the time-dependent character of location information, the ULPC component calculates the location probability regularly, i.e. every time it gets new location information events from the Context Provider. For instance, a BT device is sensed every five seconds when it is in the neighborhood of a BT dongle. This results in an update of the location probability of the user. However, the RFID authentication method in particular is much more time-sensitive as it has a very accurate location quality and requires an explicit act of the user, i.e., swiping his RFID badge in front of the reader. This means that the RFID location probability drops very rapidly in time or, in other words, the coverage area of the RFID sensor becomes larger depending on the mobility of the user (this is illustrated in Fig. 4).

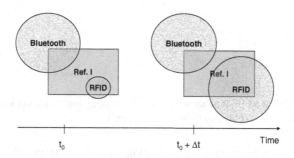

Fig. 4. Changing of RFID cell area in time

With the dropping probability, the level of authentication will drop in time as well, and subsequently the user will automatically lose authorization to resources that require a higher level of authentication than is. The application may then ask the user to upgrade his level of authentication by for instance swiping his RFID badge again.

The decay function that might be applied to the position probability depends amongst others on the mechanism that is used to determine the location of the user and the mobility of the user: the bigger the coverage area of the sensor, the less fast will the user move to another coverage area and thus the slower the authentication level will drop. In order to be able to fuse location information from different sensors the ULPC has to determine, based on timestamps, the time intervals between the most recent location update of a sensor and the locations cached from previous sensors and recalculate their coverage area based upon the mobility of the user. Though we assumed in our model the user to be moving with an average speed of 5 km/hour in the office building (see Fig. 5), this mobility pattern might be sensor and application dependent. In our calculations we assumed the following simple model to describe the mobility of the user:

$$v_{average} * p_{mobility} = v_{effective} \tag{10}$$

With $v_{average}$ representing the average velocity of the user (i.e. 5 km/hour), $p_{mobility}$ the chance that the user will walk away (e.g. 4%) and $v_{effective}$ the effective velocity of the user (i.e. 5.6 cm/second). $p_{mobility}$ strongly depends on the type of sensor and its location. As this is just a simple approach, obviously more research needs to be done here to determine a correct mobility pattern for each sensor type (see also [13]).

The increment of P_m in Fig. 5 is explained by the increase of overlap of the RFID cell with the area of interest I. After the turning point, the expanding RFID cell will have relatively less overlap with I.

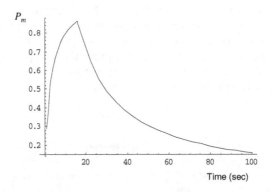

Fig. 5. Dropping of the RFID probability level in time assuming a user velocity of 5 km/hour. All other parameters were similar to those used for Fig. 1.

During application interaction more location information from new sensors may be obtained. This will result in a new level of authentication and subsequent access rights. The ID-level display of the Buddy Spotter application thus changes dynamically in time and so do the access rights the user has.

5.2 Trustworthiness

Trust plays an important role during the exchange of context information. We assume that the users trust the CMF; they have given consent to the CMF to collect their context information. Regarding the privacy of the user towards the application there is no difference compared to the potential loss of privacy when using traditional authentication services as username/password or security token. The application only obtains a probability value that the user is at a certain location. Furthermore, buddies are allowed to specify their privacy policies in an easy and efficient manner: they only have to set the proper required ID-level. Allowing users to control their personal information is an active area of research. Though many solutions have been proposed, most of them fail in offering user friendliness (e.g. [14]). We belief our parameterized approach offers the user a usable approach to control access to personal information.

5.3 Usability

Usability of security is an extremely important element of IT-security. Direct user involvement is often required in a security service. Two forms of involvement can be distinguished: action and conclusion [14]. An action involves the user to explicitly enter his username/password or swipe his RFID-badge. A security conclusion allows the users to observe some relevant security evidence regarding the security state of the system. For instance, the closed padlock at the bottom of the browser is a security conclusion. Usability principles related to security actions and conclusions are typically expressed in terms of user understanding, knowledge, mental and physical load, and willingness [14]. Our location based authentication approach fulfils several of these principles. Regarding authentication actions we strive to minimize user involvement as much as possible since that is the basic starting point of our approach. In our case the user only has to swipe his RFID badge in front of the reader to nevertheless obtain a relatively high level of authentication after fusing the information with other sources in a transparent manner. Moreover, authentication is also possible without the use of RFID. If for instance BT and WLAN are used, the user is authenticated without having to perform an explicit act.

The ID-level display informs the user about his authentication level and also shows the means by which this level has been achieved (e.g. BT, RFID, WLAN). Our first experiences indicate that users appreciate this information. In particular the benefit of being authenticated in a minimal-intrusive way is appealing. Further tests however are needed to optimize the user experience.

One could argue about the meaningfulness of the ID-level: what does an authentication level of 75% mean? The actual point of discussion here is about parameterizing security and how useful this is. "If you can not measure it, you can not improve it" (1883) — one of Lord Kelvin's famous quotations that may be very applicable to our adaptive security approach. In order to be able to measure the strength of security functions one must first parameterize them. We already mentioned that this is not easy because it involves making the performance of the security function measurable. However, a standardized reference framework that is required for this purpose is lacking. Objective and subjective notions regarding security levels are often mixed making it hard to come to such a reference framework.

We don't claim that our solution is better than other, existing solutions that have proven their usefulness already in practice for many years. Location sensitive authentication may prove useful in situations that require minimal intrusive and flexible authentication. Such situations are for instance in a hospital where medical personnel frequently has to enter credentials in order to access medical information

[15], emergency situations where access to medical information may be needed on an ad-hoc basis, or ubiquitous computing environments. In other situations, the use of context information can be very well used as an additional parameter to enhance the level of traditional authentication measures, for instance, by combining username/password with location or calendar information prior to granting access.

6 Related Work

A key element of our work consists of making security adaptive. Though several adaptive approaches for security have been described [1], [2]), the use of context information as a security adaptation parameter hasn't been considered.

To realize adaptation, we parameterize authentication. We do that on the basis of the probability that the user is at a certain location. Similarly, Ganger proposed the concept of authentication confidences as another approach to parameterize security [3]. Authentication confidences refine the current yes-or-no authentication decisions, allowing systems to cleanly provide partial access rights to authenticated users whose identities are suspect. The proposed solution direction exists of a combination of different authentication technologies. In a similar context, Noble and Corner propose a transient authentication model. In this model, a user wears a small hardware token that constantly authenticates the user to other devices over a short-range wireless link [4]. Covington et al. describe how to parameterize the authentication function [5]. Levin et al. proposed a Quality of Security Service mechanism for modulating and provisioning of predictable security service levels to users [6]. We observe that in most cases the levels of security are relatively static and pre-defined and that there is no relationship with the situational context as a means to determine the actual level of security in a dynamic and flexible manner.

The use of context information for security purposes is not new. In 2003, Leo Marcus introduced and described the logical foundations of the adaptive security infrastructure concept that also takes the environment into account [7]. Our work builds upon these foundations. Similarly, Kouadri et al. proposed a conceptual model for context-based authorizations tuning. This model offers a fine-grained control over access to a protected resource, based on a set of user's and environment state and information [8]. In [9], location and velocity information is used as a means to allow train travelers access to services offered in the train. Hager investigated methods to determine appropriate security protocols for specific wireless network applications [16]. The specific problem being addressed was that there are tradeoffs between security, performance and efficiency among current and proposed security protocols and that these tradeoffs are influenced by the constrained network capacity and limited mobile nodes (i.e. the context). Yee and Korba propose a context-aware security policy agent that is responsible for selecting security services and mechanisms for mobile Internet services according to the user's preferences, power of the mobile device and location [17]. Furthermore, security policy negotiation between the service provider and consumer is described by Yee and Korba as well [18]. An approach to building security services for context-aware environments with a strong focus on the design of security services that incorporate the use of security-relevant "context" to provide flexible access control and policy enforcement is described in

[19]. This approach is based on the concept of context-dependent roles. The related work on context aware security focuses in general on using context information for authorization purposes while we focus on context-aware user authentication.

We determine the probability that the user is in a certain area of interest. Our probability approach resembles that of [13] that describes a middleware approach for probabilistic location determination in general. There are other probability approaches for sensor data fusion but they often have a different goal. Abowd et al. describe in a Location Service that fuses sensor information using a fairly straightforward temporal and heuristic algorithm for the purpose of customized communication [20]. Bohn and Vogt [21] use a probabilistic positioning service that employs an available ubiquitous computing infrastructure for the localization of mobile devices. Data from these sources are transformed independently of each other into an abstract representation of location estimates. By means of a probabilistic fusion process, these estimates are then combined into a single position value.

7 Conclusions

The transparent nature of pervasive and ubiquitous computing environments where context information is used to enhance service experience motivates the need for security functionality that will be transparent, customized, and non-intrusive. The context sensitive adaptive security paradigm allows for adaptation of the security depending on a set of relevant information collected from the dynamic environment and the preferences and capabilities of the interacting entities, i.e. the context. As the environment evolves, the context changes and so should security in order to dynamically cope with new requirements. We argue that security services, like authentication and access control, can be made less intrusive, more intelligent, and able to adapt to the rapidly changing contexts of the environment. To validate this argument we show that by fusing various sources of location information that are available over time, the confidence in the user identity associated to the sensed devices can be increased considerably. The outcome of the location fusion and reasoning process is a value that expresses the probability that the user is at a certain location. This probability is used as a measure not only to authenticate the user based on location information but to parameterize the authentication level as well making it thereby much more adaptive to changing situational circumstances. In particular the heterogeneity of the sensed personal devices strongly contributes to the enhancement and robustness of the location-based authentication. A user is less likely to lose two or more personal devices at the same time. Furthermore, face recognition technology and calendar information could be used as additional, independent measures that help to identify the user based upon his location. For instance a web cam can identify a user at a certain location and Outlook Calendar may tell the ULPC that the user is out of office or in a certain meeting room. Future work will focus on such extensions as well as improvements in the algorithms used and user experience validation.

Acknowledgments. This research has been supported by the Dutch Freeband Communication Research Program (AWARENESS project) under contract BSIK 03025.

References

1. Schneck, P.A., Schwan, K.: Dynamic Authentication for High-Performance Networked Applications. In: Proc. of the 6th International Workshop on Quality of Service (IWQoS 1998) Napa, California, USA, pp. 127–136 (1998)
2. Ryutov, T., Zhou, R., Neumann, C., Leithead, T., Seamons, K.E.: Adaptive Trust Negotiation and Access Control. In: SACMAT 2005. Proc. of the ACM Symposium on Access Control Models and Technologies, Stockholm, Sweden, pp. 139–146. ACM Press, New York (2005)
3. Ganger, G.B.: Authentication Confidences. In: Proc. of the Eighth Workshop on Hot Topics in Operating Systems (HotOS-VII 2001), Elmau/Oberbayern, Germany, p. 169 (2001)
4. Noble, B.D., Corner, M.D.: The Case for Transient Authentication. In: Proc. of the 10th ACM SIGOPS European Workshop, Saint-Emilion, France, pp. 24–29. ACM Press, New York (2002)
5. Covington, M.J., Ahamad, M., Essa, I., Venkateswaran, H.: Parameterized Authentication. In: Samarati, P., Ryan, P.Y A, Gollmann, D., Molva, R. (eds.) ESORICS 2004. LNCS, vol. 3193, pp. 276–292. Springer, Heidelberg (2004)
6. Levin, T.E., Irvine, C.E., Spyropoulou, E.: Quality of Security Service: Adaptive Security. The Handbook of Information Security. In: Threats, Vulnerabilities, Prevention, Detection and Management, vol. III, John Wiley & Sons, Inc, Chichester (2005)
7. Marcus, L.: Local and Global Requirements in an Adaptive Security Infrastructure. In: International Workshop on Requirements for High Assurance Systems (RHAS), Monterey Bay, California (2003)
8. Kouadri Mostéfaoui, G., Brézillon, P.: A Generic Framework for Context-Based Distributed Authorizations. In: Blackburn, P., Ghidini, C., Turner, R.M., Giunchiglia, F. (eds.) CONTEXT 2003. LNCS, vol. 2680, pp. 204–217. Springer, Heidelberg (2003)
9. Hulsebosch, R.J., Salden, A.H., Bargh, M.S., Ebben, P.W.G., Reitsma, J.: Context sensitive access control. In: SACMAT 2005. Proc. of the tenth ACM symposium on Access control models and technologies, Stockholm, Sweden, pp. 111–119. ACM Press, New York (2005)
10. Ranganathan, A., Al-Muhtadi, J., Campbell, R.H.: Reasoning About Uncertain Contexts in Pervasive Computing Environments. In: Pervasive Computing, vol. 3(2), pp. 62–70. IEEE, Los Alamitos (2004)
11. Belovin, S.M.: On the Brittleness of Software and the Infeasibility of Security Metrics. IEEE Security and Privacy 4(4) (2006)
12. van Kranenburg, H., Bargh, M.S., Iacob, S., Peddemors, A.: A Context Management Framework for Supporting Context Aware Distributed Applications. IEEE Communications Magazine 44(8), 67–74 (2006)
13. Ranganathan, A., Al-Muhtadi, J., Chetan, S., Campbell, R., Mickunas, M.D.: MiddleWhere: A Middleware for Location Awareness in Ubiquitous Computing Applications. In: Jacobsen, H.-A. (ed.) Middleware 2004. LNCS, vol. 3231, pp. 397–416. Springer, Heidelberg (2004)
14. Jøsang, A., AlZomai, M., Suriadi, S.: Usability and Privacy in Identity Management Architectures. In: Brankovic, L., Steketee, C. (eds.) Proc. Fifth Australasian Information Security Workshop Privacy Enhancing Technologies (AISW 2007), Ballarat, Australia, pp. 143–152 (2007)
15. Bardram, J.: The trouble with login: on usability and computer security in ubiquitous computing. Personal and Ubiquitous Computing 9(6), 357–367 (2005)

16. Hager, C.T.R.: Context Aware and Adaptive Security for Wireless Networks. PhD thesis, Virginia Polytechnic Institute and State University (2004)
17. Yee, G., Korba, L.: Context-Aware Security Policy Agent for Mobile Internet Services. In: Proc. of the 2005 IFIP International Conference on Intelligence in Communication Systems, Montréal, Québec, Canada, pp. 249–260 (2005)
18. Yee, G., Korba, L.: Negotiated Security Policies for E-Services and Web Services. In: ICWS 2005. Proc. of the 2005 IEEE International Conference on Web Services, San Diego, California, pp. 605–612. IEEE Computer Society Press, Los Alamitos (2005)
19. Covington, M.J., Fogla, P., Zhan, Z., Ahamad, M.: A Context-Aware Security Architecture for Emerging Applications. In: ACSAC 2002. Proc. of the 18th Annual Computer Security Applications Conference, Las Vegas, Nevada, pp. 249–258 (2002)
20. Abowd, G.D., Battestini, A., O'Connell, T.: The Location Service: A Framework for Handling Multiple Location Sensing Technologies (2002), http://www.awarehome.gatech.edu/publications/location_service.pdf
21. Bohn, J., Vogt, H.: Robust Probabilistic Positioning Based on High-Level Sensor-Fusion and Map Knowledge. Technical Report No. 421, ETH Zurich (2003)

A Sensor Placement Approach for the Monitoring of Indoor Scenes

Pierre David, Vincent Idasiak, and Frédéric Kratz

Laboratoire Vision et Robotique, UPRES EA 2078, 88 Boulevard Lahitolle,
18020 Bourges, France
{pierre.david, vincent.idasiak, frederic.kratz}@ensi-bourges.fr

Abstract. Within the framework of a French project, which aims at developing a new human presence sensor, we intend to design a sensor system simulator. During the establishment of the requirements of that new sensor we raised that the mission of a global scene survey could only be performed by a collection of several systems using very diverse technologies. This article presents the development of a method for the placement of multi-technology and multi-sensor systems. The considered environments are room or set of rooms in office buildings or individual homes. We will explain how we managed to represent the use of different sensors considering their various environments. Then, the way of exploiting these models using genetic algorithms is discussed. Those models are oriented for finding system placement and therefore for helping sensor networks deployment.

Keywords: Simulator, sensor network, genetic algorithm, sensor placement.

1 Introduction

The project in which we are involved is called Capthom. It was set up to design a new low-cost human presence detector characterized by a high reliability. The development of new human presence sensors is currently needed by research projects on energy management (ERGDOM [1]) and on the medical monitoring of the elderly (GERHOME [2], PROSAFE [3], SOPRANO [4]). It is of primary importance in this type of industrial project to be able to present simulations to evaluate the future capacities of the system, both in functional and dysfunctional terms. We thus wish to create a simulation software capable of testing the various research considered for the development of the system. This kind of simulator could also be used in the deployment phase of the systems as a placement tool. The goal of such a use is to find the best way of installing a sensor network considering an established problem composed of detection objectives, detection conditions and a defined scene. During the creation of the tool and of the models the state of art led us to consider mainly camera placement works. We were therefore inspired by this way of modeling and we tried to generalize it to all types of sensors. The study of sensor placement can be very profitable in terms of money savings by limiting the number of needed systems; moreover, an intelligent placement dramatically improves the performances and also

G. Kortuem et al. (Eds.): EuroSSC 2007, LNCS 4793, pp. 110–125, 2007.

insures a growth in the lifetime of the components. We decided to characterize a sensor system by a special attribute, which we called the efficient zone, that is to say the zone in which a considered sensor is able to catch and interpret its targeted physical flow. We must also be able to manage scenarios of multiple use as well as very different environments. Hence it was necessary to make the creation and characterization of new environments available. We thus defined a representation of the scenes of use and a way of modeling their characteristics. So we defined two models, one for the sensors and the other for their environment, meeting and merging them as a whole model representing the problem of sensor selection and placement. We will describe in the next paragraphs the type of problem that we tried to solve and model, as well as the model that we set up to describe the scene. We will insist on the model of the scene and the elements characterizing the environment. We will then detail a method of modeling the sensors before presenting our way of solving the problem. We will show in this last part the adaptability and the various parameter settings of this method.

2 Definition of Our Problem and First Step in Modeling

Our first work was carried out to optimize and evaluate the placement of the future Capthom sensors. We followed the technique developed in the problems of camera placement. These problems intervene in many fields: photogrammetry, video surveillance, camera management for interfaces with virtual worlds and simulation of cinematographic shot design. Our topic, which is sensor exploitation, seems to be closer to video surveillance. In paragraph 2.1 we will present a review of the problems treated for camera placement. However we will first briefly explain our problem. The main goal that we pursued in this job was to be able to purpose the best composition of sensor network to fulfill desired objectives in a given scene. This problem involves three fundamental notions: the scene, the sensor network and the objectives definition. The scene concerns both the physical geometry of the place and its utilization. The sensor network must be designed with a representation of their efficiency, functioning and behavior. The objectives must also be clearly defined in terms of desired monitoring, efficiency as well as reliability. Our approach should therefore be applied to the design of multi-sensor installations monitoring indoor areas where any kind of physical flows are present.

2.1 Problems of Camera Placement

The basic approaches in this field of research are employed in virtual reality simulators. In the case of [5] and [6] the matter is to provide the user some help for the placement and the control of cameras in 3D environments. The first aims at browsing comfortably in a software tool and at observing a target object. For [6] the objective is to make it possible to simulate the efficiency of the placement of a camera network in a scene representing a real case (a surgical operation). In [5] screenings are computed with an adaptation of the hemi-cube algorithm [7]. The virtual environment proposed in [6] provides the user with a decision-making aid in the placement of a camera network, by providing him information concerning the coverage given by the cameras and also by putting forward the resolution of the cameras on the various zones of the image. The second class of problem is very close to the field of cinematography and tackles the issue of browsing in virtual environments. The topic of this kind of work is the way by which the visualized scenes are presented on the screen. Some approaches are similar to the realization of storyboards or to the virtual rehearsal of shots for the cinema [8]. Others are directed towards the human machine interfaces [9], [10], [11], [12]. They thus mainly utilize computations that use the internal models of cameras to find the place of the objects observed in the image plan. The position of points in images is also the main interest of work in photogrammetry. Within this framework these techniques are used to set up networks of cameras allowing a very precise measurement of the objects either with very complex geometry or whose size forbid the more classical methods of measurement. Those cases are the measurement of industrial pieces [13], [14], or of buildings of complex architecture [15]. Finally, the third field that uses placement and camera control method is the localization of objects or humans. The objective in the first case is rather close to the concerns seen in photogrammetry. It aims at carrying out a measurement as precise as possible. For video surveillance, some goals are common to the case of cinematography, because one wishes to have some objects in the shot with a sufficient resolution, but the organization of the shot is left completely free [16]. Actually, in this work the coverage provided by the cameras is the priority. All those cases of placement methods and camera control have specificities, nevertheless we note that they are all guided by their final goal. That leads the researchers to set up strategies using constraints optimization. Depending on the fields, those constraints are extremely diverse, but they are generally translated into objectives on the parameters describing the cameras.

2.2 Definition and Model of the Studied Scenes

In our problem we consider that the study is undertaken on a scene. According to the requirements of the Capthom project the scenes are either tertiary buildings or a set of rooms, in which we want to carry out a control of the energy consumption. A scene describes a zone and its environment that one wants to supervise. A scene is thus a set of element describing the physical aspects coupled with the objectives researched in term of collected information. The shapes of the rooms must hence be described as well as the position and shape of the furniture. The disturbances of physical flow must also be indicated, as well as the elements describing the mission of the system. The mission can be defined by a zoning of the scene indicating various priorities for various parts of the spaces. However, the description of the scene should not be limited to the zones to be observed, but it should also take into account some excluded zones. These kinds of areas are for example the space behind a window or behind an opened door. In the future we would like to add the utilization that humans make of the room to conduct very precise scenarios concordant with real use. The obstacles are zones in which a given flow cannot pass through, therefore they can be material like solid objects or immaterial like electromagnetic flows.

We thus have defined several types of zones to specify a scene. There are initially the internal zone or zone to be covered and the one considered as external. This definition may seem to be naive but demonstrates its utility if we want to consider the space seen behind a door, a window or any other kind of opening. One can then define the zones of material or immaterial obstacle by defining the edges of the latter. This allows to represent the geometry of the scene. A third type of zone used is the disturbing zone, which do not stop flows, but degrade them or modify them. They are utilized to model that certain phenomena which appears in houses, like the displacement of air mass of different temperatures or the radiations emitted by various equipments, can have effects that are far from being negligible for the sensors. To summarize, there are then two principal types of zones, on the one hand the zones to be observed and on the other hand the zones not to be supervised. Using levels of priority between the zones to be observed and to be avoided can graduate this binary definition.

To model those scenes we decided to start from the basis utilized in [16] that we have enriched to fulfill our multiple needs. In our models we chose to represent the scene by a list of points whose value indicates their nature (belonging or not to an obstacle). Each element of this vector corresponds to a point of the scene. The points of the scene are thus numbered in a given order. This vector, noted thereafter **Scene**, is carrying much information and can become a vector of couple of values, this is a modification of the model used in [16] that enrich is power of representation. We describe in this vector all the zones inherent to the mission. The points to be supervised are coded with a 1 in the element that carries their index and with a 0 if they do not represent an objective to be covered. The example in Fig. 1 presents a simple case where the interior of the red zone represents the points to be supervised.

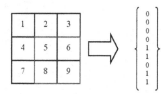

Fig. 1. Creation of the description vector of a scene

The construction of the vector representing the scene is done in two stages. The first part is the mathematical description of the scene by primitives: the vertices, the edges and the polygons. Then, the scene is sampled in the form of a vector whose elements carry information on the nature of a point belonging to the scene. We can obviously influence the smoothness of the discretization by increasing the number of points represented in the vector of the scene. The description vector of the scene is built by tests of membership to the polygons. For each point belonging to the discretization we check that it rests with the points to be observed or with those to avoid. This enables us to obtain the digital model of the scene that we will exploit to develop optimization algorithms. To exploit this digital model, we also set up an adapted model of the systems of sensors. We defined a model common to any class of sensor that we numerically adapted to the scene model for evaluating and creating construction solutions for sensor system adapted to particular problems.

2.3 Definition and Model of Sensor Systems

2.3.1 General Model

Before describing a system of sensors as a whole we tried to represent a sensor whatever its nature and the physical flow it measures in a universal way. According to us a sensor is characterized by various parameters that are the zone it covers, the precision of its measurement through this zone, its placement and the flow or perturbations to which it is sensitive. The covered zone can be very variable between various sensors. For a camera, this zone is the field of view, for a contact sensor this zone is reduced to a point. It thus appeared interesting for us to represent a sensor by its efficient zone i.e. the zone in which it can provide information on the flow it measures. Moreover, to represent the precision of the taken measurements this efficient zone must also be a spatial distribution of the reliability and measuring accuracy. For all the types of sensors we can consider that the efficient zone emanates from the sensitive cell of the system. To build these efficient zones and to adapt to various geometries of efficient zones we chose the method of ray tracing in the construction of the detection polygons of the sensors.

We give in Fig. 2 the example of the efficient zone concerning a camera. We can note that this type of efficient zone is quite similar for ultrasonic sensors and pyroelectric infrared systems.

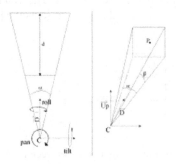

Fig. 2. Standard parameters of a camera, efficient zone

We understand on the preceding figure (Fig. 2) that the efficient zone of a camera-like sensor can be defined by three main parameters for a three-dimensional description. Those parameters are the azimuth α and latitude β angles as well as the depth of field d; concerning this last parameter, two options can be developed. At first sight, we can only consider the most distanced element that we can observe by taking only a depth distance. But we can also consider points that we cannot be analyzed because they are too close from the sensor, by considering two distances: one for the start of the effective zone and another for its end.

This kind of profile could also fit the effective zone of ultrasonic or pyroelectric infrared sensor. Indeed, an ultrasonic transducer is emitting on a zone where an object is detected if it reflects a detectable energy to the ultrasound receiver. The efficient zone for this type of sensor is then defined by the opening angle of the ultrasound emitter (that leads to a conic approximation of the detection polygon), and by the maximal distance from which an object could reflect a significant energy. For the pyroelectric infrared sensor the basic technology is very different but its efficient zone is very close to the case of the camera and the ultrasonic sensor. The pyroelectric infrared sensors currently used for human motion detection are not localization sensors, like camera or ultrasonic transducer, in the sense that they are not able to give the position of the detected objects. This type of sensor only gives a binary measure of what it monitors, that is the presence or not of an object in movement that emits an infrared radiation. Concretely, the pyroelectric infrared sensors utilized by most industries for human detection [17] are composed of a passive pyroelectric infrared sensor component set up behind a Fresnel lens. This installation is used to create a distributed set of detection lobes across the monitored room. The sensor detects intrusion when a body that emits infrared radiation crosses at least two lobes consecutively. Physically the sensor covers the room with different rays as expressed in Fig. 3. In fact the real efficient zone is not only the ray where the sensor catches infrared signals but actually the whole environment between the first and the last ray. This zone is therefore comparable to a camera detection polygon characterized in the two-dimensional value the azimuth and the depth of field. This zone can be seen as the space in which the sensor detects the phenomenon that it tracks.

By using a ray-tracing algorithm, we then obtained a list of points representing the vertices of the detection polygon for the sensor considered. Indeed, each vertex is the intersection point with the first obstacle met and whose distance with the center of the sensor is lower than the depth of field. An algorithm implemented in ©Matlab enabled us to obtain detection polygons sufficiently precise toward the scene scale considered, Fig. 4 presents a polygon made up of 102 vertices.

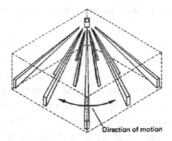

Fig. 3. Infrared rays used by an infrared movement detector (figure from [17])

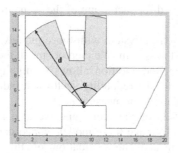

Fig. 4. A detection polygon

2.3.2 Modeling Reliability and Measurement Confidence

This process enabled us to obtain the polygon representing the efficient zone of a sensor. We used this element as a basic to detail the characteristics of the considered sensor. Indeed, by considering the laws giving the precision relative to the distance of a point, we can graduate the detection polygon with the reliability of measurements. These laws, meaning the measurement accuracy, depend on the quality of the sensor and can also be classified by mission. For a camera, we can express that the precision of information relates to the number of pixels per millimeter. However, we can also complement this information by giving the minimal resolution necessary for various applications. We will thus be able to say if the information collected at various spots of the scene is sufficient to carry out face recognition or activity recognition. We used in experimentations an ultrasonic sensor made by Polaroid whose precision decreases linearly. If we choose to represent this sensor considering that it can sweep an angle of 90 degrees with a depth of field of 12 meters, we obtain the following profile (Fig. 5).

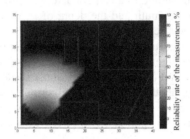

Fig. 5. Example of distribution of spatial measurement reliability

We represented the scene by a vector of points. We identically present the detection polygon in a vector where all the points of the scene are considered. In this way the vectors of the sensors and those of the scene are comparable. It is then easy to find the zones covered by the sensors as well as their characteristics. We chose to represent the sensors by two distinct but complementary vectors. The first collects the points that are in the detection polygon and carries information on the type of zone to which they belong. These zones can be as we presented in the preceding part, the zones to be seen, closed areas or zones of various priority levels. The second vector is used to describe the reliability of measurements at each point of the scene provided by the sensor. We then registered for each point of the scene the measuring accuracy that the sensor offered.

In order to represent a sensor system with those models, we added the descriptive vectors of each element of the system. We then obtain all the points "observed" by the system. For some of them, the addition allows to highlight redundancies in their observation. It is the same for the reliability vector that indicates the combinations of sensor that increase the reliability of measurement.

We now have the models of the scene and we also know how to model very different sensors. We tried to make these models concordant to be able to model the whole placement problem and the evaluation of the feasibility of the scene monitoring.

3 Problem Modeling

The problem that we pose is to be able to propose and assess sensor systems for human presence detection, in a scene proposed by an external user. We considered the following data as the input data of the problem: the shape, the furnishing and the zoning of the scene as well as the positions and orientations possible for the sensors usable in the scene. Therefore we had a finished number of sensors usable and a finished set of position for this. Consequently, we obtained a finished set of solution to the problem. The problem thus consists in finding the best solution in the space suggested. In [16] the procedure is quite similar. Nevertheless their research is only turned towards obtaining solution ensuring 100% of coverage of a room. Moreover, this work is limited to the use of cameras and not of other types of monitoring systems. We thus largely adapted the way of proceeding of this work to adapt to our needs. We largely widened the models employed and increased their representation capacity. Our contribution is to have generalized the models for any type of presence sensor and to put forward parameters much more complex than a simple covering of a room. We think in particular about the addition of the reliability vector characterizing the sensor systems.

The installations in real environment present a lot of specificities that could not be solved by a solution guaranteeing only a total coverage of the room as in [16]. Indeed the real establishments of sensor systems often mean the use of limited means as the presence of specific technical difficulties that are sometimes unforeseeable. Moreover, the priorities of installation cannot be reduced to simple research of zones to cover. To carry out the resolution of such requirements we chose to use genetic algorithms offering the flexibility and the effectiveness that were necessary to our approach. In this part we will explain the mathematical formulation of the problem. Then, we will discuss the methods usable for its resolution.

3.1 Mathematical Formulation

In order to solve the problem of sensor placement we wished to place ourselves in the traditional case of linear programming problem under constraints. I.e. we wished to formulate the problem in the form of formula (1). Function **f** is the cost function to minimize, A is the system matrix, b the constraints and x the selectable parameters whose upper and lower bounds are ub and lb.

Minimize $f(x) = c^T x$

$$\text{With respect to} \qquad A \times x > b \tag{1}$$

$$ub \geq x \geq lb$$

The construction of the models previously presented is clearly directed to this end. In fact, in this context, a point is seen by the capture system if its value in the detection vector is positive. We also note that this point must be observed if a positive value is allotted to him in the description vector of the scene. Finally, to represent a sensor system we saw that it consisted in adding the detection vectors of each component. This is what we can do by using a vector of selection. This vector is the one that makes it possible to indicate which components of the basic list are used to constitute the system. Its number of rows is the number of sensors that can be installed and carries a 1 for the element representing a selected component. In formula (1) the "b" vector is thus the description vector of the scene named before **Scene**, the "x" vector is the selection one. Matrix "A", of dimension Numbers of Points × Number of possible sensors, is built by concatenating on the right all the detection vectors of the sensors installables in the same order as the one used for the selection vector. We thus can compare for each point of the scene if it appears in the points seen by the system, i.e. if $A \times x > b$. The process can be schematized in the following way (Fig. 6).

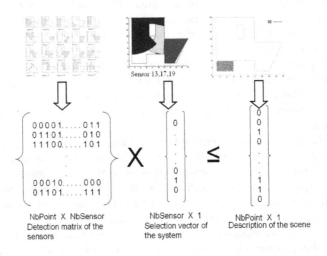

Fig. 6. Mathematical model of the placement problem

In order to pose the linear programming problem we still have to express a cost function **f** to be minimized. This function can be the means to introduce a real cost, related to the price of the material employed, the installation or the energy consumption. It can also represent the quality of the material or another more subjective evaluation. The final function can balance these various aspects to form a multi-criteria function that will be adapted according to the policy of desired installation. This function gives for each possible sensor the total cost of its installation and of its exploitation; the various parameters must be parameterized according to the desired priorities. For example, the priority can be, a low energy consumption that would result in assigning a very high cost to the systems consuming a lot of energy. On the contrary, if one simply wishes to minimize the number of systems used it is enough to assign a unit weight to all the sensors. To solve the optimization problem previously posed, there are many possible ways that we will present now.

3.2 Resolving Method

Many works on camera placement used a large variety of different optimization methods. Our work is inspired from those that have followed the 0-1 canonical model of optimization [16] to solve the problem formulated in (1). But work of Olague proposed the use of genetic algorithms that are multi-cellular [13] or "Parisian" [14]. The genetic algorithms are known for their effectiveness to solve very complex and non-linear problems, whose form of the solution set is badly known, or when the problem is difficult to formalize by traditional methods. This kind of algorithm functions like a research in parallel, which functions on the principles of natural selection, that also allows the avoidance of local optimum. In our case, the variable to be optimized is the selection vector. This vector represents the system. We thus optimized the selection of the sensors. This vector is in a simple case a binary vector, it thus lends itself particularly well to the use of optimization of 0-1 canonical models like [16] done. To solve the problem, we initially used two distinct methods: the Branch and Bound algorithm to have a comparison base with [16] and the genetic algorithms.

3.2.1 Use of the Branch and Bound Algorithm

We have adapted the format of our input data to use this method. We thus wrote all the vectors as binary vectors. This induces a huge simplification of the problem because we can consider only the problem of covering 100% of the scene and avoiding areas. The Branch and Bound method is a generic method of resolution of optimization problems, and more particularly of combinative or discrete optimization. This method makes it possible to solve complex problems by enumerating the solutions in an eligible set. However, it has of a reduction mechanism of exploration by the progressive evaluation of the solutions considered. In the Branch and Bound method we divide the basic problem in a collection of simpler sub-problems covering the totality of the root problem. The Bound function makes it possible to limit exploration, by trying to test the utility to develop or not a branch of the tree representing the total research. This is done when it is feasible to easily calculate the optimums of the sub-problems. If for instance a local minimum is higher than a solution previously calculated the branch is then abandoned. The use of the Branch and Bound method enabled us to validate our models and implementations that were made with ©MATLAB. Moreover, we then succeeded in obtaining results on the discovery of placement solution ensuring 100% of room monitoring with a minimal number of sensors. We tested it on the scene described hereafter for which a version of Branch and Bound coded with ©MATLAB gives the following results (Fig. 7).

The use of this algorithm enabled us to get our first results for the placement of the system with three main constraints which were: to cover the entire room, not to supervise the prohibited points and to be limited for the number of sensors, their characteristics and the place where they could be located. We did this to validate models for the type of scene that we wish to study. However, for the establishment of our simulation and placement software we wish to go further in the type of constraints considered. We also wish to be able to easily have partial solutions if a solution perfectly respecting all the constraints does not exist. For these various reasons, we chose to explore a new way for the optimization of the systems of sensors construction.

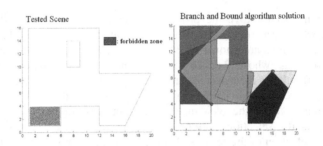

Fig. 7. Branch and Bound algorithm solution

3.2.2 Use of Genetic Algorithm

For our problem, we used a binary genetic algorithm justified by the object of the research, which is the sensor selection vector of the total system. The genetic algorithm and their functioning have been well defined in [18]. Many parameters can influence the results of such research. Nevertheless, we currently choose to focus on one main aspect. We mainly worked on the fitness function to carry out various optimization policies. These policies attempt to exploit the maximum of information that the models can carry. We used a selection function, based on geometrical law, whose probability of selection is proportional to the adaptation of the individual. This has been done to preserve a partly random selection and thus to avoid the local optimum.

The variety of the scenarios under consideration for the system and the variety of the priorities of each user make extremely interesting the possibility of proposing various configurations for the same scene. The real deployment of the solutions and the unpredictability related to the real case force to have multiple solutions to the same problem referring to a given scene. We have been able to fill this need for adaptability in the generation thanks to the use of these algorithms. Indeed, this type of algorithm is shown to be very flexible and effective in the search for solution to mixed multi-criteria problems. To emphasize the effectiveness of this option, we carried out many placement tests. The great flexibility of this method is due to the facility to create fitness functions directing researches towards different solutions. We thus expressed various policies in the creation of these functions and we implemented many tests to justify and highlight the parameters to be varied as well as the values to affect to them. Thereafter, we will present these policies and the results obtained, by pointing out the variations under consideration for the various parameters.

We initially chose policies of fitness function searching an acceptable coverage rate in ratio to the cost or the heaviness of the monitoring installation. We considered four principal parameters to be fixed to obtain results satisfying our expectations. Those consist in the maximum avoidance of the excluded zones, not exceeding the maximum number of sensors desired in a solution (being able to be fixed with the results of the Branch and Bound algorithm) and the balance between the profit in the covered zone provided by a new sensor and its cost, which was never considered in the methods seen in the bibliographic study. For each policy used we will fix a percentage of minimal additional coverage that a new sensor must induce to legitimate its installation. To influence the three goals that we have just described, we built the fitness function around four parameters functioning with the addition of bonus or handicap. The genetic algorithm that we used searches a maximum value for the result of the fitness function. We describe below the fitness function that we use.

This function takes in input the A matrices (detection matrix of the sensors) and **Scene** (description vector the scene) described previously and respectively representing the visibility of the sensors and the set of points to be observed in the scene. Are also considered in input, the selection matrix X of the sensors (X is the research solution), NbPoint the number of points in the scene, MaxSens the number of sensors used in the solution computed by the Branch and Bound if it exists, NbSens the maximum number of sensors in the scene and Cov the additional value of covering that a new sensor must bring. val is the value that will be given to the fitness function for a given individual X.

The sum function is the sum of every element of a given matrix.

```
val ← 0
for i=0..NbPoint
        if Scene ( i ) > 0
                if (A × x)( i ) ≥ Scene ( i )
                val ← val + 100 / sum(Scene ) // This line is built to work with
        if (A × x)( i ) < 0                      // percentage of coverage.
                val ← val – 100              // Seeing the forbidden zone cancel
        if sum( x ) > MaxSens                    // the bonus of seeing 100% of the scene.
                val ← val – 100          // It is the same if too much sensors are used.
val ← val + Cov × ( NbSens – sum( x ) )  // The smaller the number of sensor is, the
                                         // bigger val is.
```

This fitness function gives access to equilibrium between the covered zone and the number of deployed sensors. Indeed, a solution with a more important cover rate is not better noted than a solution with fewer sensors. It is what arises from Fig. 8 that presents the topology of the fitness function presented for a value of the Cov parameter of 20% and by considering the maximum number of desired sensors as 5. The number of prohibited points is set to zero. We can thus observe a step around a number of five sensors that makes it possible to avoid any solution using a too high number of sensors. This is used to support the convergence of the genetic algorithm towards the best solution. We also notice in Fig. 8 that solutions with a less important cover rate are as well noted as a solution with 20% additional coverage and with one supplementary sensor. Besides, they are better noted than a solution with a supplementary sensor and a higher coverage rate of only 10%. This type of policy thus enables us to go further from a research of Branch and Bound type. In fact we are not searching only a single optimum based on a single criterion. Fig. 8 presents the topology of the function if all the cases represented are accessible. For instance, if for each possible number of sensors one can obtain any possible coverage rate. In real cases, the topology is only a part of the complete case. All the points illustrated below will not be accessible, which will give topologies closer to the one presented in Fig. 9.

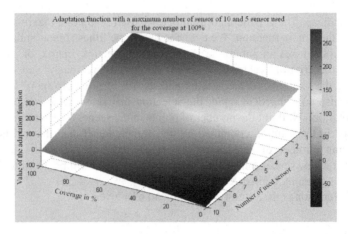

Fig. 8. Fitness function topology

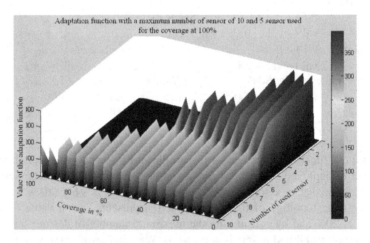

Fig. 9. Fitness function topology in a real case

In Fig. 9 the topology appears to be largely modified compared to the case when all the configurations exist. In this example, there are 20 possible sensors, considered disjoined (their detection polygon does not overlap), the user does not want to use more than 5 sensors and the various percentages of coverage that we could find are 5, 10 or 15%.

3.3 Experiment

Using the genetic algorithm is like moving on the surfaces plotted on the two previous figures and searching to find the highest point of the surface. This function was used to calculate sensor systems usable in the following scene (Fig. 10). Others types of scene, giving encouraging results, were tested but are not presented in this article.

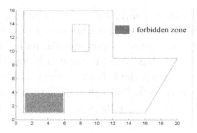

Fig. 10. The studied scene

The allowed positions for the sensor that we could employ in this scene were the following (Fig. 11). Those sensors are all camera-like systems but as explained on section 2.3.1 the polygon shape are quite the same for pyroelectric or ultrasonic component. Moreover the shape of those zones does not affect the efficiency of the method because all the polygons are used as list of points.

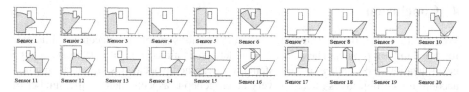

Fig. 11. Set of usable sensors

A solution ensuring 100% of coverage of the scene has been found by using a Branch and Bound algorithm and is presented in Fig. 7.

Various values of the Cov parameter were tested. We thus carried out tests for values of 5, 10, 20, 30, 40 and 50% of additional covers necessary to justify the installation of an additional sensor. The following table synthesizes the obtained results. We show on each line a test carried out with a different value of Cov.

Table 1. Configuration for the scene with coverage-centered policies

Cov	Number of utilized sensors	Rate of coverage in %	Final value of the fitness function	Index of the selected sensors			Obtaining rate of the solution for 10 tries*
5%	3	95.9815	17598	13	17	19	100%
10%	3	95.9815	26598	13	17	19	100%
20%	2	85.7805	44578		13	19	100%
30%	2	85.7805	62578		13	19	100%
40%	1	49.4590	76946			19	100%
50%	0	0	95000		NULL		100%

* For 4000 generations of 100 individuals

The use of this policy enabled us to highlight alternative solutions to the solution comprising 100% of coverage rate. These solutions are more adapted to the problem of this scene. Indeed we obtain configurations with high cover rates (95,98% and 85,78%) and a number of sensors used less important (3 even 2 compared with the 5 of the first solution). We notice in this table that the addition of sensor n°17 brings 10,2% of additional cover. It is thus added to the configurations when Cov is only set to 5 or 10%, which strictly respects the behavior awaited for our policy. It is the same for sensor 13 that represents a profit lower than 40% of additional coverage rate. The results are presented in a graphic way in the following figure (Fig. 12).

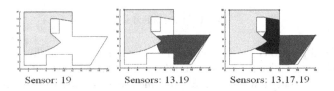

Sensor: 19   Sensors: 13,19   Sensors: 13,17,19

Fig. 12. Solution obtained with genetic algorithms

4 Conclusion

We have presented our first investigation in the establishment of a placement simulator for the Capthom project. The principal characteristics aimed for this system are a huge generalization in the choice of the systems. We searched placement solutions taking into account mixed objectives related to very heterogeneous aspects such as: the number of systems, the redundancy of measurements, the search for particular points or the electric consumption of the system. This has been made through the concordance of the scene and the sensor system models. The models were made up to recreate a classical optimization problem. Then, the problem has been represented in the form of a computable problem with the use of genetic algorithm in the ©Matlab software. We described how to solve this problem by the use of genetic algorithms. We presented a construction of the fitness function by providing an easily reusable and modifiable framework. We indicated the parameters to be modified according to the required objectives. Moreover, this implementation is shown to be very flexible and can be a basis for the design of a final simulator. Indeed, it allowed us to easily represent various phenomena as we showed with the use of a zoning of the scene and the calculation of the distribution of reliability through the scene. We presented results about policies balancing cost and coverage; others have been obtained for the management of redundancies between sensors involving progress in the system reliability. This method is to be utilized in design time. The future developments will have to validate the real application of the Capthom project. More than a simple placement aid simulator, we aim at creating a system model for sensor network but also a software that will validate the use of such systems in well-defined utilization scenarios of the scene. It will be a very interesting point to be able to consider changing environments perhaps by creating a furniture map containing probability of presence of the furnishings.

Acknowledgments. We specially want to thank all our partners involved in the Capthom project. This work was realized with the financial help of the Regional Council of the Centre and the French Industry Ministry within the framework of the Capthom project of the Competitiveness Pole S^2E^2, www.s2e2.fr.

References

1. Benard, F., Barthel, L., Campo, E., Esteve, D., Raffray, P., Delaunay, P.: Etude d'un Gestionnaire d'Energie de Concept ERGDOM. Rappots de contrat, convention EDF/ADEME n 02.04.134 (2004)
2. Zouba, N., Bremond, F., Thonnat, M., Vu, V.T.: Multi-sensors Analysis for Everyday Activity Monitoring. In: SETIT 2007 (March 25-29, 2007)
3. Chan, M., Campo, E., Esteve, D.: PROSAFE: a multisensory remote monitoring system for the elderly or the handicapped. In: 1st International Conference On Smart homes and health Telematics (ICOST 2003) Paris, pp. 89–95 (2003)
4. European project SOPRANO. Information Society Technologies (IST) public deliverable IST – 2006 – 045212 (2007)
5. Philips, C.B., Badler, N., Granieri, J.: Automatic viewing control for 3D direct manipulation. In: Proceedings of the 1992 Symposium on Interactive 3D Graphics, pp. 71–74 (1992)
6. State, A., Welch, G., Ilie, A.: An Interactive Camera Placement and Visibility Simulator for Image-Based VR Applications Stereoscopic Displays and Virtual Reality Systems. Proceedings of the SPIE 6055, 640–651 (2006)
7. Cohen, M.F., Greenberg, D.P.: The Hemi-cube a radiosity solution for complex environments. Computer Graphics 19(3) (1985)
8. Bares, W.H., Thainimit, S., McDermott, S.: A model for constraint-based camera planning. Smart Graphics. In: Papers from the 2000 AAAI Spring Symposium, Stanford, March 20-22, 2000, pp. 84–91 (2000)
9. Drucker, S., Zeltzer, D.: Intelligent Camera Control in a Virtual Environment. In: Graphics Interface 1994, pp. 190–199 (1994)
10. Drucker, S., Zeltzer, D.: Intelligent Camera Control in Graphical Environment. Ph D. thesis, Massachusetts Institute of Technology, Cambridge (1994)
11. Drucker, S., Zeltzer, D.: CamDroid: A System for Implementing Intelligent Camera Control. In: 1995 Symposium on Interactive 3D Graphics, pp. 139–144 (1995)
12. Bares, W., Lester, J.: Intelligent Multi-Shot Visualization Interfaces for Dynamic 3D Worlds. In: IUI-99, Proceedings of the 1999 International Conference on Intelligent User Interfaces, Los Angeles, pp. 119–126 (1999)
13. Olague, G.: Automated photogrammetric network design using genetic algorithms. Photogramm. Eng. Remote Sensing 68(5), 423–431 (2002)
14. Dunn, E., Olague, G., Lutton, E.: Parisian Camera Placement for Vision Metrology. Pattern Recognition Letters 27, 1209–1219 (2006)
15. Saadat, S.M., Samdzadegan, F., Azizi, A., Hahn, M.: Camera Placement for Network Desig In Vision Metrology based On Fuzzy Inference System. XXth ISPRS Congress, July 12-23, 2004. Istanbul, Turkey (2004)
16. Erdem, U.M., Sclaroff, S.: Automated Camera Layout to Satisfy Task-Specific and Floor Plan-Specific Coverage Requirements. Computer Vision and Image Understanding 103, 156–169 (2006)
17. Gobeau, J.F.: Détecteurs de mouvement à infrarouge passif. Cinquièmes rencontres capteurs Capteurs 2006, Pôle Capteurs automatismes (2006), http://www.bourges.univ-orleans.fr/pole_capteur
18. Holland, J.H.: Adaptation in Natural and Artificial Systems. University of Michigan Press, Ann Arbor (1975)

Recognition of User Activity Sequences Using Distributed Event Detection

Oliver Amft, Clemens Lombriser, Thomas Stiefmeier, and Gerhard Tröster

Wearable Computing Lab., ETH Zurich, Switzerland
{amft,lombriser,stiefmeier,troester}@ife.ee.ethz.ch
http://www.wearable.ethz.ch

Abstract. We describe and evaluate a distributed architecture for the online recognition of user activity sequences. In a lower layer, simple heterogeneous atomic activities were recognised on multiple on-body and environmental sensor-detector nodes. The atomic activities were grouped in detection events, depending on the detector location. In a second layer, the recognition of composite activities was performed by an integrator. The approach minimises network communication by local activity aggregation at the detector nodes and transforms the temporal activity sequence into a spatial representation for simplified composite recognition. Metrics for a general description of the architecture are presented.

We evaluated the architecture in a worker assembly scenario using 12 sensor-detector nodes. An overall recall and precision of 77% and 79% was achieved for 11 different composite activities. The architecture can be scaled in the number of sensor-detectors, activity events and sequences while being adequately quantified by the presented metrics.

Keywords: Event detection, activity recognition, distributed detectors, activity sensing, inertial sensors, wireless sensor networks, smart objects.

1 Introduction

A key challenge for context-aware systems is to unobtrusively recognise complex activities from sensors, distributed in clothing and objects in the environment. Attempts to embed sensors into every-day objects has been proposed for smart environments as the PlaceLab [1]. By using such assistive systems, the user will be comforted with instant and relevant information and coaching support, e.g. advice, when certain task steps have been forgotten by the worker or tasks could be performed more efficiently.

The recognition of activities has been broadly investigated using fusion methods to combine different information sources at the data, feature, or classifier level. Choosing the right fusion approach requires an analysis of resource constraints within the recognition architecture. Wireless sensor networks impose particular limitations on computation and energy resources. Since wireless transmission is the highest power consumer on a sensor node, the required communication bandwidth should be kept as low as possible. However, concerning the processing power of a sensor node, it has been shown, e.g. for sound

G. Kortuem et al. (Eds.): EuroSSC 2007, LNCS 4793, pp. 126–141, 2007.

classification [2], that sensor nodes can provide important contributions to the recognition. Consequently, for activity recognition in sensor networks, an early aggregation of the sensor data at the node is favourable to minimise bandwidth usage. This, in turn, restricts the fusion approach in the network to the (typically discrete) classifier level.

Fusing detector ensembles using spatially distributed sensing sources helps to assess large sets of activities and more complex activity sequences, demonstrated e.g. in [3]. Such an architecture requires an abstraction model to manage and integrate the nodes in the recognition process with respect to the provided information. Bobick [4] proposed a layered representation of user activities, consisting of movements, activities and interaction or sequences of activities. This hierarchy can serve as basis for a distributed recognition architecture. The lowest recognition layer lends itself ideally for a distributed implementation on sensor nodes, since the activities are aggregated, where the sensor data originates.

In this work we use a two-layered abstraction model based on the concept that activity sequences are composed of *heterogeneous activity events*. Activity events are seen as atomic units of activities, recognised at distributed sensor-detector nodes. The *atomic activities* represent an alphabet of low-level detection entities in the recognition system. A specific set of atomic activities yields a *composite activity*. The recognition of such composite activities is performed by fusing the sensor-based detections at an integrator.

Within this scope the paper makes the following contributions:

1. We present a recognition architecture that can be distributed in a network of wireless sensors. Each sensor node contributes to the activity recognition by detecting heterogeneous activity events in continuously sensed data. The distributed detection simplifies the recognition of composite activities at the integrator. Consequently, simple histogram- and symbol-based algorithms were used for the composite recognition. Section 2 details the deployed algorithms along with descriptive metrics for size and complexity of the architecture.
2. By using domain knowledge we demonstrate that atomic activities can be grouped to detector events. This approach permits the integration of useful information, supporting the recognition of composite activities.
3. We evaluate the architecture in a car assembly scenario using 12 sensor-detector nodes to recognise 11 different composite activities. Section 3 describes the scenario. Evaluation metrics, borrowed from information retrieval, are used to fully assess the recognition performance of the architecture.

1.1   Related Work

Recent approaches to recognise interaction or sequences of activities followed the concept of a layered modelling, e.g. [5,3,6,7] as it conveniently partitions the complexity of user activities. However, the solutions differ in the activity granularity of each layer depending on the application and recognition concept. Ryoo and Aggarwal [6] used three categories starting with "atomic actions" extracted from image sequences and modelled the activity sequences using a context-free grammar. Kawanaka et al. [7] used a hierarchical architecture of interacting hidden Markov models (HMMs) to represent activities and sequences of activities. Oliver et al. [3] aimed to reduce the complexity of hierarchical HMMs by independent training of the layers. Moreover, Dynamic Bayesian networks (DBNs), being a very flexible framework for reasoning, have been applied to the recognition of human activities, e.g. [8].

Uniform HMM- and DBN-based reasoning solutions have some impractical properties for the deployment in distributed sensor networks: 1) a high computational complexity and 2) the need for a large training corpus combined with an extensive parameter search in order to tune the large amount of model parameters.

Distributed recognition in wireless sensor networks poses special requirements on classification algorithms. The main constraints are the limited processing power and the energy resources of wireless sensor nodes. The impact of these limitations have been investigated from different angles, mostly using binary classifiers computing maximum likelihood estimates. Research on distributed estimators includes the minimisation of messages needed for stable decisions [9], the effects of data quantisation and message losses [10] or coding to counteract data corruption during transmission [11]. Thiemjarus and Yang [12] present algorithms for the selection of features depending on sensor location and quality.

Detection of events and inference of composite activities in sensor networks have been addressed by middle-ware solutions like DSWare [13]. In DSWare, compound events can be signalled if a number of atomic events have been observed. A confidence value was assigned to events, indicating the decision confidence. Osmani et al. [14] created overlay networks called "Context Zones" of devices, which could contribute with events for certain activities. Detected events triggered a rule inference engine working on "Activity Maps". The activity maps model possible activities and the associated events.

The recognition approaches for wireless sensor networks described above, have the drawback that they either focus on the low-level classification of relatively simple sensor signals, or work on high-level abstracted events, for which it is not clear how they can be generated. This work aims at integrating low- and high-level activity recognition and investigating their interaction.

2   Distributed Activity Recognition Architecture

The activity recognition architecture is based on a set of networked sensor-detector nodes that perform sensing and recognition of activity events using the locally available sensor data. The detector nodes transfer every event to a central integrator node. The integrator combines the information from multiple detectors and their reported events to recognise composite activities. Figure 1 provides an overview on the recognition architecture. The operation of the detector node is detailed in Section 2.2, of the integrator node in Section 2.3.

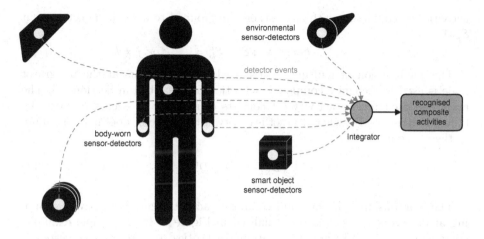

Fig. 1. Schematic of the distributed detection and classification of user activities

2.1   Activity Classification Model

The architecture follows the concept of composite activities that can be described by a finite number of atomic activities. The total set of different atomic activities $\mathcal{A}$ describes the basic detection alphabet of the architecture (Eq. 1). The set of composite activities is $\mathcal{C}$. Each composite activity $\mathcal{C}^n$ is composed of unique atomic activities from $\mathcal{A}$.

$$\mathcal{A} = \{a_1, \ldots, a_\alpha\}, \qquad \mathcal{C} = \{\mathcal{C}^1, \ldots, \mathcal{C}^\beta\}, \tag{1}$$
$$\mathcal{C}^n \subseteq \mathcal{A}, \quad 1 \leq n \leq \beta$$

The relations (Eq. 1) indicate some important metrics to describe the size and complexity of the recognition architecture. The *alphabet size* α counts the number of atomic activities used. The *composite class count* β measures the number of activity sequences that are covered by the recognition system.

The distributed detection is formed by the detector nodes and their detector events: a detector node $\mathcal{D}^i$ of the total set of detectors $\mathcal{D}$ contains at least one detector event $\mathcal{E}^i$ (Eq. 2). The *number of detector nodes* $(|\mathcal{D}|)$ and the *total number of detector events* $(\sum_{i=1}^{|\mathcal{D}|} |\mathcal{D}_i|)$ represent complexity metrics of the implemented architecture.

$$\mathcal{D}^i = \{\mathcal{E}^i_1, \ldots, \mathcal{E}^i_{\gamma_i}\}, \quad \forall i : \gamma_i \geq 1 \tag{2}$$

The primary application goals of the architecture were to use simple motion sensors and include a large activity alphabet. However, some atomic activities cannot be discriminated precisely by every detector. At the individual detector, this is observed as confusion between the atomic activities, indicating that the data patterns for the activities are not separable. Naively, the confused activities would be omitted from the detection for this node. In our approach, the affected activities are grouped to one event of the detector: for each detector D^i, atomic activities a_j conflicting with each other, are grouped to a single detector event $\mathcal{E}_j^i$ (Eq. 3).

$$\mathcal{E}_j^i \subseteq \mathcal{A}, \quad where \quad \forall i : \mathcal{E}_j^i \cap \mathcal{E}_k^i = \emptyset, \quad for \; j \neq k \tag{3}$$

The combination of multiple, distributed event detectors at the integrator node is used to recognise composite activities, as described in Section 2.3. The event-based composite activity $\hat{C}^n$ consists of a subset of events reported by different detectors $\mathcal{D}^i$, where the set is empty, if the detector does not contribute to the recognition (Eq. 4).

$$\hat{C}^n = \bigcup_i \mathcal{D}_n^i, \quad \forall i : \mathcal{D}_n^i \subseteq \mathcal{D}^i \tag{4}$$

The relationship of atomic and composite activities and the activity grouping at detectors is exemplarily visualised in Figure 2. In the simple example, composite activity $\mathcal{C}^1$ consists of atomic activities $a_1 \ldots a_3$ and corresponds to picking a tool (a_1), using it to manipulate an object (a_2) and returning it (a_3).

Each of the detectors $D^1 \ldots D^3$ uses at least one locally acquired sensor channel and recognises a subset of the atomic activities. In the presented example, acceleration signals from the objects and the user's wrist were used. The event sets for the detectors are: $D^1 = \{\mathcal{E}_1^1\}$, $D^2 = \{\mathcal{E}_1^2, \mathcal{E}_2^2, \mathcal{E}_3^2\}$, and $D^3 = \{\mathcal{E}_1^3\}$. While the events for D^1 and D^2 consist of one atomic activity each, for D^3 two activities are grouped $\mathcal{E}_1^3 = \{a_1, a_3\}$.

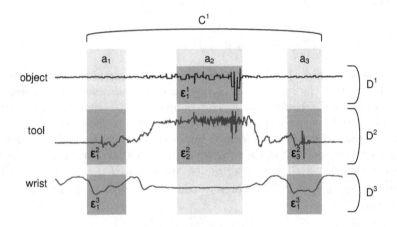

Fig. 2. Relationship of atomic and composite activities and the event grouping at detectors in an object manipulation task. The composite activity C^1 consist of three atomic activities: picking a tool (a_1) using it to manipulate an object (a_2) and returning it (a_3). Each of the detectors $D^1 \ldots D^3$ recognises a subset of the atomic activities, grouped in activity events. Acceleration signals from the users wrist and the involved objects are shown. Please see related text for details.

2.2   Detector Node Operation

The task of a detector node is to spot activity patterns in the continuous stream of data from the locally acquired sensor data. The detection step at the individual network node is introduced to minimise the transmission bandwidth requirements: instead of raw data or intermediate data compression, only the event detection result is transmitted.

The event recognition performance is a critical parameter for successful detection of composite activities. The most vital aspects for the recognition are: 1) the extraction of valid atomic activities from the embedding continuous data (spotting task) and 2) the disambiguation of different events spotted by the detector (classification task). Both must be achieved within the limited resources of the node. The first problem relates to the embedding of the sporadically occurring user activities into a large set of unknown activities, the NULL class. Due to the large variability of the unknown user activities, the NULL class cannot be modelled reliably. For the detector nodes, a feature similarity search algorithm, based on our previous work [15] was adapted to spot relevant activity events. For the disambiguation of different events, a comparison fusion of the similarity search result was used [16]. Both procedures are briefly summarised below.

Feature Similarity Search. The search for potential events was performed by determining the similarity of a data section to a pre-computed pattern of an detector event. Similarity was compared using features computed for a data section when the last sample within a constant-sized window was received. The window size and the pre-computed event pattern were determined in a training step for each detector event $\mathcal{E}^i_j$. The window was moved at a step size of 0.25 s. Every detector $\mathcal{D}^i$ implemented its own feature set F^i.

For the similarity analysis between the feature vectors, the Euclidean distance was used. A threshold on the distance, obtained during training, was applied to omit unlikely sections. The distance of a reported event to the trained pattern was normalised using the distance threshold.

The advantage of this algorithm is that it works as a binary classifier to separate one detector event from the remaining data. Several instances of the feature similarity search were used to spot different events independently.

The computational complexity of the feature similarity search scales with the frequently used sliding window approach, e.g. for sound [2]. Using the detector event notation, the complexity of the sliding window is $O(l(\mathcal{E}^i_j))$, where $\mathcal{E}^i_j$ is the considered event and $l(\mathcal{E}^i_j)$ describes the length of the event in processing units, e.g. data samples. The mean complexity of a detector $\mathcal{D}^i$ is shown in Eq. 5. The complexity is determined by the mean length of all events and the size of the feature set F^i computed for the detector. This result is considered as upper bound, depending on the efficiency of the feature implementation.

$$O\left(\frac{1}{|\mathcal{D}^i|} \sum_{j=1}^{|\mathcal{D}^i|} l(\mathcal{E}^i_j) \cdot |F^i| \right) \tag{5}$$

The independent search instances derive events that can overlap in the time-domain. A filtering was applied to resolve these conflicts. The filter retained events according to the minimal normalised distance among all overlapping events in a buffer. These events were released from the buffer after a timeout.

2.3   Integrator Node Operation

Similar challenges regarding the detection and classification of composite activities at the integrator node exist, compared to the detector nodes: 1) spotting of relevant activity sequences in continuous data and 2) classifying these sequences. Therefore an analog approach to the detector node operation was taken. However, the analysis at the integrator node is performed using activity events, requiring methods adapted for the type of input. An advantage, compared to the detector node, is that the integrator algorithms run at a massively reduced input rate.

Two different distance algorithms were investigated for the recognition, using the combined input stream from all detectors: 1) an approximate string matching by converting the detector events to characters and comparison to a template string of the activity and 2) a comparison of the event occurrence frequencies to a histogram of the activity. Both approaches require training observations from each composite activity, as detailed below.

Distances were computed between a search window of previously received events and the trained pattern at each new event in the input stream. For the composite activity detection, a variable-sized search window was used. By using a distance threshold, unlikely sections were omitted. Both, the search bounds and the threshold were determined from training instances.

As string matching approach, the *edit distance* [17] was used. The strings were generated by mapping each event type to a unique character. The edit distance returned the minimum number of edit operations (insertions, deletions, substitutions) required, to convert the string inside the search window into the training template of each composite activity. The template itself was found by computing the minimal edit distance among all training instances. This corresponds to the minimum linkage distance of the training instances. The complexity of this algorithm is $O(|\mathcal{C}^n|)$ on continuous character input, where $|.|$ is the number of events in each composite activity $\mathcal{C}^n$.

For the event frequencies approach, an *event histogram* was build for each composite activity. The histogram was initialised with the annotated event occurrences for each composite activity. The histogram was adapted by the event occurrences observed in training instances. A distance between this training histogram and the histogram determined from the search window was computed by summing the absolute differences in each event bin.

If the histograms were identical, the distance is zero. This would indicate that the events would ideally resemble the training distribution. Contrary, a larger distance indicates a different shape of the histogram. For this algorithm, the complexity for testing a composite activity is $O(\sum_{i=1}^{|\mathcal{D}|} |\mathcal{D}^i|)$. The complexity is governed by the total number of different events used in the system.

The computational complexity at the integrator node is determined by the detection algorithm and the window procedure for each composite activity. Eq. 6 shows derived bounds with ordered factors. For simplicity of the notation, the adaptive window size was replaced with the average size of the composite activities, $\frac{1}{\beta}\sum_{n=1}^{\beta}|\mathcal{C}^n|$.

$$Edit : O\left(\frac{1}{\beta}\sum_{n=1}^{\beta}|\mathcal{C}^n|\cdot\sum_{n=1}^{\beta}|\mathcal{C}^n|\right), \qquad Hist : O\left(\sum_{i=1}^{|\mathcal{D}|}|\mathcal{D}_i|\cdot\sum_{i=1}^{\beta}|\mathcal{C}^n|\right) \qquad (6)$$

The first factor in each equation indicates the difference between the two algorithms: while the first factor for the edit distance scales with the number of events included in each set $\mathcal{C}^n$, the factor for the event histogram method scales with the total number of different events used in the system.

3 Evaluation

For the evaluation of the proposed architecture, we selected the recognition of car assembly activities. In order to measure the performance of the event detectors and the integrator recognition, detection metrics, commonly used in information retrieval, were introduced. Finally, the feasibility of the architecture was assessed with the metrics.

3.1 Car Assembly Scenario

A car body, installed in a lab environment, was used to record assembly and testing activities, as they have been observed and annotated beforehand in a car mass production facility. Sensors were worn by the worker and attached to different tools and parts of the car. In total, 12 sensors were used to acquire 3D-acceleration from 47 atomic activities in 11 composite activities. Figure 3 shows a worker during a door attachment activity. The activities are summarised in Tab. 1.

Nine wireless sensor motes were used to record motion of different car parts included in the activities (front light, braking light, front and back doors, the hood and trunk door) as well as tools used for the assembly work (two cordless automatic screwdrivers and a socket wrench, see Figure 3). Three wired sensors were attached to a jacket at the wrist position of both lower arms and the upper back. The body-worn sensors and the tool sensors (screwdrivers and socket wrench) were recorded at 50 Hz, the remaining sensors were sampled at 20 Hz. Two workers each completed 10 repetitions of all composite activities wearing the jacket. A manual annotation of the activities was performed.

A total of 49 different detector events were derived from the atomic activities for all sensor-detector nodes. Tab. 2 presents the mapping along with the sensor locations for reference.

Besides the relevant composite activities, the users performed five additional activities three times during each repetition to enrich the data set diversity,

Fig. 3. Left: worker mounting the front door of the car. Right: socket wrench equipped with acceleration sensing wireless nodes.

Table 1. Composite activities C^n and the associated atomic activities

Composite activity	Atomic Activities	Description
Mount front door (C^1)	pickup door (a_1), attach door (a_2), fix screws by hand (a_3), pickup socket wrench (a_4), use socket wrench (a_5), return socket wrench (a_6), close door (a_7)	Mount front door of the car and fix screws with a socket wrench.
Mount back door (C^2)	pickup door (a_8), attach door (a_9), fix screws by hand (a_{10}), pickup socket wrench (a_{11}), use socket wrench (a_{12}), return socket wrench (a_{13}), close door (a_{14})	Mount back door of the car and fix screws with a socket wrench.
Test front door (C^3)	open (a_{15}), teeter (a_{16}), close (a_{17})	Test front door hinges.
Test back door (C^4)	open (a_{18}), teeter (a_{19}), close (a_{20})	Test back door hinges.
Test trunk (C^5)	open (a_{21}), teeter (a_{22}), close (a_{23})	Test trunk door.
Mount brake light (C^6)	fetch light (a_{24}), insert light (a_{25}), screw brake light (a_{26})	Install middle brake light at the trunk door.
Test hood (C^7)	open (a_{27}), teeter (a_{28}), close (a_{29})	Test the hinges of the hood.
Mount hood rod (C^8)	fetch rod (a_{30}), open hood (a_{31}), install rod (a_{32})	Open the hood and install a supporting rod to keep it open.
Mount water tank (C^9)	fetch tank (a_{33}), hand screw tank (a_{34}), pickup driver (a_{35}), screw (a_{36}), return driver (a_{37})	Install a water tank in the hood room using screwdriver 1.
Mount bar (C^{10})	fetch bar (a_{38}), hand screw bar (a_{39}), pickup driver (a_{40}), screw (a_{41}), return driver (a_{42})	Mount a bar under the hood using screwdriver 2.
Mount light (C^{11})	fetch light (a_{43}), install light (a_{44}), pickup screwdriver (a_{45}), fix screw (a_{46}), return screwdriver (a_{47})	Install left front light using screwdriver 2.

including writing on a notepad, working at a computer, drinking, tieing shoes and scratching the head. The activities were not further considered in the evaluation.

Table 2. Location of the sensor-detectors $\mathcal{D}^i$ and the 49 corresponding detector events $\mathcal{E}^i_j$. The detector events were derived by grouping atomic activities for each sensor-detector node.

Sensor location	Detector events
Right lower arm ($\mathcal{D}^1$)	$\mathcal{E}^1_1 = \{a_{27}, a_{21}\}$, $\mathcal{E}^1_2 = \{a_{28}, a_{22}\}$, $\mathcal{E}^1_3 = \{a_{29}, a_{23}\}$, $\mathcal{E}^1_4 = \{a_{30}\}$, $\mathcal{E}^1_5 = \{a_{32}\}$, $\mathcal{E}^1_6 = \{a_{34}, a_{39}, a_3, a_{10}\}$, $\mathcal{E}^1_7 = \{a_{35}, a_{40}, a_{45}, a_{37}, a_{42}, a_{47}, a_4, a_6, a_{11}, a_{13}\}$, $\mathcal{E}^1_8 = \{a_{43}\}$, $\mathcal{E}^1_9 = \{a_1, a_8\}$, $\mathcal{E}^1_{10} = \{a_5, a_{12}\}$
Left lower arm ($\mathcal{D}^2$)	$\mathcal{E}^2_1 = \{a_{27}, a_{31}, a_{21}\}$, $\mathcal{E}^2_2 = \{a_{28}, a_{22}\}$, $\mathcal{E}^2_3 = \{a_{29}, a_{23}\}$, $\mathcal{E}^2_4 = \{a_{32}\}$, $\mathcal{E}^2_5 = \{a_1, a_8\}$, $\mathcal{E}^2_6 = \{a_7, a_{14}, a_{17}, a_{20}\}$, $\mathcal{E}^2_7 = \{a_{15}, a_{18}\}$, $\mathcal{E}^2_8 = \{a_{16}, a_{19}\}$, $\mathcal{E}^2_9 = \{a_{25}\}$
Upper back ($\mathcal{D}^3$)	$\mathcal{E}^3_1 = \{a_{34}, a_{36}\}$, $\mathcal{E}^3_2 = \{a_{35}, a_{37}, a_{40}, a_{42}, a_{45}, a_{47}, a_1, a_4, a_6, a_8, a_{11}, a_{13}\}$, $\mathcal{E}^3_3 = \{a_{46}\}$
Front light ($\mathcal{D}^4$)	$\mathcal{E}^4_1 = \{a_{43}\}$, $\mathcal{E}^4_2 = \{a_{46}\}$
Brake light ($\mathcal{D}^5$)	$\mathcal{E}^5_1 = \{a_{24}\}$, $\mathcal{E}^5_2 = \{a_{25}\}$
Hood ($\mathcal{D}^6$)	$\mathcal{E}^6_1 = \{a_{27}, a_{31}\}$, $\mathcal{E}^6_2 = \{a_{28}\}$, $\mathcal{E}^6_3 = \{a_{29}\}$
Trunk ($\mathcal{D}^7$)	$\mathcal{E}^7_1 = \{a_{21}\}$, $\mathcal{E}^7_2 = \{a_{22}\}$, $\mathcal{E}^7_3 = \{a_{23}\}$
Screwdriver 1 ($\mathcal{D}^8$)	$\mathcal{E}^8_1 = \{a_{45}\}$, $\mathcal{E}^8_2 = \{a_{46}\}$, $\mathcal{E}^8_3 = \{a_{47}\}$
Screwdriver 2 ($\mathcal{D}^9$)	$\mathcal{E}^9_1 = \{a_{35}, a_{40}\}$, $\mathcal{E}^9_2 = \{a_{36}, a_{41}\}$, $\mathcal{E}^9_3 = \{a_{37}, a_{42}\}$
Front door ($\mathcal{D}^{10}$)	$\mathcal{E}^{10}_1 = \{a_1\}$, $\mathcal{E}^{10}_2 = \{a_2\}$, $\mathcal{E}^{10}_3 = \{a_{15}\}$, $\mathcal{E}^{10}_4 = \{a_{16}\}$
Back door ($\mathcal{D}^{11}$)	$\mathcal{E}^{11}_1 = \{a_8\}$, $\mathcal{E}^{11}_2 = \{a_9\}$, $\mathcal{E}^{11}_3 = \{a_{18}\}$, $\mathcal{E}^{11}_4 = \{a_{19}\}$
Socket wrench ($\mathcal{D}^{12}$)	$\mathcal{E}^{12}_1 = \{a_4, a_{11}\}$, $\mathcal{E}^{12}_2 = \{a_5, a_{12}\}$, $\mathcal{E}^{12}_3 = \{a_6, a_{13}\}$

In total, 4.8 hours of data were recorded, containing 2.3 hours of relevant composite activities. The data from sensors mounted on the car parts and tools were transmitted wirelessly to a root sensor node. No special synchronisation procedure was applied, since the delay of the data was found to be negligible for our purposes. In the recording sessions, an average message loss of 6.7% was observed. For some sessions the loss rate was 16.4%.

3.2 Analysis Procedure

Simple time-domain features were derived from the three acceleration signals available and optimised for each detector, such as signal sum, mean, sum of differences, maximum and minimum. The features were computed for the entire signal section of each activity and partitions of it. By using partitions of the signal section, the temporal structure within an atomic activity was captured.

A four-fold cross-validation was applied on the data set to select training and validation set for the event detection and composite activity evaluation. For this purpose the data set was partitioned into four sections. For each cross-validation iteration three data sections were used for training and one for validation. In this way, each section was used once for testing.

The detection performance was analysed with the metrics *Precision* and *Recall*, commonly used in information retrieval (Eq. 7). *Relevant activities* correspond to the actually conducted and manually annotated activities, *retrieved activities* are all objects returned by the algorithms and *recognised activities* represent the correctly returned activities.

$$Recall = \frac{Recognised\ activities}{Relevant\ activities}, \qquad Precision = \frac{Recognised\ activities}{Retrieved\ activities} \qquad (7)$$

To identify an activity as correctly returned by the algorithms, time-domain overlap check between the annotation and returned event was used.

3.3   Detector Node Performance

The recognition performance was analysed for all 12 detectors and its respective detector events. Figure 4 presents the overall performance of each detector. The best results were achieved for the detector $\mathcal{D}^7$ attached to the trunk. Figure 5 exemplarily shows the performance of the individual events for the socket wrench detector $(\mathcal{D}^{12})$.

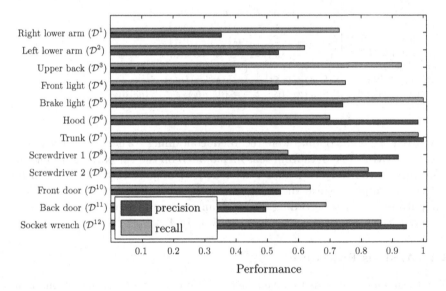

Fig. 4. Overall recognition performance of the 12 detectors. (Best performance is found towards high precision and high recall.)

Although the right and left arm detectors $(\mathcal{D}^1$ and $\mathcal{D}^2)$ contained many event types (ten and nine respectively), a meaningful discrimination was achieved for both. The front and brake light detection performance $(\mathcal{D}^4, \mathcal{D}^5)$ suffered from highly variable movements during the mounting activity. For the upper back detector $(\mathcal{D}^3)$, often event durations did not match with the annotation of the instances. The annotation was not optimised for events from the upper back detector. The weak detection at the front and back door were influenced by low amplitude of the acceleration signals during the door opening and closing activities. Finally, the data loss of all wireless sensors $(\mathcal{D}^3-\mathcal{D}^{12})$ observed during the recordings degraded the performance results for these detectors.

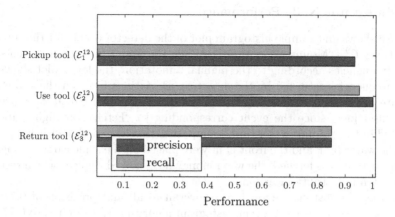

Fig. 5. Detector event recognition performance. This example shows the socket wrench detector ($\mathcal{D}^{11}$).

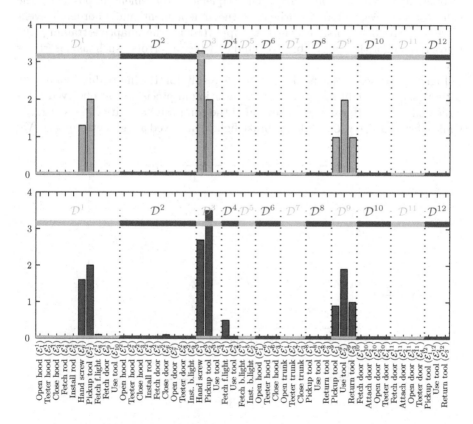

Fig. 6. Histogram plot of the detector events for the composite activity $\mathcal{C}^9$ ("Mount water tank"). Upper plot: event frequencies according to the manual annotation. Lower plot: event frequencies returned by the detectors.

3.4   Integrator Node Performance

Figure 6 shows an example histogram plot of the detector events for the composite activity $\mathcal{C}^9$ ("Mount water tank"). While the upper histogram presents the event frequencies according to the manual annotation, the lower plot shows the event frequencies returned by the detectors. Deviations between the two charts indicate insertion or deletion errors of the detectors. The bar at $\mathcal{E}_1^4$ represent event insertions, since the event corresponding to "Fetch front light" are not included in the "Mount water tank" composite. (In the experiment it happened, that the water tank and the front light were placed on the same carriage waggon. When the tank was fetched, the waggon may have moved slightly, which resulted in the fetching activity insertion.)

For the edit distance algorithm an overall recall and precision of 0.77 and 0.26 was achieved. For the event histogram algorithm the results are 0.77 and 0.79. The event histogram based approach clearly outperformed the edit distance in precision while both algorithms return a similar number of correctly recognised activities. The strict event sequence requirement imposed by the edit distance, was often disturbed by the independent and unordered event reports of the detectors. In contrast, supported by the adaptive training, the histogram algorithm was able to handle detector errors, such as insertions better.

Figure 7 presents the final recognition result for the individual composite activities. For the edit distance, a large variation in precision was observed. This indicated a low performance for several of the composite activity classes, e.g. $\mathcal{C}^{11}$ while a few other, e.g. $\mathcal{C}^6$ ("Mount brake light") achieved a high performance. We

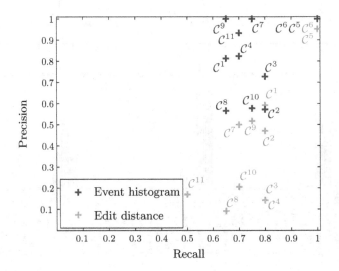

Fig. 7. Recognition performance of the composite activities in the car assembly scenario. Result for two event fusion methods are shown: edit distance and event histograms.

attributed this result to the different training and event sequence requirements of the algorithms, as discussed before.

4   Discussion and Conclusion

In this work, an online architecture for the distributed recognition of complex activity sequences was introduced. The architecture was chosen to reflect a two-layer abstraction of user activities. In the first layer, a distributed event aggregation from atomic activities was performed. In the second layer, activity sequences were recognised from the events. Our work was driven by the observation, that distributed detectors, at the user's body or the surrounding environment, sense activities in a location-dependent granularity. As a consequence, neither each detector can spot every activity in continuous sensor data nor uniquely discriminate all activities.

As a solution, an intermediate grouping of atomic activities to detector-specific events was used. With this approach, the information from each detector was adequately included in the second layer recognition, instead of naively omitting less-specific detectors. By performing the grouping manually, available domain knowledge on the relation of activity events was integrated (as detailed in the example of Section 2.1, Figure 2). Moreover, the grouping simplified the detector nodes, since less events had to be discriminated. Further work will aim at automating the manual grouping of atomic activities, while still incorporating the domain knowledge.

Moreover, the grouping allowed the transformation of temporal activity sequences into a spatial recognition model. The purely spatial recognition algorithm based on event histograms, showed even better performance for the composite activity detection than the strictly sequential edit distance. This result shows, that the distributed solution strongly simplified the complex task of activity sequence recognition.

Data loss due to wireless transmission of the sensor nodes using a standard MAC layer peaked at more than 16% in the car assembly scenario. This data loss influenced the achieved recognition performance in this work. By performing the recognition at the node directly, sensor data loss is avoided.

In the present work, a central integration and recognition of composite activities was used. However, the recognition based on the histogram algorithm could be solved by a distributed implementation of wireless sensor networks. This solution would increase redundancy in the recognition architecture.

The complexity of the detection step scales with the length of the event and the number of features used. The computational requirements are further decreased by sharing intermediate results during feature processing. The complexity of the single sliding window algorithm is a lower bound for this optimisation. Notably, the sliding window approach was even used on sensor nodes at data rates of more than 4 kHz, for spectral audio data processing [2]. Given these relations, we are confident that the distributed architecture is feasible for an online implementation on low-power sensor nodes. Similarly, we believe that the simple histogram algorithm used for the event-processing at the integrator could be processed on a wireless sensor node.

The influence of recognition errors by detector nodes is often underestimated. In some works, different error types, e.g. insertion errors, were not considered at all [14,13]. The results of the architecture evaluation showed that these errors cannot be avoided for activity recognition. Hence, theoretic or simulation-based investigations should consider them more carefully.

For detector nodes that communicate recognition results to the network, the detection metric *precision* indicates the node's bandwidth requirements. A low precision corresponds to a high number of insertion errors and increases the number of required event transmissions above the relevant events.

The car assembly scenario was selected based on the need for novel worker support systems. Such assistive systems track the task steps and report, whether tasks have been executed precisely enough. A clear emphasis of the work was to implement systems, that help the user in increasing task performance, e.g. if the tightening of screws for a car part was not detected, the worker could be asked to verify that the part was attached correctly. Moreover, using this system, the worker could train tasks at an individual pace or refresh skills when working with different car models. All of these applications require monitoring of activities, that involve multiple connected task steps.

We introduced the detection architecture along with descriptive metrics for a general quantification of its size and complexity. For the presented car assembly tasks, the architecture was composed of 11 composite classes, an alphabet size of 47 atomic activities and 49 detector events originating from 12 sensor-detector nodes. Similar scenarios can be found in other application areas, e.g. in monitoring of daily activities for health care and behaviour analysis in smart homes [1]. The presented approach could be applied to these applications as well. The metrics provide a convenient way to assess size and complexity of the distributed recognition architecture.

Acknowledgements. The work was supported by the Swiss State Secretariat for Education and Research (SER) as well as the 6th European Framework Programme Integrated Projects e-SENSE, http://www.ist-e-SENSE.org, contract number 027227 and WearIT@Work, http://www.wearitatwork.com, contract number 004216.

References

1. Intille, S.S., Larson, K., Tapia, E.M., Beaudin, J.S., Kaushik, P., Nawyn, J., Rockinson, R.: Using a live-in laboratory for ubiquitous computing research. In: Fishkin, K.P., Schiele, B., Nixon, P., Quigley, A. (eds.) PERVASIVE 2006. LNCS, vol. 3968, pp. 349–365. Springer, Heidelberg (2006)
2. Stäger, M., Lukowicz, P., Tröster, G.: Implementation and evaluation of a low-power sound-based user activity recognition system. In: McIlraith, S.A., Plexousakis, D., van Harmelen, F. (eds.) ISWC 2004. LNCS, vol. 3298, Springer, Heidelberg (2004)

3. Oliver, N., Garg, A., Horvitz, E.: Layered representations for learning and inferring office activity from multiple sensory channels. Comput. Vis. Image Und 96(2), 163–180 (2004) (Special issue on event detection in video)
4. Bobick, A.: Movement, activity, and action: The role of knowledge in the perception of motion. Philos. T. Roy Soc. B. 352(1358), 1257–1265 (1997)
5. Bao, L., Intille, S.S.: Activity recognition from user-annotated acceleration data. In: Ferscha, A., Mattern, F. (eds.) PERVASIVE 2004. LNCS, vol. 3001, pp. 1–17. Springer, Heidelberg (2004)
6. Ryoo, M., Aggarwal, J.: Recognition of composite human activities through context-free grammar based representation. In: CVPR 2006: Proceedings of the IEEE Conference on Computer Vision and Pattern Recognition. vol. 2, pp. 1709–1718 (2006)
7. Kawanaka, D., Okatani, T., Deguchi, K.: Hhmm based recognition of human activity. IEICE T. Inf. Sys. E89-D(7), 2180–2185 (2006)
8. Du, Y., Chen, F., Xu, W., Li, Y.: Recognizing interaction activities using dynamic bayesian network. In: ICPR 2006: Proceedings of the 18th International Conference on Pattern Recognition. vol. 1, pp. 618–621 (2006)
9. Predd, J.B., Kulkarni, S.R., Poor, H.V.: Distributed learning in wireless sensor networks. IEEE Signal Proc. Mag. 23(4), 56–69 (2006)
10. Saligrama, V., Alanyali, M., Savas, O.: Distributed detection in sensor networks with packet losses and finite capacity links. IEEE T Signal Proces 54(11), 4118–4132 (2006)
11. Wang, T.Y., Han, Y.S, Varshney, P.K., Chen, P.N.: Distributed fault-tolerant classification in wireless sensor networks. IEEE J. Sel. Areas Comm. 23(4), 724–734 (2005)
12. Thiemjarus, S., Yang, G.Z.: An automatic sensing framework for body sensor networks. In: BodyNets 2007: Proceedings of the Second International Conference on Body Area Networks (2007)
13. Li, S., Lin, Y., Son, S.H., Stankovic, J.A., Wei, Y.: Event detection services using data service middleware in distributed sensor networks. Telecommun. Syst. 26(2), 351–368 (2004)
14. Osmani, V., Balasubramaniam, S., Botvich, D.: Self-organising object networks using context zones for distributed activity recognition. In: BodyNets 2007: Proceedings of the Second International Conference on Body Area Networks (2007)
15. Amft, O., Junker, H., Tröster, G.: Detection of eating and drinking arm gestures using inertial body-worn sensors. In: Gil, Y., Motta, E., Benjamins, V.R., Musen, M.A. (eds.) ISWC 2005. LNCS, vol. 3729, pp. 160–163. Springer, Heidelberg (2005)
16. Bannach, D., Amft, O., Kunze, K.S., Heinz, E.A., Tröster, G., Lukowicz, P.: Waving real hand gestures recorded by wearable motion sensors to a virtual car and driver in a mixed-reality parking game. In: CIG 2007: Proceedings of the IEEE Symposium on Computational Intelligence and Games, pp. 32–39. IEEE Computer Society Press, Los Alamitos (2007)
17. Navarro, G., Raffinot, M.: Flexible Pattern Matching in Strings. Cambridge University Press, New York (2002)

Behavior Detection Based on Touched Objects with Dynamic Threshold Determination Model

Hiroyuki Yamahara, Hideyuki Takada, and Hiromitsu Shimakawa

Ritsumeikan University,
1-1-1 Noji-Higashi, Kusatsu, 525-8577 Shiga, Japan
yama@de.is.ritsumei.ac.jp

Abstract. We are developing a context-aware application for use in homes, which detects high-level user behavior, such as "leaving the home" and "going to bed", and provides services according to the behavior proactively. To detect user behavior, a behavioral pattern is created by extracting frequent characteristics from the user's behavior logs acquired from sensors online, using an extraction threshold based on the criterion of frequency. Most context-aware applications need to determine such a threshold. A conventional model determines a fixed common threshold value for all users. However, the common value is improper for some users because proper values vary among users. This paper proposes a detection method of high-level behavior with a model for determining the threshold value dynamically according to individual behavioral pattern.

Keywords: Threshold, context, behavior, ambient, proactive.

1   Introduction

We aim to develop a context-aware system which provides services, such as in the following example. Imagine a situation where a user leaves the home. Usually, the user keeps windows open while the user is in the home, and the user closes the windows before leaving the home. One day, the user has left the home and has carelessly left the windows open. In order to prevent such a danger in advance, our system informs the user that the windows are open before the user leaves the home. Such a service is valuable for the user because the service not only improves user amenity but brings relief and safety. In the above example, the timing to provide a service to the user is important. If the user is informed after the user leaves the home, the user must go back into house for closing the windows. The user should be informed before going outside the house. As another example, suppose the system provides a service in a situation of coming home, and an attempted delivery notice had arrived into a home server when a user comes home. In such a case, the system recommends the user to go to pick up a package before the user sits on a sofa and gets relaxed. We refer to such services, which should be provided proactively according to user behavior, as *proactive services*. In order to provide proactive services, the system must correctly detect characteristic behavior of the user in situations of leaving the home and coming home.

G. Kortuem et al. (Eds.): EuroSSC 2007, LNCS 4793, pp. 142–158, 2007.

Some existing studies propose methods for detecting user motion, such as "walking" and "standing up", and simple actions, such as "making tea" and "brushing teeth" [1,2,3,4]. However, not these low-level behaviors but high-level behaviors, such as "leaving the home" and "coming home", need to be detected for providing proactive services. A high-level behavior is a complex behavior in which some actions are interleaved. It is difficult to provide proactive services only by detecting low-level behaviors. We are developing a system for detecting high-level behaviors [5].

Context-aware applications, including our developing system, are built based on a model that collects online sensor data, which is acquired according to user behavior, as behavior logs and matches the logs with behavioral patterns for recognition. First, such systems collect a specific amount of sample behavior logs, which show characteristics of user behavior. Next, a behavioral pattern is created with the logs in every situation to be detected. User behavior is detected by matching behavior logs, which are acquired online from current user behavior, with each behavioral pattern. These systems need a specific amount of personal behavior logs as sample behavior logs to create a behavioral pattern for recognition. Therefore, services of the systems do not get available until enough sample behavior logs have been collected. If it takes a long period to collect sample behavior logs from the user activity, the user is dissatisfied with waiting a long time. In order not to dissatisfy the user, a behavioral pattern must be created with a small number of sample behavior logs which can be collected in a short duration. Most of existing methods create a behavioral pattern based on a stochastic method such as Hidden Markov Model (HMM) [6,7]. These methods need many sample behavior logs to create a behavioral pattern. Consider the problem to create a behavioral pattern of the situation of leaving the home. Only about 30 sample behavior logs can be collected even in a month. That means these methods cannot create a behavioral pattern in a short duration. These methods are not adequate to be put into practical use. Compared with these existing methods, a system we developed previously detects user behavior, using a behavioral pattern created with only 5 sample behavior logs which can be collected within a week [5].

Our system must set threshold values, which are used for creating a behavioral pattern and for matching online sensor data with the pattern. The first threshold is an extraction threshold. A behavioral pattern is created by extracting characteristics which frequently occur in sample behavior logs. The extraction threshold is a threshold of the occurrence frequency. If an improper value is set to the extraction threshold, behavior recognition accuracy is low because the characteristics of the user are not extracted adequately. The second threshold is a detection threshold. When a user's online sensor data is matched with a behavioral pattern, if the degree of conformity is more than the detection threshold then our system detects user behavior and provides services. Naturally, an improper detection threshold makes behavior recognition accuracy low. Not only our system but also most context-aware applications require thresholds to be set for creating a behavioral pattern and for matching the pattern. To make behavior recognition accuracy high, proper threshold settings are necessary. After many sample behavior logs are collected, initial values of the thresholds can be changed into more proper values by learning with the logs. The issue of the learning is not discussed in this paper. This paper discusses, as an issue to be solved, how to set an initial threshold value that achieves high recognition accuracy under a constraint of a small number of sample behavior logs.

There are several approaches to set proper threshold values in a variety of fields. In image processing, a setting method of a threshold used for extracting a specific area from a target image has been proposed [8]. This method can be used only if both parts to be extracted and parts not to be extracted exist together in a recognition target. Our issue does not meet such a condition, because behavior recognition in this paper considers whether a current behavior log conforms to a behavioral pattern or not. This approach in image processing cannot be applied to our issue. In other approaches, Support Vector Machines and boosting has been used for text categorization [9,10], and HMM is used for speech recognition [11]. These approaches can set a proper threshold value under the assumption that they can collect and analyze many samples of recognition target or many samples of others which have similar characteristics to samples of the recognition target instead. However, there is the constraint of a small number of sample behavior logs for creating a behavioral pattern in our issue. In addition, because characteristics of high-level behavior in homes are different among individual users, behavior logs of other people other than a user cannot be used for sample behavior logs. Although these methods can be used for learning a proper threshold value after many personal behavior logs have been collected, these methods cannot be used for setting a proper initial threshold value.

It is important to set a proper threshold value initially in order not to dissatisfy a user. In the conventional model for setting a threshold value, a developer of a context-aware application or an expert of the application domain sets the initial threshold value before introducing the system to a user's actual environment. Having the system used by some test users on a trial basis, the expert analyzes relativity between change of recognition accuracy and changes in a threshold value. The threshold value is determined such that the recognition rate averaged for all test users is the highest, with receiver operating characteristic curve, precision-recall curve, and so on. The value determined is used as an initial threshold value common to all users after introduction to actual user environment. However, it is difficult to achieve high recognition accuracy with the common threshold value for all users. Proper threshold values vary with individual behavioral pattern.

This paper aims to create a behavioral pattern which can bring out higher recognition accuracy by setting more proper threshold value than the conventional model, particularly for users whose behavior is not recognized well with the conventional model. Because it is difficult to determine the threshold value with only a small number of personal sample behavior logs, we consider to utilize data from test users as in the conventional model. However, unlike the conventional model, we cannot determine threshold values directly and also cannot create a behavioral pattern with many data from test users in advance, because characteristics of high-level behavior vary with individual user, as mentioned above. This paper proposes a method for determining an extraction threshold dynamically, based on a model which derives not a threshold value itself but a rule for determining the value by analyzing test user data. When acquiring knowledge by analyzing test user data, if the knowledge is not about an attribute which has high commonality among many users, then the knowledge is not meaningful. The conventional model determines the threshold value itself by analysis. The value obtained represents knowledge acquired without separating attributes, which have low commonality, from attributes which have high commonality. By analysis focused on attributes which have high commonality, more meaningful knowledge can be acquired. As such an attribute, the proposed method focuses on the number of characteristics composing a behavioral pattern. There is a famous number known as "the magical number seven, plus or minus two [12]" in cognitive science. This hypothesis proposes that the number of items of information which a human can instantaneously handle is about seven items. This is a common number for all people. This means that humans select about seven characteristic information items by screening a lot of information in order to instantaneously grasp the situation. From another point of view, the person can evaluate a situation properly by discarding excess information and selecting only information which is minimally necessary. Consider the number of characteristics composing a behavioral pattern. If there are too many numbers, the pattern will include excess elements which were not normally characteristics. If there are too few numbers, the pattern will miss useful elements as characteristics. This property of the number of characteristics is similar to the property of the number of items for human cognition. Considering such a property, this paper assumes that there is a universally ideal number of characteristics composing a behavioral pattern, which does not depend on individuals, as in the case of human cognition system. The proposed method derives a determination rule of an extraction threshold by analyzing test user data with a focus on the number of characteristics composing a behavioral pattern. A value of the extraction threshold is dynamically determined based on the rule when creating a behavioral pattern after introducing a context-aware application to the actual user environment. The proposed method has the following advantages.

- Focusing on an attribute which has high commonality, the method acquires meaningful knowledge from test user data, from which the conventional model cannot acquire meaningful knowledge, to detect high-level behaviors.
- The method dynamically determines a threshold value for individual behavioral patterns created with a small number of sample behavior logs, using a threshold determination rule derived from test user data.
- With a proper threshold for individual behavioral pattern, the method improves the recognition accuracy for users whose recognition accuracy is low with the common threshold value.

The result of an experiment shows that the proposed method improves behavior recognition accuracy, which is less than 80% with the conventional model, of some experimental subjects more than 10%.

The remaining part of this paper is as follows. Chapter 2 describes our behavior detection system. Chapter 3 explains a model for deriving a threshold determination rule and application of the model into our detection system. Chapter 4 shows evaluation by experiment. Finally, Chapter 5 concludes this paper.

2 Behavior Detection for Proactive Services

2.1 Detection of High-Level Behavior

We consider situations of leaving the home, coming home, getting up and going to bed, as situations in which proactive services can be provided effectively. For example, suppose when getting up, our system provides a reminder service, which reminds a user of one-day-schedule and of things to be completed by the time the user leaves the home. By providing this reminder service before the user starts preparing for leaving or for having a meal just after a series of actions when the user got up, the service can support the decision of next action of the user. When going to bed, our system provides services which brings relief and safety. For example, our system informs of that the windows are not closed. We consider proactive services are valuable services which can proactively prevent repentance and danger, which the user faces when the services are not provided.

Proactive services should not be provided mistakenly when "the user gets out of bed just for going to the toilet in the middle of sleep", or when "the user goes outside house just for picking up a newspaper". High-level behaviors, such as "leaving the home" and "going to bed", cannot be correctly detected only by recognizing simple actions as in the existing methods [3,4]. We consider that a high-level behavior is a long behavior of around ten minutes. Some actions are interleaved in the high-level behavior. In addition, characteristics of the high-level behavior vary with individual user. Therefore, a behavioral pattern for detecting the high-level behavior must be created with personal behavior logs of individual user. In order not to dissatisfy the user due to long waiting for collecting personal behavior logs, we consider services must get available within a week at the latest only with a small number of personal behavior logs.

2.2 Individual Habit in Touched Objects

In order to detect high-level behaviors, we must collect data which remarkably shows characteristics of individual user behavior as behavior logs. We focus on the aspect that most people often have habitual actions in a habitual order, for not making omission of things to do, in situations such as leaving the home and going to bed. Each user has his own characteristic behavior in such specific situations. For example, in a situation of leaving the home, there can be habitual actions such as going to the toilet and taking a wallet. That means the user habitually touches the same objects every time in the same situation. The kind

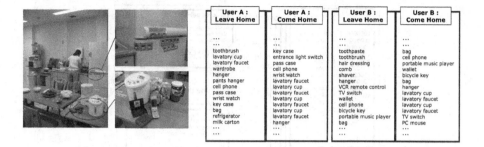

User A : Leave Home	User A : Come Home	User B : Leave Home	User B : Come Home
...	...	...	...
...	...	...	...
toothbrush	key case	toothpaste	bag
lavatory cup	entrance light switch	toothbrush	cell phone
lavatory faucet	pass case	hair dressing	portable music player
wardrobe	cell phone	comb	wallet
hanger	wrist watch	shaver	bicycle key
pants hanger	lavatory cup	hanger	bag
cell phone	lavatory faucet	VCR remote control	hanger
pass case	lavatory cup	TV switch	lavatory cup
wrist watch	lavatory faucet	wallet	lavatory faucet
key case	lavatory cup	cell phone	lavatory cup
bag	lavatory cup	bicycle key	lavatory faucet
refrigerator	lavatory faucet	portable music player	TV switch
milk carton	hanger	bag	PC mouse
...	...	...	...
...	...	...	...

Fig. 1. Objects embedded by RFID tags

Fig. 2. Examples of behavior log

of objects the user touches and their order depend on the individual user. The number of objects a user touches in situations, such as leaving the home and going to bed, is more than the number of those in other situations. In situations such as watching TV, having a meal, and reading a book, the user touches few objects, or he touches only limited kind of objects. Compared to these situations, it is obvious the user touches more objects in situations such as leaving the home. Objects the user touches indicate his intention and his behavior remarkably. The logs of touched objects are adequate for use as individual sample behavior logs.

We record histories of touched objects as behavior logs, using 13.56MHz RFID tags. As shown in Fig. 1, the tags are embedded in various objects of a living space, such as a doorknob, a wallet, or a refrigerator. Every object can be identified by unique tag-IDs stored in the tags. In contrast, a user wears a finger-ring-type RFID reader. When the user touches objects, his RFID reader reads tag-IDs of tags embedded in objects. According to the user behavior, a time series of $< tag\text{-}ID, \; timestamp >$ is recorded in a database as the behavior log of the user. Fig. 2 shows actual behavior logs recorded by our system. The table shows behavior logs of two users in situations of leaving the home and coming home. For example, in the situation of leaving the home, the habitual actions of user A are different from those of user B. Looking at the log, it is inferred that user A brushed his teeth, changed his clothes, picked up some portable commodities, and brought out a milk carton from the refrigerator. It is inferred that user B brushed his teeth, set his hair, operated a VCR and then picked up some portable commodities. These behavior logs show that kind of touched objects and their order are different among individual users even in a same situation. Similarly, comparing each user's situation of leaving the home to that of coming home, it is found that a user touches different kinds of objects or touches the same objects in a different order in different situations.

2.3   Behavior Detection with Ordered Pairs

In order to detect high-level behavior, we create a behavioral pattern represented by a set of *ordered pairs*, which show the order relation among touched objects, with histories of touched objects as sample behavior logs.

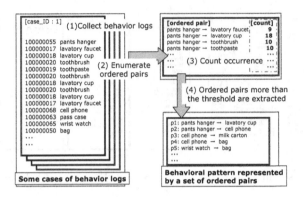

Fig. 3. How to create a behavioral pattern

The flow to create a behavioral pattern is shown in Fig. 3, with an example of a behavioral pattern in the situation of leaving the home. Generally, existing methods based on probabilistic models, such as HMM, create a behavioral pattern with high recognition accuracy using both behavior logs of the situation of leaving the home and logs of situations other than the situation of leaving the home as sample behavior logs. Consider our problem that a behavioral pattern must be created with a small number of sample behavior logs. Even behavior logs of leaving the home cannot be collected frequently. We can not expect to collect behavior logs of other situations which are adequate to make recognition accuracy high. Therefore, a behavioral pattern must be created only with behavior logs of leaving the home.

First, behavior logs of w cases are collected as sample behavior logs. The number of sample behavior logs w for creating a behavioral pattern is referred to as *window size*. The time length t_l of a sample behavior log is fixed. If m objects are sequentially touched in a behavior log l, then l is represented as a conjunction $\{o_1, o_2, \dots, o_i, \dots, o_m\}$, where, $o_{i-1} \neq o_i (1 < i \leq m)$. Second, all ordered pairs between two objects are enumerated from all collected sample behavior logs. If object o_j is touched after object o_i is touched, then the ordered pair p is represented as $\{o_i \rightarrow o_j\}$, which includes the case of $o_i = o_j$. For example, ordered pairs enumerated from a behavior log $\{o_1, o_2, o_3\}$ are $p_1 : \{o_1 \rightarrow o_2\} p_2 : \{o_1 \rightarrow o_3\} p_3 : \{o_2 \rightarrow o_3\}$. Next, the occurrence of all ordered pairs is counted up as occurrence count. The occurrence count means not the number of times that each ordered pair occurred in a sample behavior log, but the number of sample behavior logs including each ordered pair. For example, if an ordered pair occurs in all sample behavior logs, the occurrence count of the ordered pair is w. Finally, the ordered pairs where the ratio of the occurrence count to w is more than an extraction threshold $e\%$ are extracted as a behavioral pattern π.

The behavioral pattern π, which is created in advance, is matched with the current behavior log of time length t_l, which is acquired online from current user behavior. At the time when more than a detection threshold d % of ordered pairs, which compose the behavioral pattern π, exist in the behavior log, user behavior of leaving the home is detected.

For example, ordered pairs, such as $\{toothpaste \rightarrow toothbrush\}$, indicate the user's habitual actions, such as "brushing teeth". Ordered pairs, such as $\{toothpaste \rightarrow pants\ hanger\}$, indicates habitual order of the user actions, such as "the user wears pants after brushing his teeth". The behavioral pattern of a set of ordered pairs can represent the user's habitual actions and their order. Some existing studies create a behavioral pattern of a Baysian Network (BN) for recognizing simple actions [3,4]. To create a BN which has no cyclic path and can finish probabilistic reasoning in real-time, its network must be determined by hand. However, it is difficult to determine a network by hand so that the network can represent high-level behaviors in which some actions are intricately interleaved. Compared to the method using a BN, our detection method can extract characteristics of user behavior from such a complex behavior because our method uses an ordered pair, which is the smallest unit of order, to represent a behavioral pattern. There is an existing method which detects behavior in one situation with many behavioral patterns, which is created using time series association rule [13]. However, we need to observe user behavior for a long period in order to detect high-level behaviors. As a result, too many behavioral patterns are created by this existing method. Although this method needs to select only good behavioral patterns from many patterns, it is difficult to select good patterns under the constraint of a small number of sample behavior logs. Compared to this method, our detection method avoids comparing the quality among too many behavioral patterns, by potentially representing a variety of behavioral patterns with a set of ordered pairs.

2.4   Difficulty of Setting Threshold Values on Behavior Detection

We previously conducted an experiment in which we detected user behavior in situations of leaving the home, coming home, getting up, and going to bed, using our detection method. The recognition accuracy is evaluated both with *true-positive rate (TPR)* and with *true-negative rate (TNR)*. TPR shows the rate at which behavior logs in a specific situation, which logs are referred to as *true cases*, are correctly detected with a behavioral pattern of the situation. TNR shows the rate at which behavior logs in situations other than the specific situation, which logs are referred to as *false cases*, are correctly neglected with the behavioral pattern of the situation. It is preferable that both TPR and TNR are high. As a result, the recognition rates of some subjects were more than 90%. Meanwhile, the recognition rates of a few users were low rates of less than 80%.

The main cause of these differences is that the extraction threshold and the detection threshold are pre-determined values common to all users. By calculating *half total true rate (HTTR)*, which is an average between TPR and TNR, these threshold values were determined such that HTTR averaged for all users is maximum. After many sample behavior logs are collected, the recognition accuracy can be improved by learning of a behavioral pattern with the logs. However, we should solve the problem that there are differences of recognition

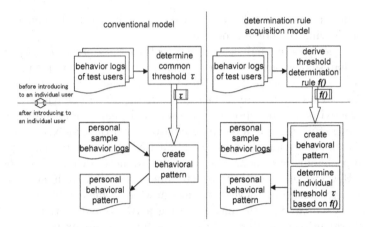

Fig. 4. Determination rule acquisition model from behavior logs of test users

rate among users depending on initial threshold values. It is necessary to improve the recognition accuracy of users, whose recognition rates are low with the common threshold values, by setting proper threshold values for individuals.

3   Dynamic Threshold Determination

3.1   Threshold Determination Rule Acquisition Model

We consider determining a threshold value dynamically for individual behavioral pattern in order to set a proper value to the threshold. For that purpose, unlike the conventional model which uses a fixed common threshold value, this paper proposes a model which acquires a rule to individually determine the threshold value for each behavioral pattern from the data of test users. The conventional model is illustrated on the left side of Fig. 4 and the threshold determination rule acquisition model, which we propose, is illustrated on the right side of Fig. 4. The horizontal center line shows a partition of the two phases for introducing a context-aware application to actual user environment. The upper portion is the development phase, before introducing the system to the actual environments of individual users. The lower side is the operation phase, after introducing the system. As shown in Fig. 4, the conventional model determines a common threshold value at the developement phase. First, the model collects behavior logs of test users. Next, for every test user, the model repeatedly creates a behavioral pattern with the logs, while matching the logs with the pattern. Analyzing the result of recognition accuracy, the model determines the threshold value with which recognition rate averaged for all test users is the highest. At the operation phase, the model creates an individual behavioral pattern with personal behavior logs. The threshold value is common irrespective of users. However, because a proper value for a threshold varies with the individual behavioral pattern of each user, behavior recognition accuracy of some users may be low with the common value.

To dynamically determine a proper threshold value for individuals, it is preferrable to acquire knowledge from personal behavior logs of individual user. However, it is difficult to determine a proper threshold value only with a small number of personal behavior logs. Therefore, the proposed model dynamically determines a threshold value by using both knowledge acquired by analysis of test user data and knowledge acquired from personal behavior logs. First, our model collects sample behavior logs of test users. Second, our model repeatedly creates a behavioral pattern with the logs and matches the logs with the pattern, for every test user. Next, our model analyzes the correlation between a threshold value and the recognition accuracy. If the threshold value is directly determined by analysis, the same problem occurs as in the conventional model. Our model derives not a threshold value itself but a rule f for determining the value by analysis. The threshold value is not determined at the developement phase. At the operation phase, the threshold value τ is determined for individual behavioral pattern by combining the rule f and knowledge acquired from a small number of personal behavior logs. If an analyst derives a rule f by focusing on an attribute, which does not depend on individual users, and personal behavior logs are used to consider attributes, which depend on individual users, then our model can determine a proper threshold value.

3.2   Effectivity of Dynamic Determination of Extraction Threshold

We apply the proposed model to our behavior detection system. The system has the extraction threshold and the detection threshold, which are described in Chapter 2.3. Primarily, it is important to set a proper value to the extraction threshold. In this paper, we consider a method for determining the value of extraction threshold dynamically.

The number of ordered pairs composing a behavioral pattern changes according to change of the extraction threshold, and affects the quality of the extracted behavioral patterns. It is preferable that a behavioral pattern includes many ordered pairs which are characteristics of user behavior in true cases. At the same time, the pattern should include few ordered pairs which can be characteristics of user behavior in false cases. If a behavioral pattern is composed of too few ordered pairs due to setting the extraction threshold high, then the behavioral pattern may not include some ordered pairs which should be normally included as user characteristics. On the other hand, if a behavioral pattern is composed of too many ordered pairs due to setting the extraction threshold low, then the behavioral pattern may include excessive ordered pairs which are not normally user characteristics. In particular, such fluctuation is a sensitive problem under the constraint of a small number of sample behavior logs. Suppose an improper value is set to the extraction threshold. It is impossible to extract ordered pairs adequately without excesses and shortages. Accordingly, recognition accuracy is low because differences between true cases and false cases are small when matching those cases with the behavioral pattern created with such ordered pairs. A proper extraction threshold sharpens differences between true cases and false cases. Consequently, recognition accuracy becomes high.

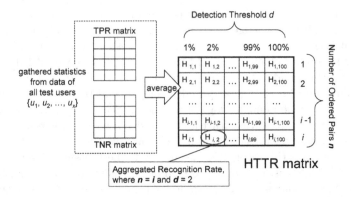

Fig. 5. TPR matrix and TNR matrix on statistics of test user data

3.3   Rating of Extraction Threshold by Statistics

Based on the threshold determination rule acquisition model, we derive a determination rule for setting the extraction threshold from data of test users. As mentioned above, the number of ordered pairs n affects the quality of behavioral patterns. The property of "the number of characteristics used for recognition", such as the number of ordered pairs, is similar to a cognitive property of human. "The magical number seven, plus or minus two [12]" in cognitive science proposes the hypothesis which indicates that humans select about seven characteristic information items by screening a lot of information in order to instantaneously grasp the situation. Consider the number of ordered pairs n. In both the case of excess ordered pairs and the case of insufficient ordered pairs, recognition accuracy is low. This property of the number of ordered pairs is similar to the property of the number of items for human cognition. Therefore, this paper assumes that there is a universally ideal number of ordered pairs, which does not depend on individuals, as in the human cognition system. In the issue of behavior detection, attributes such as kind of objects and their order have little commonality among users. It is difficult to derive a meaningful rule directly from these attributes. We attempt to derive a determination rule for the extraction threshold by evaluating the threshold value with a focus on the number, which has high commonality, of ordered pairs.

With an example of a behavioral pattern of a user v in the situation of leaving the home, we describe the proposed method which determines the threshold value dynamically. Before creating a behavioral pattern of user v, the threshold determination rule f is derived from behavior logs of x test users at the development phase. First, the proposed method executes the following procedure for every test user. The window size w is a given value which is common to all users.

1. Collect behavior logs in the situation of leaving the home as true cases, and also collect behavior logs in situations other than that as false cases.
2. Select w true cases as sample behavior logs.

3. Create w behavioral patterns with each setting of the extraction threshold value $e = 100 \times 1/w$, $100 \times 2/w$, ..., $100 \times w/w$, using the w true cases.
4. With all settings of the detection threshold d from 1% to 100%, match all true cases and all false cases with the w behavioral patterns.
5. Repeat step 2 to step 4 k times.

Second, TPR and TNR are calculated by gathering statistics on all results of the matching. As shown in Fig. 5, matrixes for the statistics of the rates are formed. The matrixes show the recognition rate with each number n of ordered pairs and each setting of the detection threshold. When a maximum number of ordered pairs is i in all created behavioral patterns, each matrix forms $i \times 100$ matrix. Finally, an HTTR matrix is formed. Each element H in the HTTR matrix is calculated by averaging each element in the TPR matrix and in the TNR matrix. Results of each row are respectively calculated with different numbers of statistical data. In the process of statistics, the method records the number of statistical data leading to results of each row of the HTTR matrix. Because there are w settings of the extraction threshold per behavioral pattern, the total number of statistical data is $w \times k \times x$. Each row of the HTTR matrix is rated with a rating score. The rating score s_i of the ith row is calculated as follows.

$$s_i = \ln(p(i)) \times \max_j(H_{i,j})$$

$\max_j(H_{i,j})$ means the maximum value in 100 elements of the ith row. $p(i)$ is the proportion of the number of statistical data used for the ith row to the total number of statistical data $w \times k \times x$. $\ln(p(i))$ is a coefficient for adding the reliability of statistics to the rating score. This method gives a higher rating score to rows using more statistical data. Next, these rows are equally divided into c clusters, such as cluster 1:{row 1, row 2, row 3}, cluster 2:{row 4, row 5, row 6}, ... The rating score of a cluster is calculated by averaging rating scores of all rows in the cluster. The value of c is empirically set to a proper number by an analyst. We assume that there is an ideal number of ordered pairs. However, because the number of ordered pairs composing a behavioral pattern depends on the number of ordered pairs occurring in sample behavior logs of individual user, one ideal number is not always identified using statistics of test user data. Therefore, this method attempts to find, not one ideal number, but "how much number is good roughly", by calculating rating scores of clusters. These rating scores correspond to the threshold determination rule. That is, when a behavioral pattern is created after introducing the behavior detection system to actual environment of user v, the extraction threshold is determined such that the behavioral pattern is composed of the number, which corresponds to as high rated cluster as possible, of ordered pairs.

4   Evaluation

4.1   Experiment

This paper describes an experiment to verify the efficacy of the proposed method. The experiment sets the time length t_l of a behavior log to 10 minutes. Before the experiment, we conducted a questionnaire survey for 2 weeks. In the questionnaire, subjects recorded the complete details about kind of objects the subjects touched and their order in 4 situations of leaving the home, coming home, getting up, and going to bed every day. With the questionnaire results, we could confirm that many people respectively touch different objects or touch objects in different orders, in different situations. After that, we experimentally embedded the RFID system described in Chapter 3 into the living space. RFID tags are embedded in many household goods such as kitchen gas stove, kitchen sink, and electric appliances, in every spaces such as living, kitchen, entrance, and so on. In such a space, we collected behavior logs of actual objects which subjects touched in the 4 situations respectively. The logs acquired online from subjects' behavior are stored in a database. We collected 70 behavior logs per subject.

To compare the proposed method, which dynamically determines the extraction threshold, with the method using the conventional model, which sets a fixed common value to the threshold, the experiment calculates TPR and TNR for behaviors of individual subjects in the 4 situations by repeatedly creating a behavioral pattern and matching behavior logs with the pattern, using behavior logs in the database. Here, true case means behavior logs of each situation, where a behavioral pattern is created, and false case means behavior logs of situations other than the situation of true case. The window size w is set to 5 in the experiment.

First, a threshold determination rule for the proposed method was derived by the calculations described in Chapter 3.3 with behavior logs of 8 subjects. In the experiment, rows in an HTTR matrix are divided into 100 clusters. Basically, each cluster includes three rows. But there are a few exceptions. Rows from the first row to the fifth row are included in a cluster which is rated as the second place from bottom, because they are empirically too small number as sample behavior logs. In addition, all of rows following the 300th row are included in the cluster same as the 300th row, whose cluster is rated as last place. Next, the following procedure was executed in order to calculate individual behavior recognition accuracy with 8 subjects. In this experiment, user behavior in each situation must be correctly detected in ten minutes, the time length t_l.

1. Select 5 sample behavior logs from true cases and create a behavioral pattern with the logs, based on the extraction threshold.
2. Select other 1 behavior log from true cases, and match the log with the behavioral pattern.
3. Match all behavior logs of false cases with the behavioral pattern, with all settings of the detection threshold d from 1% to 100%.
4. Repeat 100 times from step 1 to step 3, with a new behavioral pattern which is created by selecting new combination of 5 true cases every time.

Here, TPR is calculated based on cross validation. However, we limit the number of sample behavior logs used for creating a behavioral pattern to 5, which can be collected within a week. TNR is calculated by matching all false cases with all created behavioral patterns. The extraction threshold is determined when creating a behavioral pattern in step 1 using the threshold determination rule described

Table 1. Result of "Leaving the Home"

subject	TPR(%)	TNR(%)*
A	99	91.94
B	95	88.36
C (#1)	89 (+18)	92.84
D (#2)	94 (- 6)	98
E	99	99.68
F	100	95.04
G	99	96.6
H (#2)	88 (-10)	91.14

*TNR is rounded off
in the 3rd decimal place

Table 2. Result of "Coming Home"

subject	TPR(%)	TNR(%)*
A	91	95.25
B	99	99.38
C (#1,#2)	90 (+14)	84.88 (-9.13)
D (#1)	98 (+13)	98.8
E	98	99.5
F	100	100
G	100	99.78
H	100	100

*TNR is rounded off
in the 3rd decimal place

above. In this way, TPR, TNR and HTTR of all subjects are calculated for the case in which the extraction threshold is dynamically determined. After that, these rates in the case of using a fixed common value as the extraction threshold are calculated by similar steps. In that case, the extraction threshold is fixed to 80% in step 1 such that recognition accuracy is the highest. Although TPR, TNR and HTTR are calculated with all settings of the detection threshold from 1% to 100%, the results of the two methods are compared using TPR and TNR on a detection threshold with which HTTR of each method is the highest per subject.

A user touches less number and less kinds of objects, in situations other than the 4 situations to be detected in this experiment. Therefore the proposed method, which focus on kind of objects the user touches and the order of the objects, can distinguish among the 4 situations and other situations easily. Previously, we conducted an experiment in which we recognized behavior logs including behavior logs of situations other than the 4 situations. Only up to 7% of ordered pairs, which compose individual behavioral pattern, occurred in situations other than the 4 situations. This result showed that user behavior in situations other than the 4 situations has no chance to be mistakenly detected by the proposed method. With this result in mind, we evaluate the recognition accuracy only with the 4 situations in the experiment of this paper. This means we evaluate our behavior detection method under more difficult conditions.

4.2 Discussion

Based on the result of the t-test, the experiment results are evaluated with the idea that difference of more than 5% is a statistically-significant difference

Table 3. Result of "Getting Up"

subject	TPR(%)	TNR(%)*
A	96	96.2
B	84	82.48
(#2)	(- 6)	(-14.3)
C	75	96.23
(#1)	(+11)	(+12.52)
D	100	89.91
(#2)		(-9.98)
E	97	59.38
(#3)	(+31)	(-27.13)
F	96	91.45
(#2)		(-8.23)
G	100	99.98
H	59	93.6
(#3)	(-22)	(+30.22)

*TNR is rounded off
in the 3rd decimal place

Table 4. Result of "Going to Bed"

subject	TPR(%)	TNR(%)*
A	76	74.44
B	93	70.88
C	95	99.98
D	91	95.94
(#1)	(+15)	
E	47	85.68
(#1)	(+12)	
F	99	97.92
G	100	98.84
H	97	93.92
(#1)	(+15)	

*TNR is rounded off
in the 3rd decimal place

between the proposed method and the method using the conventional model. As a result of the experiment, recognition rates in the proposed method are shown from Table 1 to Table 4. The tables respectively show the results of leaving the home, coming home, getting up, and going to bed. Each table shows the TPR and the TNR by the proposed method. In addition, the difference between the proposed method and the method using the conventional model is shown in parenthesis under each value. If the value is a positive value, then the proposed method has increased the rate. The differences which are less than a statistically-significant difference are not shown.

Looking at TPR and TNR in the tables, notable results are grouped into 3 groups from #1 to #3. Group number is written under subject name in each table. In group #1, TPR or TNR have increased with the proposed method. In each situation, there is at least 1 subject whose TPR or TNR have increased with the proposed method. Particularly, subject C of Table 1, subject C of Table 3, subject D and E of Table 4 have significantly increased. Their rates have increased more than 10% with the proposed method from low rates which are less than 80%. In group #2, TPR or TNR have decreased with the proposed method. However, even after decreasing, the rates can keep more than 80% for all subjects in group #2. Considering that our detection method must be introduced into a variety of user environments, the detection method must achieve high recognition accuracy stably for behaviors of as many users as possible. The detection method should not be effective on only a portion of users. In the experiment, the proposed method has decreased the rates of some subjects whose recognition rates are very high with the method using the conventional model. This decrease is not ideal result. However, the proposed method has increased significantly the rates of some subjects whose recognition rates are low with the method using the conventional model. This result shows the proposed method can achieve stabler behavior detection than the method using the conventional model. Overall, the result of the experiment means the recognition accuracy can be improved by determining a better value of the extraction threshold with the proposed method. The result has proved the proposed method is effective. Exceptionally, the proposed method is not effective on subjects of group #3. About their TPR and TNR, one rate has increased and the other has decreased, based on just a basic relation of trade-off.

5 Conclusion

This paper proposed a detection system of high-level behavior, such as "leaving the home", and proposed also a method for dynamically determining threshold values, which should be set in order to introduce the system to a variety of user environments. An experiment has proved our method is effective. Our method improved the recognition rate of subjects, whose rates were low with the common threshold value, more than 10%. The present recognition rate is not enough practical. In the future, we will attempt to achieve higher recognition rate by combining the present method with other informations such as position of users. In addition, we will evaluate our method by introducing more user environments.

References

1. Barbič, J., et al.: Segmenting motion capture data into distinct behaviors. In: Proc. Graphics Interface 2004, pp. 185–194 (2004)
2. Moore, D.J., et al.: Exploiting human actions and object context for recognition tasks. In: Proc. ICCV 1999, pp. 80–86 (1999)
3. Patterson, D.J., et al.: Fine-grained activity recognition by aggregating abstract object usage. In: Gil, Y., Motta, E., Benjamins, V.R., Musen, M.A. (eds.) ISWC 2005. LNCS, vol. 3729, pp. 44–51. Springer, Heidelberg (2005)
4. Wang, S., et al.: Common sense based joint training of human activity recognizers. In: Proc. IJCAI 2007, pp. 2237–2242 (2007)
5. Yamahara, H., et al.: An individual behavioral pattern to provide ubiquitous service in intelligent space. WSEAS Transactions on Systems 6(3), 562–569 (2007)
6. Aoki, S., et al.: Learning and recognizing behavioral patterns using position and posture of human body and its application to detection of irregular state. The Journal of IEICE J87-D-II(5), 1083–1093 (2004)
7. Kidd, C.D., et al.: The aware home: A living laboratory for ubiquitous computing research. In: Streitz, N.A., Hartkopf, V. (eds.) CoBuild 1999. LNCS, vol. 1670, pp. 191–198. Springer, Heidelberg (1999)
8. Kimura, Y., et al.: New threshold setting method for the extraction of facial areas and the recognition of facial expressions. In: Proc. CCECE 2006, pp. 1984–1987 (2006)

9. Shanahan, J.G., et al.: Boosting support vector machines for text classi cation through parameter-free threshold relaxation. In: Proc. CIKM 2003, pp. 247–254 (2003)

10. Cai, L., et al.: Text categorization by boosting automatically extracted concepts. In: Proc. SIGIR 2003, pp. 182–189 (2003)

11. Asami, T., et al.: A stream-weight and threshold estimation method using adaboost for multi-stream speaker veri cation. In: Proc. ICASSP 2006. vol. 5, pp. 1081–1084 (2006)

12. Miller, G.A.: The magical number seven, plus or minus two: Some limits on our capacity for processing information. The Psychological Review 63(3), 81–97 (1956)

13. Mori, T., et al.: Behavior prediction based on daily-life record database in distributed sensing space. In: Proc. IROS 2005, pp. 1833–1839 (2005)

Towards Mood Based Mobile Services and Applications

A. Gluhak[1], M. Presser[1], L. Zhu[1], S. Esfandiyari[1], and S. Kupschick[2]

[1] Center for Communication Systems Research, The University of Surrey,
Guilford, GU2 7XH, United Kingdom
{a.gluhak, m.presser, l.zhu, s.esfandiyari}@surrey.ac.uk
[2] Human Factors Consult,
Köpenicker Straße 325, Haus 40, 12555 Berlin, Germany
stefan.kupschick@human-factors.de

Abstract. The introduction of mood as context of a mobile user opens up many opportunities for the design of novel context-aware services and applications. This paper presents the first prototype of a mobile system platform that is able to derive the mood of a person and make it available as a contextual building block to mobile services and application. The mood is derived based on physiological signals captured by a body sensor network. As a proof-of-concept application a simple mood based messaging service has been developed on top of the platform.

Keywords: Wireless body senor networks, context-awareness, data fusion, mobile services, mood based services.

1 Introduction

The introduction of context-awareness to the mobile communications world has paved the way for a new class of highly adaptive and personalized mobile services and applications. The idea behind such context-aware services and applications is the use of knowledge about the user's current context and associated preferences in the application logic, in order to provide enhanced user experience, tailored to the situation and preference of a user. Much of the research in recent years has concentrated on the means to derive necessary context information of a user, its processing and use in application frameworks and the development of novel context-aware services and applications that make use of the available context information. As a result a variety of different mobile services and applications have emerged, the majority of those exploiting location information or offline context (manual input of information) such as preferences or the calendar of a user.

Recent advances in wireless sensor networks (WSN) have created new opportunities in the way that context information can be captured. Using WSNs a greater diversity of contextual information can be obtained for static or mobile users. Despite advances in processing and reasoning of sensor data a variety of unsolved challenges still remain. Examples are the capturing of complex contextual information such as the mood or emotions of a person.

G. Kortuem et al. (Eds.): EuroSSC 2007, LNCS 4793, pp. 159–174, 2007.

This paper presents first steps towards the provision of mood based mobile services and applications documenting research undertaken in the IST e-SENSE project [1]. A portable system is presented that allows the capturing of the mood of a mobile user and making this context information available to mobile applications on the platform. The system is based on a body sensor network, wirelessly connected to a PDA class device, which serves as a gateway to the service platform. As part of the platform an online mood detection algorithm has been implemented on the device that interprets the physiological parameters captured by the body sensor network. A simple mood based messaging service has been realized on top of this platform. To our knowledge this is the first completely mobile platform that is able to derive the mood of a person and integrate it into a mobile solution. Although the initial results are promising, a lot of research challenges need to be solved, in order to reliably detect a variety of mood states of a mobile person. We believe that our initial experiences described in this paper will encourage other researchers to get engaged in this complex but exciting area of research.

The reminder of the paper is structured as follows. Section 2 briefly surveys related work in the area, while section 3 provides the motivation behind a mood based messaging service as context aware applications for mobile users. Section 4 gives an overview of the system and the design of its different components. Section 5 describes background information of the measurement campaigns conducted for the development of the mood detection algorithm and provides details of its implementation. The developed mood based messaging application and experimental setup is described in section 6, followed by a discussion of initial experiments and experiences. Final remarks are given in section 8.

2   Related Work

The mood of a person is important contextual information for many human interactions. Knowledge about the mood or in the extreme case emotional state of a person enhances cognition. Machines, computing systems, their applications and even persons with autism or Asperger's Syndrome do not have a natural ability to recognize or convey mood and emotional states. This section gives a brief summary of the work so far in the commercial and research communities for mood models applicable to machine mood recognition systems and what information these systems use and how they obtain it as well as their application.

2.1 Emotion Models

Current theories from psychology on emotions can be grouped into theories that focus on how emotions arise and how they are perceived, and theories focusing on how observed emotions could be categorized or structured. Since theoretical aspects on how emotions arise, when and how they are perceived, and which biological mechanisms induce them are less important for systems to recognize emotions, these approaches will not be reviewed in this paper. Please refer to the respective literature for further information [12][13][14].

Among the theories for categorizing or structuring emotions, two main theories are currently established: a discrete approach, claiming the existence of universal 'basic emotions' [15] [16] and a dimensional approach, assuming the existence of two or more major dimensions which are able to describe different emotions and to distinguish between them [17]. There is still controversy on the matter of which approach is the one that best captures the structure of emotion even though attempts have been made to conflate the two [18].

Discrete emotion theories claim the existence of historically evolved basic emotions, which are universal and can therefore be found in all cultures. Several psychologists have suggested a different number of these, but there has been considerable agreement on the following six: anger, disgust, fear, happiness, sadness and surprise. Several arguments for the existence of these categories have been provided, like distinct universal facial signals, presence in other primates etc.

Dimensional emotion theories use dimensions rather than discrete categories to describe the structure of emotions. According to a dimensional view, all emotions are characterised by their valence and arousal (see Figure 1). Valence is defined by its two poles negative/bad and positive/good, whereas the two poles of the arousal dimension are sleepy/calm for very low arousal and aroused/excited for very high arousal. These two dimensions are not claimed to be the only dimensions or to be sufficient to differentiate equally between all emotions, but they have proven to be the two main ones, accounting for most of the variance observed [17].

2.2 Mood Recognition Systems

Mood systems can be categorized into systems using manual mechanisms to input mood states, audio/visual recognition systems, in particular facial recognition for video and key word or voice carrier frequency for audio systems, physiological body sensor systems, and other systems that use specific information such as pressure sensors on seats to infer e.g. movement or posture.

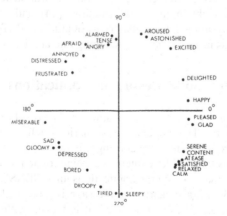

Fig. 1. A Circumplex model of affect (taken from Russell, 1980)

Example applications that use mood obtained by manual input mechanisms are already commercially available. In particular the iPod application Moody [3] uses manual input to tag tracks with mood information. Similarly Nokia in the context of the MOBILIFE project [4] developed a ContextWatcher application [5] that also tags mood information to content; in the case of mood this is also manual.

Systems that rely on video processing seem to be the most reliable and mature autonomous systems based on one type of input, although direct comparison of video processing systems and other systems do not exist in the literature. However these systems are unsuitable for mobile applications as they rely on facial pattern recognition [5] or a video camera infrastructure for movement pattern recognition [7]. The majority of the facial expression mechanisms are based on Ekman's work called Facial Action Coding System (FACTS) [8]. It is the most popular standard currently used to systematically categorize the physical expression of emotions. The affective computing lab at MIT for example uses facial expression recognition methods to provide human computer interfaces of computers with a notion of the users' mood [9]. Alternatively systems using an electromyogram (EMG) can also be used to obtain facial expressions [10], these systems do not rely on video inputs, however use similar techniques to evaluate facial movement patterns to derive mood.

Systems that use audio input such as the voice carrier frequency have shown promising results and are applicable to mobile scenarios. [11] describes such an example of using speech recognition to establish a notion of mood. The application described in the paper is related to the car environment during which the driver interacts with the computer system of the car. Audio based system requires a certain speech or sound input of significant length to be reliable, which only makes them applicable for conversational scenarios or voice interactions with computers.

Physiological body sensor networks have mainly been developed for medical applications. These systems can also be used for machine mood recognition as physiological parameters are unbiased due to the nature of the signal (a person is usually not able to influence physiological parameters); however physiological reactions can be very difficult to measure in mobile environments and are not straight forward to interpret to obtain mood information. [24] and [25] present example systems measuring physiological parameters to obtain mood. Typically only one or two parameters, such as heart rate and/or breathing rate are used.

3 Use of Mood in Mobile Messaging Applications

As a case study we use the e-SENSE application scenarios [31] to illustrate the use of mood for mobile users. The application scenarios include *push like messaging applications* that generate content according to the mood of users and *pull like messaging applications* that deliver content according to the mood of users.

Examples of push like messaging applications in e-SENSE are security related such as the *Danger Warning* scenario – a message is generated according to certain mood states of a user, e.g. if fear is detected, it is tagged with location information and sent to appropriate contacts (e.g. local authorities or family member). A less critical application of the push messaging is blogging. The e-SENSE scenarios also describe an extensive application space of enhanced and automated blogging using sensory information, such as mood. A diary is generated automatically containing location and mood information, which then can later be updated by the user with pictures, movie clips and text.

So far the content based on the user's mood was generated at the user; on the other hand, pull like messaging applications use mood information to provide appropriate content and application logic. For instance the e-SENSE scenario application *Happy Messaging* provides certain content defined by the user profile according to a person's mood, e.g. sending pictures to cheer the user up.

e-SENSE generated 26 application scenarios of which many address mood as a central context building block for applications. A subset of the scenarios was also used to build audio visual demonstrations which can be found at www.ist-e-sense.org.

4 System Platform

This section provides an overview of the developed system platform. The system platform consist of a wireless body sensor network (BSN), worn at the mobile user and the personal mobile device of a user. The BSN is connected to mobile device, providing the captured physiological parameters to algorithms and applications that execute on the system platform. In the following an overview of the hardware and software architecture of the platform is provided.

4.1 Hardware Architecture

The BSN is based on three SensiNode [32] Micro.2420 wireless sensor nodes. Like many other Motes, the Micro.2420 is designed around a MSP430 micro controller and provides an IEEE 802.15.4 radio in 2.4 GHz ISM band. Two of the three sensor nodes are directly worn at the body of a person, each attached with two sensor probes and respective signal conditioning boards (see Figure 2). The third sensor node is attached to the USB port of the mobile device acting is a bridge to the 802.15.4 network.

The BSN provides four different sensor sources to extract a variety of different physiological parameters of the mobile user. The four different sensor probes attached to the sensor nodes of the BSN are:

- ECG electrodes with snap leads: used as a source to derive the heart rate and its variability of a person
- Piezo Respiratory Belt Transducer: used as a source to derive the breathing rate of a person
- Finger electrodes: used as a source to measure the electro dermal response or activity of a person
- Skin temperature sensor: used as a source to measure skin temperature.

Two custom signal conditioning boards have been developed for the different sensor probes. The first signal conditioning board hosts necessary circuitry for ECG and breathing rate sensor. For the ECG signal, an ASICs is used that is specifically designed for the amplification and conditioning of bio-potential signals by one of our partners [27]. The respiratory belt requires additional amplification due to the low output voltage of the piezo-electric signal source. The second signal conditioning board hosts a measurement bridge for the skin resistance of a person and an input for the skin temperature sensor. Both signal conditioning boards provide the conditioned signals as analogue output to the AD converters of the micro controller on the sensor nodes. Adequate drivers that allow a configuration and interaction with the signaling boards had to be developed.

As a mobile user device a Nokia 770 Internet Tablet has been selected. There are two major motivations for the selection of the device. The N770 runs a Linux 2.6.16 kernel, allowing easy porting of Linux based tools that are used for the communication with the sensor network. The second motivation is the configurable USB host functionality, to connect the bridging sensor node directly via USB. Most PDAs don't offer a USB host controller, and are only able to operate as slaves. The N770 can be configured to operate as a USB host device. Appropriate FTDI drivers for USB-serial communication can be easily cross-compiled for the platform.

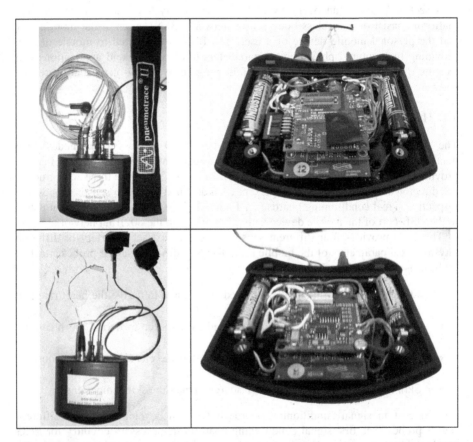

Fig. 2. BSN components: The figure shows outside and inside view of the developed BSN nodes for ECG/breathing node (above) and EDA/skin temperature node (below)

Although the N770 can act as a USB host device, it does not provide any output power over the USB interface. The sensor node has been integrated together with an external battery in an extension cover that can be attached to the N770, as shown in Figure 3.

Fig. 3. N770 setup with extension cover for sensor node and battery

4.2 Software Architecture

The two sensor nodes of the BSN with the sensor probes run a light-weight publish/subscribe middleware on top of a 6LowPAN network protocol stack [32] and FreeRTOS operating system. The middleware allows application entities on the N770 to discover available sensor information and subscribe to it at different sampling rates. The subscriptions are forwarded to the respective nodes, and once received a node starts publishing sensor information at the requested rates. The other sensor node that is attached to the N770 runs the same protocol stack, however, instead of the middleware, only a simple bridging application. All source code on the sensor node has been developed entirely in C.

On the N770, a mood recognition entity has been developed. The mood recognition entity communicates with the BSN via the nRouted process. The nRouted is a serial forward like application that allows application processes to connect via a TCP socket and routes packets to and from the sensor network via the USB serial interface. The mood recognition entity subscribes to the different sensor information types it requires. The middleware in the BSN ensures that respective sensor information is forwarded to the N770. Applications such as our mood based messaging service are then able to obtain real-time physiological parameters and a mood state vector from the mood recognition entity. Details of the mood recognition entity and the mood based messaging service can be found in section 5 and 6 respectively.

5 Mood Detection Algorithm

This section describes the development of the mood detection algorithm. The first part describes details of a measurement campaign that has been undertaken, in order to collect sufficient physiological data traces for the development. Then details on the offline analysis of the data sets are provided. Furthermore a first online version of the mood detection algorithm developed for the system platform is presented.

5.1 Measurement Campaigns

The test sample consisted of 40 subjects (27 male) with a mean age of 30 and an age range from 22 to 54, recruited from the Center for Communication Systems Research CCSR and Surrey's School of Management, who voluntarily took part in the experiment and were not rewarded. The sample was multi-cultured with representatives from 16 different countries participating.

The experiment was performed at CCSR at the University of Surrey in Guildford, England over a period of five days. The subjects sat in a separated windowless test room in 2m distance of a 2.4m x 1.35m screen. Subjects were equipped with the HealthLab System [28] for physiological measurement, which included a chest belt for respiration measurement, ECG and EDA electrodes, EMG electrodes for facial muscle measurement, a neck microphone for voice parameters and a wristband for measurement of skin temperature. Procedures were explained to them and it was made clear that they had to watch short films and afterwards state how they had felt in doing so. The investigators observed the trial via a glass pane from a separate room.

5.2 Offline Analysis and Algorithm Development

During the experiments data was recorded from the HealthLab System and monitored in order to check the correct fit of the sensors and the correctness of the data. There were separate data records for each film and person, which contained the data sets of all sensors. The received data had to pass through various pre-processing routines before the required parameters were extracted.

Electrocardiogram (ECG): An important feature which can be extracted from the ECG is the heart rate variability (HRV). HRV is a measure of variations in the heart rate and is usually calculated by analyzing the time series of beat-to-beat intervals. The raw ECG signal has to pass through various pre-processing and verification routines to ensure the accuracy of the resulting array of R-peaks. Following parameters are directly derived from the calculated time series of R-peaks:

- *Heart_rate:* The average number of heart beats per minute. Different studies indicated fairly strong correlations with valence [19][20].
- *RMSSD:* The root mean square of the standard deviation.
- *pNN-50:* The percentage of differences between adjacent beat-to-beat intervals that are bigger then 50 msec.

As mental stress increases, the variability of the heart rate decreases and thus, the *RMSSD* and the *pNN-50* should theoretically decrease as well. The two parameters should also correlate significantly [29].

A frequency domain analysis is performed to investigate effects of the sympathetic and parasympathetic nervous system. A common frequency domain method is the application of the discrete Fourier transformation to the beat-to-beat interval time series. This analysis expresses the amount of variation for different frequencies. Several frequency bands of interest have been defined in humans [30]:

- *High Frequency band* (HF) between 0.15 and 0.4 Hz. HF is driven by respiration and appears to derive mainly from vagal activity.
- *Low Frequency band* (LF) between 0.04 and 0.15 Hz. LF derives from both vagal and sympathetic activity and has been hypothesized to reflect the delay in the baroreceptor loop.
- The *LF/HF ratio* describes the relation between the activating (sympathetic) and deactivating (parasympathetic) influence of the autonomic nervous system [21]. Therefore, rising values could indicate an increased activity level.

Breathing rate: After baseline and shock removal the peaks of the single breaths are detected and the resulting parameter *Breathing rate* is computed. It describes the average number of breaths per minute during a film.

Electrodermal activity (EDA): Changes in electrodermal activity are thought to reflect dilations and secretions of the eccrine sweat glands. They are innervated by the sympathetic nervous system, but not by the parasympathetic nervous system. Electrodermal activity is often associated with levels of emotional arousal [22]. At large, electrodermal activity consists of two components. The tonic component is a low frequency baseline conductivity level. The phasic component is of higher frequency. The tonic component of the data can be easily computed by applying a long-term average filter (about 30 secs) to the data. Statistically meaningful tonic parameters are:

- *EDA_changes:* The total number of turning points of the signal. We assume that smaller numbers could imply a low level of arousal.
- *EDA_increase:* The percentage of increasing data of the low-pass filtered data curve. This is a long-term measure and can be interpreted as inversely proportional with arousal.

The phasic component, which basically consists of short term reactions, is determined by a pattern-recognition algorithm, which separates the detected 'responses' from the rest of the data. A 'response' is a short term variation of the signal and has to meet certain criteria: The amplitude is larger than 3 kOhm and the duration of the decrease is between 0.5 and 5 seconds [24]. Based on that following phasic parameters are extracted:

- *EDA_responses:* The total number of responses. Larger numbers could be a sign of a higher level of arousal.
- *EDA_slope:* The average slope of an EDA-response. Higher slopes could indicate a higher intensity of a reaction and thus correlate with arousal.

Electromyogram (EMG): This is a measure of electrical potentials generated by the muscles in activation and rest. *Zygomatic* EMG, which collects muscle activity at the corners of the mouth and *corrugator* EMG, which measures muscle activity of the eyebrows, were used in this study. Very strong correlations with the valence dimension are described in [23]. For both EMG datasets, identical parameters were extracted:

- *EMG_reactions:* The total number of typical contractions of the respective muscle. The number increases with rising facial activity, for example laughing.
- *EMG_activity:* The percentage of the data in which any activity is measured.

Skin temperature: The measured temperature evolves inversely proportional to the *skin resistance*. Two parameters were extracted from the signal:

- *TEMP_mean:* The mean value of the smoothed signal.
- *TEMP_rise:* The gradient of the linear regression function of the signal. Positive values imply a long-term increase of the skin temperature during the film and therefore a lowered arousal level.

The development of the mood-detection algorithms aims at assigning the aggregated physiological data to a specific sector of the coordinate system, which is defined by the two axes 'valence' and 'arousal'. Offline analysis of features extracted from the collected data sets has shown good detection rates of up to 80% for arousal and 70% for valence based on 5 features [32] using classifiers based on Support Vector Machines (SVM). ECG, EDA, Breathing rate and Skin Temperature have provided the strongest indications for the level of arousal and have therefore been utilized in the online algorithm described below.

5.3 Online Algorithm

The design of the online version of the algorithms proved more challenging as incoming data has to be continuously processed in real time on the computational constraint devices (the BSN nodes and the Nokia N770). Consequently the capabilities of the implemented mood recognition algorithm is limited to the detection of the level of arousal, interpreted as two emotional states, namely *relaxed* and *activated*.

Figure 4 gives an overview of the signal processing chains utilized for the extraction of desired features for the mood recognition algorithm. The measured analogue signals at the sensor probes are conditioned in the sensor processing boards at the sensor nodes, before being sampled at different rates. The ECG signal is sampled at 250Hz, while the other remaining signals are sampled at 20Hz. Digital pre-filtering takes place on the sensor node before transmission to the Nokia N770. Further processing steps for the extraction of required features of the physiological signals take place in the Nokia N770.

The most complex processing chain in terms of algorithmic and computational complexity is the *ECG processing chain*. A peak detection algorithm has been implemented with automatic threshold detection, in order to extract the RR peak distances of the heart beat. Additional mechanisms increased the reliability of peak detection in presence of noise artifacts in the signal or packet loss.

Several features are computed from the RR peak distances: the heart rate, RMSSD and PNN-50. Furthermore a spectral analysis is performed to identify high and low frequency components and their ratio. For the spectral analysis a 256 FFT is utilized.

The *EDA processing chain* considers both the short term (ST) phasic components in form of skin resistance response (SRR), and the long term (LT) tonic components in form of the skin resistance level (SRL).

The *skin temperature processing chain* computes short and long term development of the skin temperature over time in form of the mean and temperature rise.

All physiological data is continuously provided to the processing chains and buffered for further processing. The processing chain is invoked every 5 seconds on the currently buffered data set and necessary features are extracted and provided to the mood detection algorithm, which currently only detects the level of arousal. The interval of 5 seconds provided a good tradeoff between required resolution accuracy and computational effort on the resource limited system.

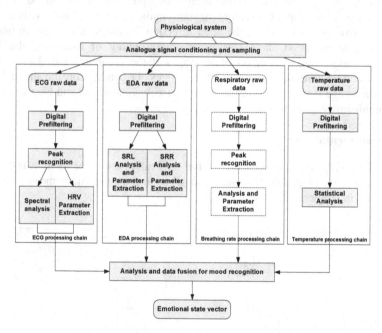

Fig. 4. Signal processing chains for physiological data

6   Mood Based Messaging Application

In order to demonstrate the utilization of the system platform by a mobile context-aware service, a mood based messaging application has been developed. In the following subsections the mood based messaging application and experimental setup are briefly presented, followed by a discussion of initial experiments.

6.1 Overview of the Mood Based Messaging Application

The mood based messaging application (MMA) provides a simple peer to peer text based chat and messaging service that allows the exchange of messages with one or more parties. The message exchange is realized via a multicast channel, to which all interested communication parties are able to subscribe. Figure 5 shows the main window of the MMA. Like a normal messaging application, the MMA allows manual entry and exchange of messages as well as displaying of sent and received text messages. In addition the MMA interacts with the mood entity that has been implemented as part of the service environment of the system, in order to obtain physiological parameters and a mood state vector of the attached person.

On the right hand side, different physiological parameters are displayed, that have been extracted from the original sensor information of the BSN. This includes heart rate, its variability expressed as pnn50 and power spectrum density of LH and HF components and their ratio, long term and short term trend of the EDA and the skin temperature. On the bottom of the figure the intensity of the mood vector is represented as a bar graph. Currently two states namely "relaxed" and "activated" are shown. A growing bar in one or the other direction indicates an increase in intensity in the recognized mood state. Besides manual sending of messages, the MMA provides an automatic generation of messages if the computed mood state vector reaches a certain threshold in either of the states. Depending on the mood state a different message is sent. This message can be freely customized and tagged with e.g. location information or other sensor data.

6.2 Experimental Setup

In the following the experimental setup typically used for demonstrations is described. The developed BSN system is attached to the test subject, measuring ECG, EDA and skin temperature.

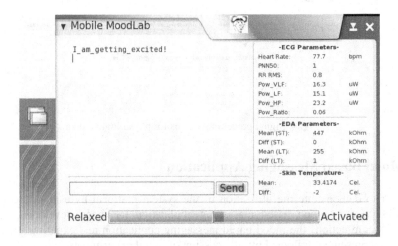

Fig. 5. Main window of the mobile mood based messaging application

The nodes wirelessly transmit required sensor information, which is received by the sensor node attached to the mobile PDA device.In addition the time variations of the real-time physiological parameters are visualized on a laptop in an oscilloscope type application. The Laptop also has a sensor node attached that listens on the same channel in promiscuous mode.

The most difficult part of the experiment is the controlled induction of a certain mood state. After investigation of different audio visual methods, a scenario based only on audio has been selected. For the test case the test subject wears noise cancellation headphones and listens to a carefully edited music track. The track starts with a 2 min long period of silence, which is used to record a physiological baseline of a person. Then a period of classical music begins that puts the test subject into a more relaxed mood state. After a certain period of time, the music unexpectedly changes to an extreme heavy metal track. As a consequence a transition in the test subject's mood state from a "relaxed" to an "activated" state is expected.

6.3 Discussion

Figure 6 shows an example of the calculated mood vector that has been recorded during the audio experiment. During the silence period a baseline of the person is initially recorded. The test subject seemed slightly activated in anticipation of the music to start. After the classical music started the mood vector decreased constantly indicating relaxation of the person.

Towards the end of the classical music period the mood vector returned to zero, indicating that the relaxation phase of the test subject ended reentering a neutral state. The sudden change to the heavy metal music is notable at the following increase of the mood vector, indicating a change to an activated state of the test subject. The initial activation eases off over time as the test subject is getting used to the music. The experiment showed a similar development of the mood vector with different participants for the first execution of the experiment. A reproduction of the test results on the same test subject using the same audio material showed a greater variance of the mood vector over time, as the test subject quickly got used to audio material as they remembered the music transitions. Especially the relaxation phase became shorter, as the test subject activated when anticipating the start of the heavy metal song. Also the use of the music depends on the subjective perception of a test subject, as not all found for example classical music relaxing. Inducing a desired mood state during an experiment proves very difficult to control and substantial research is still required to develop methodologies and techniques that allow the induction of mood states in a reliable fashion, while considering the individual characteristics of human nature.   As expected, the application automatically generated messages when a certain threshold of the mood vector is exceeded. In our case the thresholds has been set to 0.2 to indicate an activated state of a person and -0.2 to indicate a relaxed state. A corresponding mobile device then received a message notifying the user that the corresponding party (in this case the test subject) is now relaxed or excited. Although the messaging application is very basic, it should simply serve as a proof-of-concept.

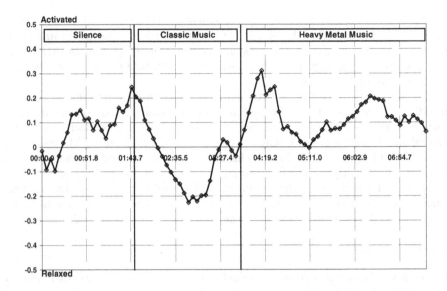

Fig. 6. Mood vector during the audio experiment

More complex applications as described in section 3 can be realized as further steps. It should be noted that although the whole system is wearable and thus designed for the use of a mobile user, actual measurement campaigns and the evaluation have been performed on stationary test subjects. The system still requires evaluation on mobile test subjects and the mood algorithm may require further adaptation in order to ensure adequate state detection.

7 Conclusions

This paper reports initial work undertaken towards the provision of mood based mobile services and applications. A completely mobile platform that allows the execution of mood based messaging services is presented. The platform is based on a body sensor network and PDA class device. Although the current application and capability of the platform are relatively simple and limited to the detection of activation, it demonstrates that mood based mobile services may become a reality in near future. As next steps we are planning to include features derived from the breathing strap into the mood detection algorithm and, in addition, enhance the algorithm to compute the level of valence. A combination of activation and valence together with the additional features will allow the classification of more emotional states and with higher reliability. In order to enhance the diversity of application space, we are currently integrating the mobile platform into an IMS based instant messaging and presence service.

A second measurement campaign will be conducted, this time, directly involving the developed mobile platform. We hope to derive more insights on the reliability of the mobile platform, especially with the involvement of mobile users.

Acknowledgement

This paper describes work undertaken in the context of the e-SENSE project, "Capturing Ambient Intelligence for Mobile Communications through Wireless Sensor Networks" (www.ist-e-SENSE.org). The authors would like to acknowledge the contributions of their colleagues from the e-SENSE Consortium.

References

[1] IST FP6 project e-SENSE, http://www.ist-e-sense.org/
[2] Dey, A.K., Abowd, G.D.: Towards a better understanding of context and context-awareness. In: Proceedings of the Workshop on the What, Who, Where, When and How of Context-Awareness, CHI 2000 Conference on Human Factors in Computer Systems, ACM Press, New York (2000)
[3] iPod Moody: http://www.apple.com/downloads/macosx/ipod_itunes/moody.html
[4] IST FP6 project MOBILIFE, http://www.ist-mobilife.org/
[5] Koolwaaij, J., Tarlano, A., Luther, M., Nurmi, P., Mrohs, B., Battestini, A., Vaidya, R.: Context Watcher - Sharing Context Information in Everyday Life, Web Technologies, Applications, and Services, WTAS 2006, Calgary, Canada (2006)
[6] Cohen, I., Sebe, N., Garg, A., Lew, M.S., Huang, T.S.: Facial expression Recognition from Video Sequences. Computer Vision and Image Understanding 91(1-2), 160–187 (2003)
[7] Coulson, M.: Attributing Emotion to Static Body Postures: Recognition Accuracy, Confusions, and Viewpoint Dependence. In: Journal of Nonverbal Behavior, vol. 28(2), pp. 117–139(23). Springer, Heidelberg (2004)
[8] Ekman, P., Friesen, W.: Facial Action Coding System. Consulting Physiologists Press, Palo Alto (1977)
[9] Scheirer, J., Fernandez, R., Picard, R.W.: Expression glasses: a wearable device for facial expression recognition. In: Conference on Human Factors in Computing Systems, Pittsburgh, Pennsylvania, pp. 262–263 (1999) ISBN:1-58113-158-5
[10] Branco, P., Encarnação, L.M.: Affective computing for behavior-based UI adaptation. In: Proceedings of the Workshop on Behavior-based User Interface Customization, IUI 2004, Madeira, Portugal (January 13-16, 2004)
[11] Jones, C.M., Jonsson, I.-M.: Automatic recognition of affective cues in the speech of car drivers to allow appropriate responses. In: Proceedings of the 19th conference of the computer-human interaction special interest group (CHISIG), Canberra, Australia (2005) ISBN:1-59593-222-4
[12] Scherer, K.R., Ceschi, G.: Lost luggage emotion: A field study of emotion-antecedent appraisal. Motivation and Emotion 21, 211–235 (1997)
[13] Frijda, N.: The emotions, Studies in Emotion and Social Interaction. Cambridge University Press, New York (1986)
[14] Roseman, I.J., Antoniou, A.A., Jose, P.E.: Appraisal determinants of emotions: constructing a more accurate and comprehensive theory. Cognition and Emotion 10(3), 241–277 (1996)
[15] Plutchik, R.: A general psychoevolutionary theory of emotion. In: Emotion: Theory, Research, and Experience: Theories of Emotion, vol. 1, pp. 3–33. Academic Press, New York (1980)
[16] Ekman, P.: An argument for basic emotions. Cognition and Emotion 6 (3/4) (1992)

[17] Russell, J.A.: A circumplex model of affect. Journal of Personality and Social Psychology 39, 1161–1178 (1980)

[18] Feldman Barrett, L., Russell, J.A.: Independence and bipolarity in the structure of current affect. Journal of Personality and Social Psychology 74(4), 967–984 (1998)

[19] Prkachin, K.M., Williams-Avery, R.M., Zwaala, C., Mills, D.E.: Cardiovascular changes during induced emotion: an application of Lang's theory of emotional imagery. Journal of Psychosomatic Research 47(3), 255–267 (1999)

[20] Neumann, S.A., Waldstein, S.R.: Similar patterns of cardiovascular response during emotional activation as a function of affective valence and arousal and gender. Journal of Psychosomatic Research 50, 245–253 (2001)

[21] Malik, M., Bigger, J., Camm, A., Kleiger, R.: Heart rate variability - Standards of measurement, physiological interpretation, and clinical use. European Heart Journal 17, 354–381 (1996)

[22] Cook, E.W., Lang, P.J.: Affective judgement and psychophysiological response. Dimensional covariation in the evaluation of pictorial stimuli. Journal of psychophysiology 3, 51–64 (1989)

[23] Bradley, M., Greenwald, M.K., Hamm, A.O.: Affective picture processing. In: The Structure of Emotion, pp. 48–65. Hogrefe & Huber Publishers, Toronto (1993)

[24] Roedema, T.M., Simons, R.F.: Emotion-processing deficit in alexithymia. Psychophysiology 36, 379–387 (1999)

[25] Wang, H., Prendinger, H., Igarashi, T.: Communicating Emotions in Online Chat Using Physiological Sensors and Animated Text. In: CHI 2004, Vienna, Austria, April 24-29, 2004, ACM, New York (2004)

[26] Bickmore, T., Schulman, D.: The Comforting Presence of Relational Agents. In: CHI 2006, April 22-27, 2006, Montreal, Canada (2006)

[27] Yazicioglu, R.F., Merken, P., Puers, R., Van Hoof, C.: A 60μW 60 nV/√Hz Readout Front-End for Portable Biopotential Acquisition Systems. In: IEEE International Solid-State Circuit Conference (ISSCC 2006), February 4-9, 2006, San Francisco Marriott, CA, USA (2006)

[28] Koralewski Industrie-Elektronik oHG, http://www.koralewski.de/

[29] Mietus, J.E.: From Variance to pNNX, Harvard Medical School Boston (2006)

[30] Malliani, A., Pagani, M., Lombardi, F., Cerutti, S.: Cardiovascular neural regulation explored in the frequency domain. Circulation 84, 1482–1492 (1991)

[31] D1.2.1 - Scenarios and Audio Visual Concepts, e-SENSE project deliverable (September 2006)

[32] Lichtenstein, A., Oehme, A., Kupschick, S.: Emotions in Ambient Intelligence - deriving affective states from physiological data. In: Affect and Emotion in Human-Computer Interaction, Springer, Heidelberg (to appear, 2007)

[33] Sensinode Ltd: http://www.sensinode.com

Recognising Activities of Daily Life Using Hierarchical Plans

Usman Naeem, John Bigham, and Jinfu Wang

Department of Electronic Engineering, Queen Mary University of London,
Mile End Road, London, United Kingdom, E1 4NS
{usman.naeem, john.bigham, jinfu.wang}@elec.qmul.ac.uk

Abstract. The introduction of the smart home has been seen as a way of allowing elderly people to lead an independent life for longer, making sure they remain safe and in touch with their social and care communities. The assistance could be in the form of helping with everyday tasks, e.g. notifying them when the milk in the fridge will be finished or institute safeguards to mitigate risks. In order to achieve this effectively we must know what the elderly person is doing at any given time. This paper describes a tiered approach to deal with recognition of activities that addresses the problem of missing sensor events that can occur while a task is being carried out.

Keywords: Smart Homes, Elderly Care, Hierarchal Activities of Daily Life, Task Segmentation, Task Associated Sensor Events.

1 Introduction

From the turn of the last century in 1901 the life expectancy for both men and women has continued to rise in the UK, which has lead to more elderly people in society. It has become difficult for children to look after their aged parents due to increased geographical mobility with children working and living remotely from their parents, lifestyle preferences and commitments, which leads to more elderly people depending on care homes. The introduction of smart homes has been seen as a suitable mechanism to allow people the opportunity to extend safely their independent lives and so defer entry to care homes. One of the ways to establish whether an elderly person is safe or to provide relevant help is to monitor the Activities of Daily Life (ADL) that they are carrying out and provide assistance or institute safeguards in a timely manner. In this paper we illustrate our approach through simple domestic examples. In order to understand the intentions of the elderly people, they need to be monitored. However since privacy is an issue as extensive monitoring, e.g. with cameras, can be intrusive, the approach chosen depends on the introduction of more automation in the form of algorithms that are able to discriminate between different ADLs using a few sensor events as is practicable, while achieving tolerable false positives and false negatives. In the experiments described the recognition of ADLs is based on data that is collected from RFID sensors.

G. Kortuem et al. (Eds.): EuroSSC 2007, LNCS 4793, pp. 175–189, 2007.
© Springer-Verlag Berlin Heidelberg 2007

There has been significant amount of research focused on efficient and reliable ADL identification. A popular technique in detecting ADLs is 'dense sensing' [1], which collects sensor data from many objects rather than relying on visual based engines. Numerous individual objects such as a kettle are tagged with wireless sensors or transponders that transmit information to a server via an RFID reader when the object is being used or touched. The interpretation of such sensor data is relatively easy for activities that are represented by sequential models that follow a standard path of execution. However, when a task can be carried out in more than one way or if a particular sensor event is missing due to data transfer problem. For example, if a person decides not to take milk or sugar in his or her tea when they usually do, this can sometimes been seen as a missing sensor event. Hidden Markov models have been used to carry out task identification. One such approach was by Wilson [2], where episode recovery experiments where carried out and analysed by a Viterbi algorithm which was responsible determining which task is active from the sequence of sensor events. The approach was successful in carrying out unsupervised task identification; however it was not as efficient when the tasks were carried out in any order. Multiple Behavioural Hidden Markov Models [3] have also been used to carry task identification. This approach was based on the idea of creating multiple hidden Markov models for each variation of a task, in order to accommodate each variation that could be carried out for a task. The latter approach was able to carry out task identification even if a sensor event was missing as a missing sensor event was treated as an insertion, which was very much like the approach used for substituting any unexpected sequence data in DNA motifs [4]. Other approaches that have been developed in order to carry out reliable task identification and mitigate the missing sensor problem have used techniques that involve ontologies [5] and data mining techniques [6]. Ontologies have been utilised to construct reliable activity models that are able match an unknown sensor reading with a word in an ontology which is related to the sensor event. For example the sensor reading 'mug' (which is an unknown sensor event in a model that is being interpreted) could be matched to a 'cup' sensor reading in a model for making tea that uses the term 'cup'.

2   Hierarchal Activities of Daily Life

For the work in this paper the ADLs have been modelled in a hierarchical structure, which allows us to decompose the ADLs into different models. With this type of modelling ADLs can correspond to simple tasks, such as "switch on kettle", or more a complex activities such as "make breakfast". The lowest tier of the Hierarchy of Activities of Daily Life (HADL) consists of the components responsible for gathering the sensor events within the home. The second level is task identification. A task is defined as the lowest level of abstraction in the higher tier(s). It can be associated with a simple goal of the monitored individual. The process of task identification maps each sensor event to the possible tasks which are associated with the sensor event, e.g. the sugar bowl sensor can be associated with following tasks: 'Make Coffee' or 'Make Tea'. This identification can be performed by a range of processes, such as hidden Markov models, though a simpler approach is used in the experiments described. At the higher levels there are further sub-goals and goals of the person being monitored, and these are modelled using a knowledge representation language that can represent plans. Each (sub) goal corresponds to an ADL. A task can be thought of as a lowest level goal (of the monitored individual) modelled using the planning knowledge representation language. It can however be modelled by some other modelling tool, such as a hidden Markov model. The number of levels above the task identification level depends on the complexity of the task. In this way the ADLs are nested within other ADLs. Additionally the ADLs may occur in parallel with other ADLs or have other temporal constraints. These are represented in the planning language used.

For each task (a) and sensor event (b), we can assigned a probability $P[a \mid b]$. This is required when carrying out task segmentation in the task identification. The entire sensor event stream is segmented into appropriate task segments. The segmented tasks are then used to determine which ADL is currently active. When performed, a task generates sensor events, and so task association mapping and recognition is based on analysing the sensor data, while ADL recognition is based on recognising constituent tasks.

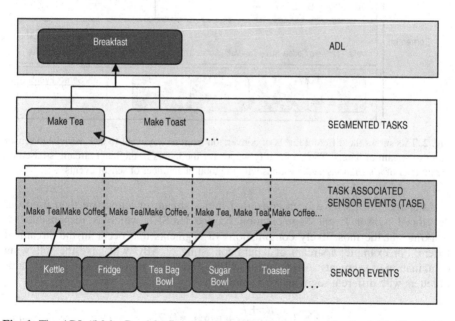

Fig. 1. The ADL "Make Breakfast" consists of a simple sequence of tasks, Make Tea, Make Toast..., but these may be in any order, or indeed be performed in parallel. The lowest tier of the HADL is the sequence of sensor events that have been detected, these sensor events are then associated with all the tasks that correspond to the sensor event, for example kettle sensor event can be associated with make tea or make coffee. These Task Associated Sensor Events (TASE) are then segmented into tasks using a statistical model which will be explained further in the paper.

3   Task Segmentation

Segmenting tasks can be carried out by simply segmenting sensor events into segments that correspond to a particular task. However this approach can sometimes generate sensor event segments that are incorrect and bear no resemblance to the task that is actually being carried out. In order to refine this problem we have manipulated the sensor events in to Task Associated Sensor Events (TASE) and developed a segmentation algorithm that is able to segments tasks efficiently. This algorithm was based on a statistical model which was created for text segmentation by Utiyama et al [2].

This method was used to find the maximum-probability segmentation of text, and does not need any training data, as it estimates probabilities from the stream of text. In the context of segmenting tasks and using the task segmentation algorithm the TASE are converted into letters so that we get a stream of letters, for example; Task(Make Tea)= letter(A), Task(Make Coffee)=letter(B), Task=(Make Toast)=letter(C)...Task(n) =letter(n).

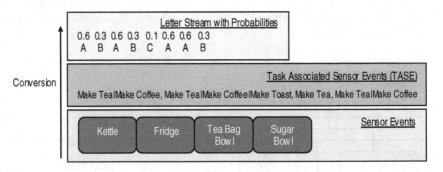

Fig. 2. This shows the different levels of conversion from sensor event to task associated sensor event to stream of letters. The probability values for the letters in the letter stream is based on the number of associations each task has with the total the number of sensor events.

After the TASEs have been converted into a stream of letters we then used our developed Sensor Event Segmentation Engine (SESE), which is responsible for working out the most likely combinations of segments that occur in the stream of letters. For example, a stream of letters consisting of ABC will have the following combination of segments: A|B|C, A|BC, AB|C, and ABC, which leads to four streams of letters with different segmentation points.

$$\sum_{j=1}^{n_i} \log \frac{n_i + k}{p+1} + \log n_i * 0.2 \tag{1}$$

Equation 1 is applied to each segment within each stream of letters, which outputs an overall cost for each stream. The stream of letters which has the lowest cost is generally close to correct segmentation or has been correctly segmented. Therefore the SESE analyses at sample of the 10 lowest cost segmented streams, which gives a good idea of which task is currently active. It is evident that on many occasions the results provided by the SESE may not be perfect in terms of accuracy, but this is where the higher tier of the HADL is used to refine the interpretation. The higher tier will be discussed in the next section.

Let AB|C be the stream of letters that algorithm 1 is going to be applied to. In relation to this stream of letters, n_i represents the length of the segment within the stream of letters, (AB)=2, (C)=1.

k represents the frequency of each letter in the stream of letters, (A)=1, (B)=2, (C)=1. n represents the length of the text stream, which is 3, and p is the prior probability assigned to each letter.

Table 1. Task Segmentation for stream AABCA with probabilities A=0.9, B=0.5, C=0.2

Cost of Stream	1st Segment	2nd Segment	3rd Segment	4th Segment	5th Segment
2.592679	A	AB	CA		
2.592679	AA	B	CA		
2.592679	AA	BC	A		
2.594235	A	A	B	CA	
2.594235	A	A	BC	A	
2.594235	A	AB	C	A	
2.594235	AA	B	C	A	
2.617737	A	A	B	C	A
2.733313	A	A	BCA		
2.733313	A	ABC	A		
2.733313	AAB	C	A		
2.761272	AA	BCA			
2.761272	AAB	CA			
3.011331	A	ABCA			
3.011331	AABC	A			
3.423917	AABCA				

The task segmentation in Table 1 shows the cost of each stream with different segments, with the lowest cost shaded in orange, while the other shaded sections form the sample of the 10 lowest cost streams. From the table it is clearly evident that the segmentation carried out gives a clear indication of what task might be currently active. For example Tasks like A have been segmented correctly, as well as that this technique provides the high level with more alternatives when mapping these tasks with the ADL plans.

Whatever the method used for task identification, the next step is to use the modelling of the possible goals and sub-goals of the individual to assist the interpretation. This is now described.

4   High Level Activity Recognition

The aim is to support recognition of tasks through feedback from beliefs held about ADLs. Initial task recognition is as has been described in previous sections, while the next steps in recognition through the hierarchy of constituent ADLs. The ADLs are represented in a hierarchical plan representation language. While a common way of representing and modelling high level behaviour is workflows, which are typically modelled using an augmented Petri Net [3]. Workflows are often used to model business processes. However, workflows are too prescriptive in their ordering and in their way of representing combinations of activities when trying to model the legitimate variations in human activity associated with a set of goals, so a richer knowledge representation language has been chosen. The language that has been used for the recognition of the activities as well as the elderly person's intentions is Asbru [4]. The Asbru language is a process representation language, which has similarities to workflow modelling. The roots of Asbru are in the modelling of medical protocols, which can be complex. It is hoped that this language will prove to be a suitable representation language to model behaviours of the monitored subjects.

Asbru is a task-specific and intention-oriented plan representation language initially designed to model clinical guidelines. Asbru was developed as a part of the Asgaard project to represent clinical guidelines and protocols in XML. Asbru has the capability to represent the clinical protocols as skeletal plans, which can be instantiated for each patient that requires a specific treatment. These skeletal plans are a useful guide for physicians when monitoring patients on a treatment protocol [5]. Asbru has many features which allow each skeletal plan to be flexible and to work with multiple skeletal plans. These plans in Asbru have been used to represent ADL and sub-activities within an ADL, e.g. Prepare Breakfast is an ADL, and a sub-activity of this ADL is to enter the kitchen.

In Asbru when a goal is achieved the plan is labelled as executed. When the pre-conditions of an ADL have been met then the ADL is classified as being executed. For example, for the goal 'eat egg' to start execution a pre-condition could be that the goal 'make egg' should be labelled as executed. Additionally an ADL can be classified as mandatory or optional. If an ADL has sub-goals that are classified as mandatory then these sub-goals must be executed before the ADL is labelled as executed. If optional then the sub-goal need not be executed. Sub-goals can be ordered in many ways. Common ones are sequential (in strict order), parallel (executed simultaneously), in any order (activated in any order but only one sub-goal can be executed at a time) and unordered (executed without synchronisation). The monitoring system being developed allows multiple activities to be tracked including tasks that may occur at the same time. Asbru also allows temporal intervals to be associated with goals, but that has not yet been incorporated into our monitoring system.

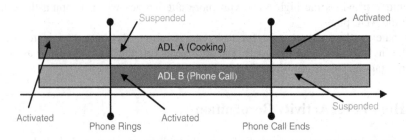

Fig. 3. If an elderly person is cooking dinner (ADL A) and the phone rings (ADL B) then the elderly picks the phone up, then with the aid of the conditions Asbru can suspend A and start ADL B. Once the elderly person is off the phone then ADL A will reactivated and ADL B will be suspended as more phone calls will come during the course of the day.

4.1 Modelling with Asbru

We briefly describe an example of how an ADL is modelled with Asbru in the higher tier of the HADL.

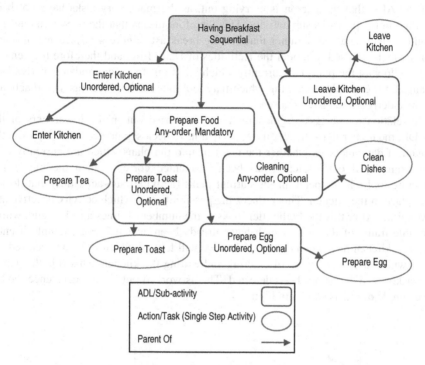

Fig. 4. Modeled example in Asbru, which is known as an ADL plan

We suppose that the following actions/tasks are detected in the lower tiers of HADL – Enter Kitchen, Prepare Toast and Clean Dishes - in this order.

At the detection of each task, the following processes will take place:

The main root ADL plan Having Breakfast is sequential, which means that the child activities within Having Breakfast will be executed in a sequential order, working its way from enter kitchen sub-activity to exit kitchen sub-activity.

When **Enter Kitchen** is detected then the sub-activity plan Enter Kitchen is set to complete, the **Enter Kitchen** is also a single step activity, which then allows the system to moves onto the sub-activity of the root ADL plan. A single step activity is a plan that cannot be decomposed any further and is called a task, which is what it is called when it is detected in the lower tiers of HADL.

The next task that may be detected is **Prepare Toast**. As this is also a single step activity then this is also set to complete, however the system could not continue to the next sub-activity of the root ADL plan as the sub activity for Prepare Food is mandatory, which means that all the child activity plans and tasks within this plan must be detected before it can proceed to the next sub-activity. As well as being mandatory, a plan may be optional, which means that a root parent activity does not need its child activities to be set to complete in order for it to move to other sub-activities.

The next task that is detected is **Clean Dishes**, this indicates the plan in Figure 4 is not the ADL that the person is carrying out, as the mandatory tasks have not been fulfilled in the previous sub activity. This therefore means that the person in question might be having a snack rather than having breakfast. Figure 4 is just one of many plans which are used to model the Activities of Daily Life, and therefore they enable us to follow all the plans concurrently which allows us to accommodate any dynamic changes, for example something which may look like having breakfast could actually be the elderly person having a snack.

In relation to the task segmentation that is carried out in the lower tiers of the HADL, the higher tier retrieves the tasks that have been segmented correctly from the stream of data and sees whether the tasks fit into the plans which are currently idle. For example, if enter kitchen has been segmented correctly then the higher tier planning tool will suspend all the current plans (activities/sub-activities) that do not take place in the kitchen. This reduces the possibilities of which activity is active at a given time. After this the higher tier looks at the number of times a task occurs within the time frame of the tasks which have already been detected. For example if enter and exit kitchen have been detected, then we will look at which task has occurred the most within that time frame of entering and exiting the kitchen. This fills the gap of the task(s) which has not been detected. This is worked out by an occurrence model, an example of this is shown in Figure 5.

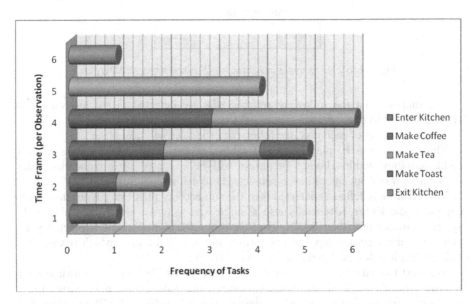

Fig. 5. The occurrence model shows the frequency of each task given the time frame, which is measured in sensor observations

In Figure 5 if enter and exit kitchen have been detected then according to the occurrence model the most likely task that could fill one or many gaps in the higher tier ADL plans is Make Tea. This is because the frequency of the task "Make Tea" is incrementing with each time frame until time frame five, which suggest that the task may have been completed. However up until time frame four it is not evident whether "Make Tea" or "Make Coffee" is being carried out by the elderly person. In this situation both tasks can be mapped to the high level plans as the planning language Asbru is capable of managing concurrent tasks and once it is evident that make tea is the correct task as shown in time frame five then task make coffee can be suspended.

5 ADL Detection Experiments

The objective of the ADL detection experiments was to determine which ADLs are active from the collected sensor data stream. The accuracy of these experiments was determined by the percentage of detection rate of identifying an ADL. For each possible plan a discrepancy count and more importantly a surprise index is computed, whenever a new task is recognised in the lower tier of the HADL. The discrepancy count simply computes the number of sensor events that are consistent with the plan being the current ADL. The surprise index is used to account for the fact that the absence of some sensor events can be more unusual than others, and quantifies this by accruing a measure of how likely a sensor event is when a task is being executed.

A discrepancy is computed whenever there is a missing mandatory action/task, such as make tea for the ADL Make Breakfast. The surprise index is the maximum of the conditional probability of a missing sub plan/activity and actions/tasks. In order to generate the detection rates for each ADL in these experiments, each ADL has been assigned a surprise index threshold. If the surprise index exceeds an ADL's surprise threshold when the ADL is actually being performed, then that is taken to mean that the ADL has not been detected correctly. For example, the ADL Make Breakfast has

Table 2. Surprise Threshold and Execution Order of the ADL/ Sub Activities, with the ADLs in bold and the sub activities in italics

ADL/ Sub Activities	Surprise Threshold	Execution Order
Breakfast	1	Sequential
Prepare Food	1.25	Any Order
Clean Dishes	1.5	Unordered
Laundry	1	Sequential
Wash Clothes	1	Sequential
Dry Clothes	1	Sequential
Put Shopping Away	1.25	Any Order
Unpack Shopping	1.25	Any Order
Prepare Meal	1.3	Any Order
Make Chicken Curry	1.25	Sequential
Make Fish & Chips	1.25	Sequential
Warm up Meal	1.25	Sequential
Clean up Kitchen	1.3	Any Order
Clean Dishes	1.5	Unordered
Dish wash Dishes	1.3	Sequential

a sequential execution order and has surprise index threshold of 2. While carrying out the experiment if a surprise index that is over 2 is found then this means that Make Breakfast is not the ADL. Table 2 shows the surprise threshold for each of the ADLs and sub activities that have been used in the experiment.

The six experiments were in a kitchen (Figure 6) and the ADLs being tested for detection are all kitchen oriented. The experiments were conducted with non-intrusive RFID transponders installed around the kitchen and on its cupboards and utensils, such as on the kettle, dishwasher, and toaster. The data generated from the transponders was collected by a RFID reader that is the size of match box and was worn on the finger of the subject conducting the experiment. For these experiments 10 adult volunteers had been recruited from the community to carry out the ADLs. The ADLs ranged from making breakfast to putting shopping away. The reason why 10 subjects were chosen is because people have different ways and ordering of carrying out a particular ADL, so there will be variability in the sensor stream.

Fig. 6. Graphical representation of the kitchen where the experiments were carried out, showing the locations of some of the sensors

The experiments are divided into two sets; one set is a 'distinctive' series of sensor data while the other set is the 'non-distinctive' series. The distinctive series makes use of sensor events where there is usually a determining sensor reading for each ADL. For example the 'fairy bottle' sensor is exclusive to the task 'washing up dishes', which makes it a distinctive sensor event which could determine if the ADL is active. On the other hand, the non-distinctive series does not make use of any sensor events which might be a distinctive when detecting an ADL. This is a harder challenge. Within the two sets of experiment there were three experiments that were conducted, which means each subject conducted six experiments in total. Table 3 shows the objective of each experiment conducted.

Table 3. Experiment Objectives

Experiment Number	Type of Experiment
1 & 2	Distinctive Series & Non- Distinctive Series – Subjects carried out 5 ADLs specified in the prescribed order provided. The tasks which were optional did not need to be carried out.
3 & 4	Distinctive Series & Non- Distinctive Series – Subjects carried out 5 ADLS in any order and were allowed to carry out the tasks within an ADL in any order. The ADLs are not interweaved.
5 & 6	Distinctive Series & Non- Distinctive Series – Subjects were allowed to carry out any 2 ADLs concurrently and in any order, e.g. make tea while putting the shopping away. Here the ADLs are interweaved.

The experiment is modelled around 5 ADLs, which consist of 25 tasks and 45 sensor events, Figure 7 shows the ADLs with their associated tasks that have been used for the experiments.

Fig. 7. The root plan is the ADL (e.g. Breakfast), the child nodes are the sub-plans/activities which are made up of tasks which are also known as single step plans

However more ADLs have been modelled as plans in Asbru, so that there are conflicting situations where one task could be a part of more than one ADL. The reason for conducting different type of experiments is to have a sufficient amount of data to test the HADL approach, which includes TASE mapping, Task Segmentation and ADL Recognition.

6   Evaluation and Results

For all of the experiments the percentage of the detection rates for each ADL was determined using the surprise index and how many times the 5 ADLs carried out for the experiments were recognised successfully.

Table 4. Distinctive Series Results for Experiments 1, 3, and 5, with the ADLs in bold and the sub activities in italics

ADL/ Sub Activities	Experiment 1 Prescribed Detection Rate [%]	Experiment 3 Random Detection Rate [%]	Experiment 5 Concurrent Detection Rate [%]
Breakfast	90	87	84
Prepare Food	93	91	86
Clean Dishes	86	83	80
Laundry	100	96	95
Wash Clothes	100	96	95
Dry Clothes	100	96	95
Put Shopping Away	95	92	89
Unpack Shopping	95	92	89
Prepare Meal	89	82	80
Make Chicken Curry	84	80	78
Make Fish & Chips	86	79	75
Warm up Meal	90	89	88
Clean up Kitchen	89	86	80
Clean Dishes	86	83	80
Dish wash Dishes	100	97	96

Table 5. Non-Distinctive Series Results for Experiments 2, 4, and 6, with the ADLs in bold and the sub activities in italics

ADL/ Sub Activities	Experiment 2 Prescribed Detection Rate [%]	Experiment 4 Random Detection Rate [%]	Experiment 6 Concurrent Detection Rate [%]
Breakfast	82	79	77
Prepare Food	85	83	79
Clean Dishes	80	77	75
Laundry	96	92	90
Wash Clothes	96	92	90
Dry Clothes	96	92	90
Put Shopping Away	89	85	84
Unpack Shopping	89	85	84
Prepare Meal	85	81	77
Make Chicken Curry	82	77	76
Make Fish & Chips	83	75	74
Warm up Meal	85	81	80
Clean up Kitchen	81	78	74
Clean Dishes	80	77	75
Dish wash Dishes	97	95	93

The results of the experiments carried out with the set of distinctive sensors (Table 4) show that ADLs like "Breakfast", "Laundry", "Put Shopping Away", "Warm up Meal" and "Dish Wash Dishes" were detected correctly on a regular basis. As well as that the detection rate percentage for these ADLs did not have a radical change when carrying out these ADLs in a random or concurrent with other ADLs. This does not mean to say that the other ADLs were not regularly detected correctly;

we just feel it was important to outline the mentioned ADLs as they are reliant on distinctive sensor events in order for them to be recognized (e.g. microwave was a distinctive sensor event for the task warm meal). The results of these particular ADLs will be compared with the experiment results for the non-distinctive series. In summary these results show that our developed hierarchical approach is capable of managing concurrent as well as randomised sensor events and tasks and most importantly to recognize which ADL is currently active.

The results from non-distinctive experiments (Table 5) show a slight decrease in the detection rate for each of the ADLs. A decrease was expected as the distinct sensor events were taken away from these set of experiments. However, the decrease that was witnessed was small, as the average of the detection rates for all the ADLs after all the experiments was 86.3%. Figure 8 shows the detection rates for the five ADLs mentioned and from this we see that it does not make a significant change to the detection of the ADLs if the distinct sensors have not been detected.

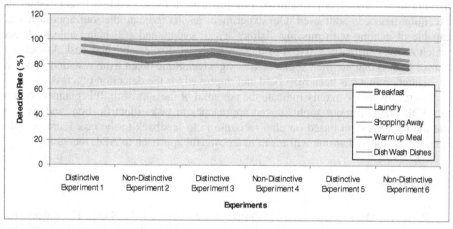

Fig. 8. Comparison of all the experiments with the ADLs that rely on distinct sensor events

The reason why our approach was able to detect ADLs without their distinct features was because of the planning capability of the higher tier. The planning capability of the representation language used was able to have all the ADLs mapped as plans which allowed our approach to be able to predict which ADL was active. The predictions were made on the basis of the events and their probabilities that had been gathered in the lower tier of our approach. Additionally, the higher tier is capable of dealing with tasks which occur in any order or are missing, as long as few tasks which are associated with ADL have occurred. Otherwise, it would be impossible to detect the ADL.

In terms of the dealing with the missing sensor events at the lower tier, the TASE was capable of dealing with this, as it provided all the possible task associations for the sensor event. Therefore if a sensor event was missing then the other sensor events which got manipulated into TASE will be able to provide some of idea of what task is active. With the aid of the task segmentation we are then able to filter out the most likely tasks that have occurred, which is then refined and mapped in plans in the higher tier of the HADL.

A limitation of this approach is that it does not take time into consideration. This is limiting as time can play a crucial part in detecting which ADL at what time of the day is active. For future enhancements the higher tier of our approach will incorporate task (and goal) durations. Also since the detection system will be aware of what time of day it then it will know the ADL plans which are usually executed around that time. In addition timing will play an important part in the lower tier of our approach, as time can be used to measure how long it takes on receiving different type of sensor events.

7 Conclusion

In this paper we described a tiered approach to interpreting sensor data when monitoring ADLs. The problem of missing sensor events and different orders of execution has been addressed. Our experiment results indicate that our approach was capable of dealing with missing distinct sensor events and still being able to detect which ADL is currently active. For the lower tier we also established Task Associated Sensor Events that are then segmented using a technique taken from research into text segmentation that has been reworked and improved for the detection of a task.

The results here can only indicate the potential of the approach. It is planned to link the ADL plans to more sophisticated approaches to tasks identification, and then to use the identification based on plan recognition to feedback to the task identification stages. It is hoped that this will result in a powerful approach to ADL recognition.

References

1. Philipose, M., Fishkin, K.P., Perkowitz, M., Patterson, D.J., Kautz, H., Hahnel, D.: Inferring activities from interactions with objects. IEEE Pervasive Computing Magazine 3(4), 50–57 (2004)
2. Wilson, D.H., Wyaat, D., Philipose, M.: Using Context History for Data Collection in the Home. In: Gellersen, H.-W., Want, R., Schmidt, A. (eds.) PERVASIVE 2005. LNCS, vol. 3468, Springer, Heidelberg (2005)
3. Naeem, U., Bigham, J.: A Comparison of Two Hidden Markov Approaches to Task Identification in the Home Environment. In: Proceedings of the 2nd International Conference on Pervasive Computing and Applications, Birmingham, United Kingdom (2007)
4. Eddy, S.R.: Profile hidden Markov models (Review). Bioinformatics, Department of Genetics, Washington University School of Medicine, USA, vol. 14, pp. 755–763 (1998)
5. Munguia-Tapia, E., Choudhury, T., Philipose, M.: Building Reliable Activity Models using Hierarchical Shrinkage and Mined Ontology. In: Proceedings of the Fourth International Conference on Pervasive Computing, pp. 17–32. Dublin, Ireland (2006)
6. Wyatt, D., Philipose, M., Choudhury, T.: Unsupervised Activity Recognition using Automatically Mined Common Sense. In: Proceedings of the Twentieth National Conference on Artificial Intelligence, pp. 21–27 (2005)

7. Utiyama, M., Isahara, H.: A Statistical Model for Domain-Independent Text Segmentation. In: Proceedings of the 39th Annual Meeting on Association for Computational Linguistics, Toulouse, France, pp. 499–506 (2001)
8. Zurawski, R., Zhou, M.: Petri Nets and Industrial Applications: A Tutorial. Industrial Electronics, IEEE Journal 41(6), 567–583 (1994)
9. Fuchsberger, C., Hunter, J., McCue, P.: Testing Asbru Guidelines and Protocols for Neonatal Intensive Care. In: Proceedings of the Tenth European Conference on Artificial Intelligence, Aberdeen, United Kingdom, pp.101–110 (2005)
10. Kosara, R., Miksch, S., Shahar, Y., Johnson, P.: AsbruView: Capturing Complex, Time-oriented Plans - Beyond Flow-Charts. In: Second Workshop on Thinking with Diagrams, Aberystwyth, United Kingdom, pp. 119–126 (1998)

GlobeCon – A Scalable Framework for Context Aware Computing

Kaiyuan Lu, Doron Nussbaum, and Jörg-Rüdiger Sack

School of Computer Science, Carleton University,
1125 Colonel By Drive, Ottawa, Ontario, Canada K1S 5B6
{klu2,nussbaum,sack}@scs.carleton.ca

Abstract. In this paper, we propose a context framework reference model that aims at supporting distributed context aggregation, processing, provision and usage in ubiquitous computing environments. Based on this reference model, we introduce our design of GlobeCon, a scalable context management framework that can be embedded into context aware systems operating in local as well as wide area networks. One of the key aspects of GlobeCon is that it aggregates context information from largely distributed, heterogeneous and disparate sources in a hierarchical manner and presents context aware systems with a unified context access interface. GlobeCon is scalable in numerical, geographical and administrative dimensions, which is achieved by three means: (i) organizing context aggregation hierarchically, (ii) effectively distributing the load of context collection and context processing (i.e., interpreting and reasoning) among local context managers, and (iii) providing a dynamic federation of the distributed directory servers.

1 Introduction

One of the main characteristics of ubiquitous computing is user-centricity where technology automatically moves into the background and adapts to people [17]. Consequently, services developed for ubiquitous computing need to be user-tailored and proactively adapt to users' needs. For this adaptation, services must have sufficient information about users and their immediate and/or possibly future contexts. As a consequence, the notion of context awareness plays an integral role in designing services and systems in the ubiquitous computing paradigm.

The term "context" has been interpreted in many different ways in the literature since it first appeared in [18]. There context was defined as location, identities of nearby people and objects and changes to those objects. One of the most widely accepted definitions, also used in this paper, is given by Dey [7], according to which "Context is any information that can be used to characterize the situation of entities that are considered relevant to the interaction between a user and an application, including the user and the application themselves". Most common entities refer to persons, places or objects.

Early context aware systems, including [1][6][18][21], were aimed at demonstrating the potential of context awareness by prototyping applications. These

G. Kortuem et al. (Eds.): EuroSSC 2007, LNCS 4793, pp. 190–206, 2007.

systems though were domain specific and tightly coupled context acquisition and management with application logic. Lately, researchers have been focusing on the architecture design of context aware systems to support the entire context processing flow [3][7][8][12]. This improves programming abstractions and aids in the development of context aware applications.

Throughout this paper, we attempt to contribute to this development by proposing a scalable context management framework for ubiquitous computing, which we call *GlobeCon*. "Scalable" in this paradigm has three dimensions as specified in [2]: numerical (i.e., the increase of workload of a single resource), geographical (i.e., the expansion of a system from concentration in a local area to a more distributed geographic pattern), and administrative (i.e., the coexistence of a variety of software/hardware platforms in a system).

1.1 Motivation

In previous work [9][10], we have designed and implemented an agent-based framework, referred to as Intelligent Map Agents (IMA), which aimed at providing access to, and support for manipulation of, spatial information. The IMA architecture is user-centric. As such, the IMA provides each user with a personalized set of functionality capabilities, interaction patterns and visualization. Hence, context plays an integral role in our IMA system. However, the usage of context in the current IMA is, at present, somewhat limited in terms of the types of context used (primarily location-based), their capabilities, and the design of context manager (embedded into personal agents). This sparked our interest and provided initial motivation for the research presented here which is our contribution towards the design of a scalable context management framework, GlobeCon. It is intended to be embedded into, or support rapid development of, context aware systems and applications operating in local as well as wide area networks.

1.2 Related Work

Growing rapidly since the Active Badge project [21], research on context aware computing has attracted a lot of attention as witnessed by a large body of literature. For comprehensive discussions, we refer the interested reader to recent survey articles [5][15][19]. We classify work in this research area as belonging to one (or more) of four main fields: designing techniques and sensors with desirable context sensing capabilities; devising context models to facilitate efficient context usage, inference and storage; building frameworks or middleware to support context aware computing; and developing context dependent applications to demonstrate the effects of context. In this paper, we concentrate our work and literature review on the third stream.

Context Toolkit [7] was one of first and influential projects that emphasized the isolation of application logic from the details of context sensing and processing. A conceptual context handling framework, based on a centralized component (i.e., the discoverer), was proposed. This framework provides context gathering, interpretation, and aggregation support for applications. CASS [8] uses a middleware approach to support context awareness for hand-held computers. The middleware resides on a centralized resource-rich server. CASS supports context acquisition, interpretation, and reasoning. CMF [14] is a stand-alone software framework designed for mobile terminals. It supports acquisition and processing of contexts in a user's surroundings, and gives the contexts to applications residing on the user's mobile device. CoBrA [3] is a broker centric agent-based infrastructure supporting context aware computing in dynamic smart spaces such as an intelligent meeting room. It has a central context broker which provides four functional modules: context acquisition, context knowledge base, context reasoning engine and context privacy control.

The above mentioned frameworks have in common a centralized component in their design, which may become a bottleneck when the number of context sources (i.e., sensors) and context consumers increases dramatically. Moreover, they do not scale well in the geographical dimension. SOCAM [12] is service-oriented context aware middleware architecture, which addresses the scalability problem with a global-wide Service Locating Service (SLS)[13]. SOCAM supports context acquisition, discovery, interpretation and context access through a set of independent services, which advertise themselves with the SLS. SCI [11], a generalized context framework, tries to address the scalability problem with a two-layer infrastructure, where the upper layer is a network overlay of partially connected nodes (named Ranges), and the lower layer concerns the contents of each Range in the overlay network. A Range provides context handling services within an area. However, the authors did not mention how to create an overlay network to support the interactions among Ranges, which in our opinion is one of the key challenges in solving the geographical scalability. Similarly, Solar [4] addresses the scalability issue using overlay networks, but does not provide a way to construct such an overlay network. In CoCo [2], the authors argued to utilize the existing infrastructures (i.e., grids, peer-to-peer networks, and Content Delivery Networks) to support scalability in a context management framework. Based on this assumption, CoCo provides an integration layer, which interfaces with heterogeneous context sources, maps the retrieved context to a standard information model, and provides it to context consumers in a unified way.

While recently some research is being carried out on the design of scalable context frameworks, a number of interesting research challenges remain to be solved, e.g., how to orchestrate the sheer mass of contexts from heterogeneous sources, how to support the efficient discovery of appropriate context information, how and where to store the vast amount of contexts, how to maintain the consistency, security and privacy of contexts in large-scale distributed environments, etc. In this paper, we attempt to answer some of the questions.

1.3   Contributions

In this paper, we present our contribution towards supporting context awareness in ubiquitous computing environments. Firstly, inspired by the philosophy of the OSI reference model [20], we propose a context framework reference model that aims at supporting distributed context aggregation, processing, provision and usage. Secondly, based on the reference model, we introduce our design of GlobeCon, a scalable context management framework that supports context aware systems operating in local as well as wide area networks. GlobeCon is different from existing context frameworks [3][8][12][14], which either operate in restricted spaces or have limited capability in network scaling. GlobeCon achieves scalability in numerical, geographical and administrative dimensions by three means: (i) organizes context aggregation in a hierarchical manner, (ii) effectively distributes the load of context collection and context processing (i.e., interpreting and reasoning) among Local Context Managers (LoCoM), and (iii) provides a dynamic federation of the distributed directory servers. GlobeCon also exhibits two additional features: (a) Dual Resource Discovery: a proximity-based discovery of raw context sources and a Universal Discovery Service (UDS) to discover the largely distributed LoCoMs which provide raw contexts as well as processed high-level contexts. This feature promotes the efficiency of discovering context sources and allows flexible context queries; (b) Context Classification: GlobeCon distinguishes *common context* and *personal context*, and applies different storage schemes on the two classes of context, which exhibit different acquisition patterns and require different levels of security and privacy protection.

1.4   Paper Outline

The remainder of this paper is organized as follows. In Section 2, we present design considerations of GlobeCon and our context classification, and propose a context framework reference model. In Section 3, we introduce our architectural design of GlobeCon. In Section 4, we discuss how to integrate GlobeCon with the IMA system and describe a health care application of GlobeCon and the IMA. In Section 5, we conclude the paper and point out our future work.

2   Scalable Context Management Framework

A context management system can be designed and implemented in a variety of ways depending on different requirements and deployment environments. GlobeCon aims at support context awareness. Hence, it needs to provide several fundamental function modules centered around context handling, so that we can relieve developers of context aware applications from the burden of having to deal with the entire context processing flow. The functional modules include:

- *Context acquisition:* collects and updates raw context from largely distributed, dynamic, heterogeneous context sources.
- *Context representation:* provides a unified data model to explicitly represent context semantically, and thus enhance context sharing.
- *Context processing:* processes the context and provides context reasoning (e.g., by combining multiple raw contexts to generate higher level contexts which usually can not be detected by sensors directly).
- *Context storage:* maintains a persistent historical context storage for later retrieval and analysis.
- *Context accessing:* supports efficient processing of plain as well as expressive queries for context information.

In addition to the functional modules, we identify a number of design considerations/issues that are of particular relevance for GlobeCon. Fig. 1 shows the concepts and approaches used in GlobeCon to address the issues.

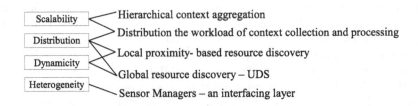

Fig. 1. Key concepts and approaches used to address the design issues

– *Scalability (C.Sca):* exists in three dimensions: numerical, geographical and administrative, as discussed in Section 1. In order to be a scalable framework, GlobeCon needs to have the ability to handle the growing demand in any of the three dimensions in a graceful manner without degrading its performance.
– *Distribution (C.Dis):* refers to the large number of raw context sources and context processing resources which are spread over large geographic areas. GlobeCon should shield the distribution from the end users.
– *Dynamicity (C.Dyn):* lies in two aspects: (i) context sources and users may be mobile and changing their states, e.g., a sensor moves from place to place and switches its status between active and inactive; (ii) the pervasive computing environment itself is highly dynamic, e.g., the traffic condition of a road network.
– *Heterogeneity (C.Htg):* arises in any system dealing with multiple data sources. GlobeCon should handle context data residing at heterogeneous and disparate sources (i.e., context sources with different interfaces and raw context data in different formats) and needs to provide these context with a unified view.

Privacy/security and fault tolerance are certainly important issues for systems like GlobeCon. However, they are not the main focus of our current work.

2.1   Our View of Context

In this section, we present our view of context, which classifies context as common and personal. *Common context* describes common places or spaces, and objects that are shared among all users or a group of users. Common contexts include physical context (e.g., light density, noise level, temperature, air pressure etc.), logical context (e.g., business processes, price, etc.), and technical context of common computing devices (e.g., processing power, memory capacity, bandwidth, latency etc.). *Personal context* describes a user's profile, physical, social and emotional situation, and objects/ entities owned by the user. Personal contexts can be further divided into domain specific context which are sets of information applied respectively to individual application domains, and general personal context which applies to most application domains (e.g., user profile, preference, the technical aspects of user's personal computing devices, etc.). As to the classification of application domains, the North American Industry Classification System (NAICS) [16] provides a guideline as a starting point.

2.2 Context Framework Reference Model

In ubiquitous computing environments the amount of context information is vast and the number of context sources, which are usually widely dispersed throughout the globe, can be very large too. To make it easier for applications to find context sources which can provide the required context, we need a solution that manages and indexes distributed context sources in a way that required context can be discovered efficiently and without (or with minimal) user interactions. Inspired by the philosophy of the OSI (Open Systems Interconnection) Reference Model [20] for computer networks, we are proposing a layered model for scalable context frameworks that provide distributed context handling support for context aware systems operating in local as well as wide area networks.

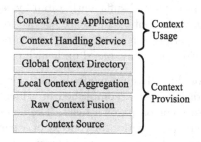

Fig. 2. Context Framework Reference Model

Our model has six layers as shown in Fig. 2: (a) context source layer, (b) raw context fusion layer, (c) local context aggregation layer, (d) global context directory layer, (e) context handling service layer, and (f) context aware application layer. Each layer built upon the one below it. The six layers together provide support for context aggregation, processing, provision and usage. At the lower "*context provision*" hierarchy, each layer offers a well defined set of context provision services. With each level of the hierarchy, users are exposed to and able to access more sophisticated contextual features.

- The context source layer comprises the ground level context sources, i.e., sensors. A context source is responsible for observing raw contexts within its sensing range and capability(ies), and reporting them to the layer above it.
- The raw context fusion layer has the tasks of interacting and gathering raw context from a number of adjacent context sources with heterogeneous APIs. Another task of this layer is to map the raw context obtained in various formats into a unified context representation, which should be compliant with a pre-specified context model.
- The local context aggregation layer collects context from raw context fusion layer, therefore its local context knowledge base covers a larger geographical

and logical area. This layer also preforms reasoning on its local context to reach high level context which is more meaningful for high level applications. The reasoning is usually based on pre-defined domain knowledge.

- The global context directory layer exists as a global registry and a matching tool, which enable context consumers to search within the context provision hierarchy for context providers having the needed context information.

The upper *"context usage"* portion of this model is concerned with the usage of context provided by the lower context provision hierarchy.

- The context aware application layer contains a variety of ubiquitous services that are context dependent and aimed at end-users. Each of such customizable service is usually developed for one particular application domain, e.g., health care, tourism, ubiquitous GIS and etc.
- The context handling service layer is a middleware layer residing between the application layer and the context provision hierarchy. It provides application domain specific, large-scale context handling functions (e.g., reasoning and aggregation) that are requested sufficiently often by services in the application layer, for example, a context inference service for a ubiquitous health care system, which performs reasoning on the context gathered from the lower context provision hierarchy, based on health care domain knowledge.

The key advantage of this reference model is that the aggregation of largely distributed context information is accomplished in a hierarchical way, thereby, each layer only needs to focus on its own tasks. The complexity of other tasks is shielded in other layers, and queries for context of a particular level of detail can be addressed directly to the corresponding layer.

3   Architectural Design of GlobeCon

Based on the reference model of Section 2.2, we introduce our architectural design of GlobeCon, a scalable context management framework. Fig. 3 depicts an overview of the multi-layered GlobeCon architecture, and gives an explicit mapping between GlobeCon and the reference model. The lowest level encompasses a variety of heterogeneous context sensors. The second level is comprised of Sensor Managers (SM) which manage their own set of sensors within their geographical vicinities. The third level contains Local Context Managers (LoCoM). Each LoCoM manages a number of SMs within their own physical or logical vicinities. A Universal Directory Service (UDS), which resides on the fourth level, provides a global registry for LoCoMs. This is followed by the context usage hierarchy. Context handling services provides larger scale context handling functionalities (e.g., reasoning and aggregation) beyond the scope of individual LoCoMs. Context aware applications are end-user oriented context dependent services. Each layer represents a fairly complex subsystem focusing on a different phase of the entire context processing flow. These layers are seamlessly integrated together to support distributed and scalable context aggregation, processing, provision and usage. In the following subsection, we elaborate on the role of each layers.

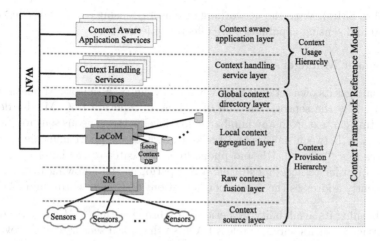

Fig. 3. An overview of GlobeCon architecture

3.1 Sensor Level

Sensors are the ground level context sources. A sensor in GlobeCon is responsible for acquiring raw context data within its sensing range and capability, connecting with and reporting the obtained context to its SM(s) via wired or wireless communication channel. In GlobeCon, a sensor proactively register itself with a SM after deployment. This is achievable by two alternative ways: (i) a sensor has pre-knowledge of which SM it should be connected with (e.g., preconfigure a sensor with this information before its deployment); and (ii) a sensor can discover available SMs by listening "*SM-advertisement*" messages within its communication range. For both of the two approaches, authentication and authorization process may necessary to ensure privacy and security.

A sensor can be either stationary (e.g., sensors mounted on the walls) or mobile (e.g., in-vehicle sensors and biosensors attached on a person). In terms of form, sensors in GlobeCon consist of software sensors and hardware sensors. A software sensor captures context directly by accessing and reading from existing software applications or services, such as obtaining CPU usage from the OS of a laptop, or accessing a user's schedule using APIs of the e-calendar on his PDA. Sophisticated software sensors are able to apply domain logic to deduce higher level context, e.g., identifying a user's availability by calculating the percentage of free time in his e-calendar. A hardware sensor is a physical device usually of small size, measuring physical aspects of context, e.g., light, noise and temperature. Modern hardware sensors are more advanced since they are typically equipped with small processor and disk spaces to perform simple on-sensor processing.

In some cases, depending on communication resources, a number of sensors can collaborate with each other to form a network and a data sink collects context from these sensors via multi-hop communication. In this scenario, the entire sensor network is considered as one logic sensor and the data sink acts as a portal to SM(s). With the networking of sensors, sensing coverage is effectively enlarged and sensing capabilities are effectively federated.

3.2 Sensor Manager

A SM manages its own set of sensors within its vicinity. A sensor-SM relationship is initiated by the sensor. The aim of SMs is to relieve a LoCoM's burden of aggregating context from any number of low level heterogeneous sensors. Acting as a gateway between sensors and LoCoMs, SMs can support multiple network interfaces (e.g., infrared, RF and phone-line) and cater to the heterogeneity of sensors. In order to accomplish this, a SM performs the following tasks, which consequently addresses the design considerations of *C.Dyn*, *C.Htg* and *C.Dis*:

- Publishing its availability by periodically broadcasting *"SM-advertisement"* messages over its wired/wireless LAN, so that local sensors can discover it.
- Initializing, configuring and synchronizing its sensors, e.g., set data sampling rate, specify data acquisition method (on-demand or event-driven).
- Monitoring and controlling the operations of its sensors, e.g., detect malfunctioning sensors and take appropriate actions (i.e., reset or de-register a sensor) to avoid inaccurate or stale context information.
- Gathering raw contexts from sensors in two possible transmission modes: push-mode in cases of periodical stream-based transmission, and pull-mode in cases of irregular requests on raw sensory readings.
- Associating obtained context with subject and sampling time, and converting the raw context in heterogeneous formats into a unified context representation in accordance with pre-specified context model.
- Detecting and resolving potential context information correlation and inconsistency existed in the obtained contexts.
- Discovering and registering with the available LoCoMs by listening *"LoCoM-advertisement"* messages within its communication range. If there are more than one available LoCoMs, a SM selects one as its manager and marks the rest as context subscribers.

Within the same physical space, there may be several SMs where each SM works for a particular application domain. For example in a hospital cafeteria where visitors come to eat, there is one SM collecting health care related contexts for visitors (e.g., temperature, humidity and heart activity), while another SM gathering dining related contexts (e.g., user's food preference). In such cases, a sensor with multiple sensing capabilities may report its context data to more than one SMs[1], as shown in Fig. 4. GlobeCon has two types of SMs: stationary sensor manger (SSM) and mobile sensor manager (MMS). Due to its mobility, a mobile SM may be out of the range of any LoCoMs. In this case, alternative schemes are devised to avoid context information loss/unavailability. For example, instead of tossing the local context away, a MSM may temporarily cache the obtained

---

[1] A sensor may report the same piece of context to multiple SMs to achieve high degree of availability of context information with the penalty of data redundancy.

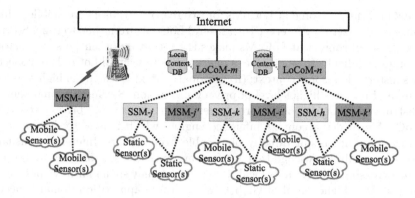

Fig. 4. An example illustrating the subsystem of LoCoM, SMs and Sensors

context locally (which requires large local disk space) and report its cache to a LoCoM when a LoCoM is available later, or a MSM may compress the context data, or it may tune sensors' sampling rate to reduce the volume of context data.

3.3   Local Context Manager

A LoCoM is responsible for aggregating context from SMs distributed within its local coverage. "Local" here indicates a logic relationship between a LoCoM and its SMs, and is not necessary limited to the proximity at the geographical level. It may imply proximity at the network level (e.g., within the same WAN) or proximity at the level of application domain(e.g., health care or tourism). LoCoMs are designed to perform computation and communication intensive functionalities (e.g., conducting context reasoning and processing potentially large numbers of context queries) within the context management processing flow, therefore, they are usually hosted by stationary device(s) with high performance computing, and high bandwidth network connections. One stream of the main tasks of a LoCoM is related to SMs management:

– Periodically broadcasting "*LoCoM-advertisement*" messages over its wired and wireless WAN so that SMs within its communication range are able to be aware of its availability.
– Handling registration requests from local SMs, and maintaining a list of registered SMs. There are two types of SM registration, where a LoCoM plays different roles respectively: SM manager or SM subscriber.
– Detecting and de-registering inactive SMs due to failure or out-of-connection. As a consequence of de-registration, a LoCoM needs to update accordingly its self-description (refer to the last paragraph in this subsection).
– Gathering context information from the registered active SMs.

As a SM manager, a LoCoM provides a local context database to store the *common context* [2] retrieved from the SMs under its management. Historical *Personal*

---

[2] For definition of *common context* and *personal context*, please refer to Section 2.1.

context may not be stored at LoCoMs for two reasons: (i) users are mobile, which implies that a user may interact with several LoCoMs from time to time. Storing users' personal context at LoCoMs makes the retrieval of their personal historical contexts difficult. In worst case, a query needs to visit all LoCoMs to collect a user's historical personal context; (ii) personal context is usually of high sensitive nature and requires higher level of privacy protection. Storing personal context at distributed LoCoMs poses more challenges on security and privacy control.

LoCoM provides a context inference engine, which is able to deduce high-level context that is not directly observable from sensors. Inference is usually based on domain knowledge that is often organized as sets of rules. A LoCoM conducts reasoning only on its local context which may span multiply application domains. At a higher level of GlobeCon, there are application domain specific context reasoning services, which conduct larger scale inference based on context information gathered from multiple LoCoMs. The main reason of having two levels of context reasoning is to effectively balance the computational load of context inference which, as demonstrated in [12], is a high computation intensive task, and becomes a bottleneck if loads are not distributed strategically.

In order for the local context to be global-wide retrievable by the services in the *context usage hierarchy*, LoCoMs register with the Universal Directory Service (UDS) and provide external access to their local context knowledge base via the Internet. First, a LoCoM generates a self-description on what kind of context information it can provide in accordance with its list of SMs. It then submits the self-description and its communication interfaces to the UDS. After knowing a LoCoM via the UDS, a context consumer (i.e., the services in the context usage hierarchy) can issue context queries directly to the LoCoM.

3.4   Universal Directory Service

The universal directory service (UDS) is at the heart of the GlobeCon framework. It contains all required information about the available LoCoMs, which can be viewed as services by UDS, and the context handling services (refer to Section 3.3.5), and how to reach and make use of them. Thus, higher level applications/services can be aware of the availability of these services via the UDS. The UDS can be viewed as one logical registry, while in real implementation and deployment, it comprises many distributed UDS servers that are federated together. The federation of UDS servers enables GlobeCon to operate in dynamic and distributed environments that address local as well as global needs.

The UDS constructs a hierarchy of the registered services. Fig. 5 shows an example of such a hierarchical service space. The top level of the hierarchy consists of LoCoMs, 3rd party information servers (e.g., an existing weather information server) and context handling services. The service space hierarchy can be extended to include context aware application services at the top layer of GlobeCon. While drilling down the hierarchy further, the classifications used by the UDS can be either locations, types of provided context, or application domains. In order to not overload any single UDS server, the tree-like hierarchical structure is divided into non-overlapping zones. The division is dynamic to

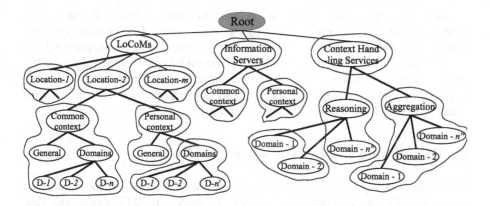

Fig. 5. Part of the hierarchical UDS service space with the division into zones

further balance the loads in cases of new service registrations and de-registrations of existing services. The advantages of this hierarchical structure are two fold. Firstly, it allows the UDS to efficiently manage the large numbers of services with a variety of types. Secondly, it certainly speeds up lookup processing for location-based and application domain based services. Because in order to find a match, a lookup request traverses only the UDS servers along one (or a few, in case of more complex queries) branch in the tree structure, instead of traversing all of the UDS servers. Therefore, this UDS design particularly addresses the design considerations of *C.Sca*, *C.Htg*, *C.Dis* and *C.Dyn*.

3.5   Context Usage Layers

The upper two layers in Fig. 3 map directly to the *context usage* hierarchy of the context framework reference model. The two layers include two groups of services: context handling services and context aware application services. The first group provides large scale, domain specific context handling services (e.g., context reasoning and context aggregation) that are beyond the service scope of individual LoCoMs. For example, they can be context inference services which perform context reasoning on context information gathered from multiple LoCoMs. One such context reasoning service usually focuses on one particular application domain. Another example can be context aggregation services, which collect context of certain type (e.g., location) or of certain event (e.g., traffic jams) from multiple LoCoMs and store the gathered context at one logic location. With the help of context aggregation services, the access to the largely distributed context sources can be transparent to the higher level applications. This group of context handling services will register themselves with the UDS.

The other group of services is a variety of ubiquitous context aware services at the application level, aiming at end users. Each of these services utilizes the context information acquired from the lower *context provision hierarchy* and provides user-oriented context aware services on a particular application domain. Even though it is not mandatory for them to register with the UDS, these services may utilize the UDS to search, discover and make use of the appropriate services from the previous group registered with UDS. Examples of such context aware specialized services are in health care, tourism, ubiquitous GIS, road emergency help, target tracking, etc. For each type of service from either of the two service groups, there may be many of them available on the Internet. They may appear as competing services and can provide the same service with different QoS parameters and different pricing.

4   Embedding GlobeCon with the IMA

GlobeCon is designed to be general purpose, we plan to show the validity of the concept by integrating GlobeCon with the IMA. The IMA, Intelligent Map Agents, is an agent-based user-centric framework. It aims at providing access to, and support for manipulation of, spatial information via a set of personalized services. In this setting, context plays an integral role in supporting user-tailored and user-adaptive features of the IMA system. A layered view of the IMA architecture is depicted in Fig. 6. By embedding GlobeCon within the IMA, we can seamlessly integrate the IMA system with the underlying sensor-rich environments, which capture surrounding contexts silently. Furthermore, GlobeCon facilitates IMA components to easily access the context acquired by the GlobeCon context aggregation hierarchy. Depending on context types, two extra components are devised to facilitate the interaction between GlobeCon and the IMA: *Personal Context Manager* and *Common Context Proxy*, as shown in Fig. 6.

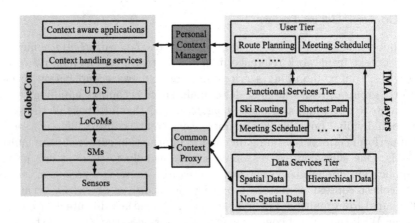

Fig. 6. Embedding GlobeCon with IMA by adding two interfacing components

Personal Context Manager (PCM) represents private and intimate knowledge about a user, and allows IMA personal agents to obtain a user's personal context. It can be viewed as services and mapped to the context handling service layer in the GlobeCon. There is one PCM for a user or a group of users depending on applications. The personal PCM for a user (or a group of users) is the subscriber for all his/her (or their) personal contexts that are captured and processed by the LoCoM, SMs and sensors subsystem. For example, in this scenario, a mobile user John, wearing a detectable ID tag (e.g., RFID tag) and other possible wearable sensors, is moving around in potentially very large geographical areas. When John is within the range of a LoCoM, the MSM, which resides on his PDA and manages his RFID and other wearable sensors, will register with the LoCoM. The LoCoM will then be aware of John's PCM (either notified by the connection request initiated by the PCM, or told by this MSM) and relay all of John's local personal context[3] to his PCM. If this MSM is out of the range of any LoCoMs, it may take alternative actions described in Section 3.3.2 and ultimately relays John's personal context captured by its sensors to John's PCM. Personal contexts that are relayed and/or stored at a PCM can then be queried and utilized by the IMA services to support customization and personalization.

Common Context Proxy (CCP) is used by the IMA to fetch common context from the GlobeCon system. For the usage of common context, the IMA services can be mapped directly to the top layer of the GlobeCon architecture shown in Fig. 3. That is, a IMA service may issue lookup queries to the UDS searching for appropriate LoCoMs or/and context handling services that hold the required common context, and subscribes context directly from them. However, letting heterogeneous IMA services to interact directly with LoCoMs and context handling services, may involve re-engineering their communication interfaces. Therefore, employing the concept of separation of concerns, CCP comes in place to provide an integration layer which bridges the smooth interaction between IMA services and the GlobeCon. In this sense, CCP can be implemented as a special IMA service that accepts context requests from other IMA services and acts upon these services to query the GlobeCon for required common context.

4.1   A Use Case in an Intelligent Health Care Application

In this section, we present a use case of the GlobeCon architecture. It is an intelligent context aware health care application as shown in Fig. 7. The following scenario demonstrates the context data/processing flow of GlobeCon, as well as the communication and interaction between the distributed services.

This application involves a number of context aware services. The services provided by the IMA system include: the *patient monitoring service* (PM) which gathers real-time physical data of patients, examines the data, determines critical conditions based on which it either issues recommendations to the patients or calls other health care related services; the *medical service* (MD) which is responsible for assigning patients to appropriate physician regardless of patients' location; the *emergency/ambulance service* (EM) that can find and dispatch the nearest available ambulance to patients' location; the *shortest path service* (SP)

---

[3] Note that the local personal context provided by a LoCoM can be provided by either John's personal wearable sensors or the stationary sensors under the management of the LoCoM (e.g., a video camera mounted on the entrance door of a grocery store).

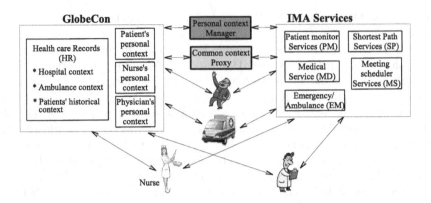

Fig. 7. An intelligent health care application using the GlobeCon framework

which takes points of a road network as input and computes shortest cost path (available in our IMA) between the given points where cost can vary e.g., from distance to travel time; and the *meeting scheduler service* (MS) which explores a number of ways for coordinating a meeting between users [9]. The GlobeCon provides the *health records service* (HR) which collects and maintains historical health care related records. The GlobeCon context provision hierarchy captures and aggregates the personal context of the patient, the physician and the nurse.

Consider a patient wearing a number of physical sensors which monitor his heart rate, blood pressure, body temperature and body humidity etc. Those sensors are managed by a MSM residing on the patient's PDA. By noticing the patient's appearance via the UDS, the PM and HR subscribe his physical context from the MSM via LoCoMs. Noticed that, whenever applicable, these physical context can be delivered to the PM and HR through LoCoMs which usually have high bandwidth connections. When the PM detects a critical condition based on the patient's current context, let's say abnormal blood pressure, which may cause serious illness if not treated properly, the PM will invoke the MD which fetches the patient's historical health record from the HR and his current physical context. Based on the patient's current context and historical health record, the MD assigns a physician to him using a pre-defined assignment algorithm. The MS will then be invoked by the MD to compute a feasible meeting time/location for the patient and the assigned physician, which takes into account the context of their meeting preference and available time acquired via the GlobeCon system. During the meeting calculation, the MS calls the SP to get shortest cost paths for the patient and the physician.

5   Conclusions and Future Work

One of the major challenges in providing scalable context management is flexibility and mobility, since the sensors, computing resources and end users are dynamic and mobile in ubiquitous computing environment. Therefore, in this paper, we propose a context framework reference model aimed at supporting distributed context aggregation, processing, provision and usage. Based on this reference model, we introduce our design of GlobeCon, a scalable context management framework that can be embedded into context aware systems operating in local as well as wide area networks.

We currently focus on the initial phase of GlobeCon prototype implementation. Then, we plan to embed GlobeCon with our existing implementation of the IMA system. By providing context aggregation and provision supports to the IMA applications, we can further demonstrate the feasibility and performance of GlobeCon from experimental point of view.

Although privacy and security are an integral part of any context management frameworks, we do not focus on these in our work. It is partially because privacy and security are self-contained research topics. Further, we plan to integrate, in the future, existing privacy and security components and policies to our framework. However, we do plan to conduct research on the privacy/security issue which in particular related to context, namely the threat of potential information leakage, which refers to unintentional disclosure of private information through knowledge mining process based on public information.

References

1. Abowd, G., Atkeson, C., Hong, J., Long, S., Kooper, R., Pinkerton, M.: Cyberguide: A mobile context-aware tour guide. ACM Wireless Networks 3, 421–433 (1997)
2. Buchholz, T., Linnhoff-Popien, C.: Towards realizing global scalability in context-Aware systems. In: 1st Int'l. Workshop on Location- and Context-Awareness (2005)
3. Chen, H., Finin, T., Joshi, A.: Semantic web in the context broker architecture. In: Proc. of IEEE Int'l. Conference on Pervasive Computing and Communications, IEEE Computer Society Press, Los Alamitos (2004)
4. Chen, G., Kotz, D.: Solar: An open platform for context-aware mobile applications. In: Proc. of 1st Int'l. Conference on Pervasive Computing, pp. 41–47 (2002)
5. Chen, G., Kotz, D.: A survey of context-aware mobile computing research. Dartmouth College Technical Report TR2000-381 (November 2000)
6. Cheverst, K., Davies, N., Mitchell, K., Friday, A.: Experiences of developing and deploying a context-aware tourist guide: the GUIDE project. In: Proc. of 6th Annual Int'l. Conference on Mobile Computing and Networking, pp. 20–31 (2000)
7. Dey, A.K., Salber, D., Abowd, G.G.: A conceptual framework and a toolkit for supporting the rapid prototyping of context-aware applications. Journal of Human-Computer Interaction (HCI) 16(2-4), 97–166 (2001)
8. Fahy, P., Clarke, S.: CASS: a middleware for mobile context-aware applications. In: Workshop on Context Awareness, MobiSys (2004)
9. Gervais, E., Liu, H., Nussbaum, D., Roh, Y.-S., Sack, J.-R., Yi, J.: Intelligent Map Agents - A ubiquitous personalized GIS. Accepted by ISPRS Journal of Photogrammetry and Remote Sensing, special issue on Distributed Geoinformatics (2007)
10. Gervais, E., Nussbaum, D., Sack, J.-R.: DynaMap: a context aware dynamic map application. presented at GISPlanet, Estoril, Lisbon, Portugal (May 2005)

11. Glassey, R., Stevenson, G., Richmond, M., Nixon, P., Terzis, S., Wang, F., Ferguson, R.I.: Towards a middleware for generalised context management. In: 1st Int'l. Workshop on Middleware for Pervasive and Ad Hoc Computing (2003)
12. Gu, T., Pung, H.K., Zhang, D.Q.: A service-oriented middleware for building context aware services. Network and Computer Applications 28(1), 1–18 (2005)
13. Gu, T., Qian, H.C., Yao, J.K., Pung, H.K.: An architecture for flexible service discovery in OCTOPUS. In: Proc. of the 12th Int'l. Conference on Computer Communications and Networks (ICCCN), Dallas, Texas (October 2003)
14. Korpipaa, P., Mantyjarvi, J., Kela, J., Keranen, H., Malm, K.J.: Managing context information in mobile devices. IEEE Pervasive Computing 2(3), 42–51 (2003)
15. Mitchell, K.: A survey of context-aware computing, Lancaster University, Technical Report (March 2002)
16. NAICS: North American Industry Classification System, http://www.naics.org
17. Satyanarayanan, M.: Pervasive computing: vision and challenges. IEEE Personal Communication Magazine 8(4), 110–117 (2001)
18. Schilit, B., Theimer, M.: Disseminating active map information to mobile hosts. IEEE Network. 8(5), 22–32 (1994)
19. Strang, T., Claudia, L.P.: A context modeling survey. In: Davies, N., Mynatt, E.D., Siio, I. (eds.) UbiComp 2004. LNCS, vol. 3205, pp. 34–41. Springer, Heidelberg (2004)
20. Tanenbaum, A.S.: Computer Networks, 3rd edn. Prentice Hall, Inc., Englewood Cliffs (1996)
21. Want, R., Hopper, A., Falcao, V., Gibbons, J.: The active badge location system. ACM Transactions on Information Systems 10(1), 91–102 (1992)

ESCAPE – An Adaptive Framework for Managing and Providing Context Information in Emergency Situations*

Hong-Linh Truong, Lukasz Juszczyk, Atif Manzoor, and Schahram Dustdar

Vitalab, Distributed Systems Group, Vienna University of Technology
{truong,juszczyk,atif.manzoor,dustdar}@infosys.tuwien.ac.at

Abstract. Supporting adaptive processes in tackling emergency situations, such as disasters, is a key issue for any emergency management system. In such situations, various teams are deployed in many sites, at the front-end or back-end of the situations, to conduct tasks handling the emergency. Context information is of paramount importance in assisting the adaptation of these tasks. However, there is a lack of middleware supporting the management of context information in emergency situations. This paper presents a novel framework that manages and provides various types of context information required for adapting processes in emergency management systems.

1 Introduction

Supporting emergency situations (e.g., disaster responses) could benefit a lot from the recent increasing availability and capability of computer devices and networks. Pervasive devices have been widely used to capture data for and to coordinate the management of tasks at disaster fields. Communication networks help bringing collected data to the back-end center, and at the back-end, high performance computing/Grid systems and data centers can conduct advanced disaster management tasks. Still, there are many challenging issues in order to utilize advantages brought by pervasive computing, mobile networks, and Grid computing for supporting the management of emergency situations.

Take the scenarios being supported by the EU WORKPAD project [29] as examples. Aiming at supporting natural disaster responses, the front-end of the WORKPAD system will assist teams of human workers by providing services that are able to adaptively enact processes used in disaster responses. On the one hand, these services are executed on mobile ad-hoc networks whose availability and performance are changed frequently. On the other hand, processes carried out in disaster scenarios are normally established on demand and changed accordingly to meet the current status of the disaster scenarios. It means that depending on the context of specific scenarios, tasks conducted during the scenarios can be altered, either in an automatic fashion or a manual one controlled by team leaders. Therefore, it is of paramount importance that these services are able

---

* This research is partially supported by the European Union through the FP6-2005-IST-5-034749 project WORKPAD.

G. Kortuem et al. (Eds.): EuroSSC 2007, LNCS 4793, pp. 207–222, 2007.

to use context information to adjust the enactment and collaboration of processes. This leads to a great demand for frameworks that can be used to provide and manage various types of context information. However, most of existing context information management middleware are targeted to indoor and small scale environments, e.g., as discussed in [8]. There is a lack of similar middleware for emergency situations. The environment in which supporting tasks for emergency situations are performed is highly dynamic and unstructured.

In this paper, we discuss ESCAPE, a peer-to-peer based context-aware framework for emergency situations with the focus on crisis situations, such as disasters, in pervasive environments. We present the design and implementation of the context management services within ESCAPE that can support multiple teams working at different sites within many responses for emergency situations to collect and share various types of context information. ESCAPE is able to manage relevant context information which is described by arbitrary XML schema and required for emergency responses, and to provide the context information to any clients. In this paper, we also illustrate early experiments of the current prototype of the framework. The ESCAPE framework is an ongoing development. Therefore, in this paper, we only focus on the discussion of the management and provisioning of context information in emergency situations at middleware layer, instead of presenting adaptation techniques and applications.

The rest of this paper is organized as follows: Section 2 discusses the requirement and motivation. Section 3 presents the related work. The architecture of the context management services is described in Section 4. We describe the management of context information in Section 5. Prototype implementation is outlined in Section 6. Experiments are illustrated in Section 7. We discuss existing issues in the framework in Section 8. We summarize the paper and discuss about the future work in Section 9.

2   Requirements and Motivation

Effective responses to an emergency situation (e.g, a natural disaster) require key information at sites where the situation occurs in order to optimize the decision making and the collaborative work of teams handling the emergency. Context information can substantially impact on responses to the situation. The key to the success of responses to an emergency is to have effective response processes which are actually established on-demand and changed rapidly, depending on the context of the situation. Such effective response processes cannot be achieved without understanding the context associated with entities inherent in the situation. Required context information related to entities at each site in the emergency situation is related to not only teams performing responses, for example, information about team member tasks, status of devices, networks, etc., but also entities affected by the emergency situation, such as victims and infrastructure. To date, context information is widely used in context-aware systems but most of them are not targeted to emergency scenarios.

In our work, we consider the support to the management of various kinds of emergency *situations* such as disasters (e.g., earthquake and forest fire). In such situations, many support *teams* of *individuals* will be deployed, as soon as possible, at *sites* (e.g., a village) where the situations occur in order to conduct situation *responses* (e.g., to rescue victims). All members of the teams will collect various types of data and perform different tasks, such as relief works or information gathering, for supporting response tasks. Within emergency situations, teams are equipped with diverse types of devices with different capabilities, such as PDAs and laptops. These devices have limited processing capacity, memory and lifetime. Moreover, the underlying network that connects these devices together is established as a mobile ad-hoc network in which usually a few devices can be able to connect to the back-end services. In addition to professional teams, teams of non-professional volunteers can also be established in a dynamic fashion. Given the current trend in mobile devices consumption, many people have their own networked PDAs and smart phones which can be easily used to support emergency situations.

Moreover, as teams in the site may perform tasks in a dangerous environment and the teams lack a strong processing power and necessary data, the front-end teams may need support from teams at the back-end. This requires us to store context information at the back-end due to several reasons. For example, teams at the back-end can use context information to perform other tasks that could not be done by the front-end teams. Furthermore, as people at the front-end are working in dangerous environments, latest location information (one type of context information) of people who are in dangerous environment can be used for, e.g., to rescue them in case they are in danger.

To support the above-mentioned scenarios, context-aware support systems for emergency situations using mobile devices and ad-hoc networks have to be developed. An indispensable part of these systems is a context management middleware which must be able to collect various context information related to the emergency situation. Such context information will be utilized at the site by multiple teams and by the back-end support teams. As the middleware has to be deployed in constrained devices, various design issues must be considered. Since devices do not have a strong communication and processing capability, the network of teams is unstructured and the operating environment is highly dynamic. As a result, context information is exchanged between various peers in a dynamic and volatile environment. Therefore, a P2P (peer-to-peer) data exchange model is more suitable.

Context management services have to exchange context information with many supporting tools, such as GIS (Geographic Information Systems) and multimedia emergency management applications. Moreover, we have to make the front-end services interoperable with the back-end (Grid-based) services which might belong to different organizations. As a result, SOA-based models and techniques will be employed in the management and dissemination of context information. In this aspect, the middleware operates on Pervasive Grid environments [17]. However, the context management framework for emergency scenarios should be flexible or be easily adapted to handle context information specified by different models at multiple levels of abstraction since context information in emergency situations is not known in advance. In this respect, the framework should support an extensible data model, e.g. XML, and query mechanism, e.g. XQuery [31] and XUpdate [33]. Then, various plug-ins used to handle specific cases, such as sensors, event notification, event condition action, etc., could be seamlessly integrated into the framework to support situation-specific scenarios.

3   Related Work

Several studies of context-aware systems have already been conducted, such as in [8]. In this paper, we concentrate our study only on existing middleware for managing context data. RCSM [35,5] is a middleware supporting context sensitive applications based on an object model. The JCAF (Java Context Awareness Framework) supports both the infrastructure and the programming framework for developing context-aware applications in Java [4,10]. JCAF is based on the peer-to-peer model but it does not support automatic discovery of peers or a super-peer. Moreover, the communication is based on Java RMI (Remote Method Invocation). The AWARENESS project [1] provides an infrastructure for developing context-aware and pro-active applications. It targets to applications in mobile networks for the health care domain. The PACE middleware [15] provides context and preference managements together with a programming toolkit and tools for assisting context-aware applications to store, access, and utilize context information managed by the middleware. PACE supports context-aware applications to make decisions based on user preferences. The GAIA project is a CORBA-based middleware supporting active space applications [26]. GAIA active space has limited and well-defined physical boundaries so GAIA is not suitable for emergency management. It is targeted to small and constrained environment such as smart homes and meeting rooms. The CARMEN middleware [11] uses CC/PP for describing metadata of user/device profiles while the Mercury middleware prototype [28] describes user, terminal, network, and service profiles using CC/PP. We observed that most of above-mentioned middleware do not support a variety of context data inherent in emergency situations. Furthermore, most of them are targeted to in door environment and do not support scenarios of collaborative teamwork in emergency situations in which front-end teams and back-end systems are connected and exchange context information through a large scale and highly dynamic environment.

Relational databases are widely used to store context information. For example, [24] stores context information about geography, people and equipments in a relational database. In [20], Location historical information is stored in a database that can be accessed using SQL. The PACE middleware provides a context management whose back-end is a relational database [15]. The Aura context information service is a distributed infrastructure which includes different information providers [22]. Context-aware applications can subscribe to middleware in order to obtain context information through subscription/notification [24,10] or can query information stored in persistent databases, e.g., in [15]. We take a different approach in which we rely on XML-based information and XQuery/XPath for accessing context information. It helps to facilitate the integration with different types of application and support the extensibility and generalization of the middleware in handling different types of context information.

Context information is also widely utilized in personal applications in pervasive computing such as electronic communication [25], in office and education use, e.g., for monitoring user interactions in the office [30], for managing presentations in meeting rooms [16], and for office assistants [34]. In [9], context information in hospitals is used in context-aware pill containers and hospital beds. [21] describes how context information can be used in logistics applications. Our work is different as we aim at supporting emergency management applications which require much diverse context information and operate in a highly dynamic environment. The ORCHESTRA [3] and OASIS [2] projects focus on disaster/crisis management, however, they do aim at supporting context-awareness.

4   Architecture of ESCAPE Context Management Services

4.1   Architectural Overview

Figure 1 presents the architecture of the ESCAPE context management framework which includes context information management services (CIMSs) and the back-end support system. Each individual's device will host an instance of the CIMS used by the individual who is responsible for collecting and managing context information related to the individual. As individuals are organized into teams, each team will establish a network of CIMSs. For each team, one CIMS whose hosting device has more powerful capability, such as the device of the team leader, will act as a super peer. This super peer will periodically gather context information available in CIMSs within its team and then push the context information to the back-end systems. Within a team, we use a simple peer structure: all peers are equal and there is no forwarding mechanism among peers[1]. Any peer which wants to obtain context information provided by another peer just directly queries or subscribes context information from the provider.

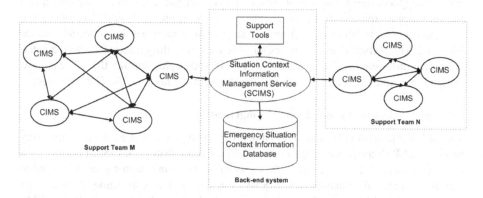

Fig. 1. Architecture of the ESCAPE context management framework for emergency situations

At the back-end, we store context information into a situation context information management service (SCIMS) which keeps all the context data related to a situation. By using this information, various support teams at the back-end can utilize rich data sources and computing services to support teams at sites. Furthermore, the context information managed by SCIMS can be used for post-situation studies. Two different teams can exchange context information by either using the back-end systems, e.g., in case the teams are not located in the same site or the network connection among

---

[1] This is only at context management middleware level. The underlying network can support multi-hop communication, depending on mobile ad-hoc network techniques employed.

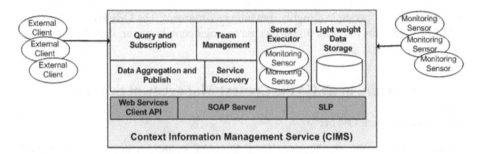

Fig. 2. Overview of the context information management service

two teams is not available, or utilizing the team leader CIMS by searching team leader devices in the network.

Figure 2 presents the architecture of a context information management service (CIMS) which is used to manage and provide context information at an individual's device. The *Web Services Client API* is used to communicate with other Web services. The *SOAP server* provides features for building services based on SOAP. *SLP* (Service Location Protocol) component supports team service advertisements and discovery. The two components *Service Discovery* and *Team Management* will be responsible for discovering other CIMSs and for managing CIMSs belonging to the same team, respectively. The *Query and Subscription* component is responsible for processing data query and subscription requests from any clients. The *Data Aggregation and Publish* component gathers context information from various peers and publishes the information to the back-end. The *Sensor Executor* is responsible for controlling internal sensors collecting context information. These sensors are considered as plug-ins of CIMS. The context information gathered at each CIMS will be stored in the *Lightweight Data Storage*.

4.2 Service Discovery and Team Management

A CIMS will publish information about itself by exploiting multicast service discovery based on SLP (Service Location Protocol)[27]. Each CIMS is described mainly by a triple (`teamID, individualID, serviceURI`) in which `teamID` is used to identify the team; all instances of CIMS within a team will have the same `teamID`. The element `individualID` identifies the individual whose device manages the CIMS whereas `serviceURI` specifies the URI of the CIMS. This triple information will be mapped into SLP advertisements. Based on that, service discovery can be performed.

A CIMS will publish its service information periodically and will keep a record of this information of all CIMSs in its team. A CIMS will check its team members regularly by pinging them. By doing so, each CIMS has an up-to-date record of all its team members. Currently, we use a configuration file to specify the intervals based on which CIMS should publish its information and check its team members presence.

4.3 Publishing and Querying Context Information

In our framework, context information will be collected by different clients and monitoring sensors (software- and hardware-based sensors). CIMSs will provide interfaces for these clients and sensors to publish and query context information. Being able to handle different kinds of context information, CIMSs will accept any type of context information that is described in XML format, without knowing the detailed representation of context information.

CIMS provides two mechanisms for sensors/clients to publish context information. We distinguish two cases: sensors execution will be and will not be controlled by CIMS. In the first case, sensors will be invoked directly by CIMS as a plugin. To support this, we develop a generic interface that sensors should implement. This interface includes three main methods named `setParameters`, `execute` and `getContextInfo`, allowing CIMS to initialize sensors, invoke sensors instances and obtain XML-based context data without considering how the sensors are implemented. In the second case, CIMS provides Web services operations for any sensors/clients to publish context information. To retrieve context information, clients have to use Web services operations provided by CIMS. Clients can also specify XQuery-based requests in searching for context information from a CIMS.

4.4 Customized Middleware Components

Supporting reconfigurable middleware components is important in pervasive environments since specific platforms, e.g., PDA or laptop, have a multitude of varying capabilities and support technologies that should be exploited differently. We identify two main components within the CIMS that should be reconfigured according to underlying platforms: *Query and Subscription* and *Lightweight Data Storage*. As we aim at supporting different kinds of context information, the context management framework does not bind to any specific representations of context information as long as the representations are in XML. Using existing tools like Xerces [7] and kSOAP2 [19], we are able to generate and process XML data in constrained devices. Nevertheless, supporting advanced XML processing functions, such as query and update of XML data with XQuery [31] and XUpdate [33], in constrained capability devices is very limited. Our *Lightweight Data Storage* is based on eXist XML database [12] when a CIMS is deployed in normal devices supporting Java SE (e.g., laptop). Otherwise, the storage will be based on a round-robin model in which context information is stored into XML files, and XQuery processing is based on MXQuery [23] which is a lightweight XQuery engine for mobile devices.

5 Management of Context Information

5.1 Context Information Level

In emergency situations, context information will be collected by and exchanged among individuals and teams within different scopes of knowledge, e.g. within the knowledge of an individual or a team at a site during a response or the whole situation. Therefore, we support five levels of knowledge in which context information is available: *individual, team, site, response* and *situation*. The *individual* level indicates the context information within the knowledge of an individual. It means that context information is either associated with or collected by an individual. The *team* level indicates context information gathered from all individuals of a team. The *site* level specifies context information collected from all teams working on the same site whereas the *response* level indicates context information gathered within a response. The *situation* level specifies all context information gathered during the situation. The five levels of knowledge provide a detailed structure of context information. Based on that, context information can be efficiently shared among teams and utilized for different purposes.

5.2 Storage and Aggregation of Context Information

A single level/place in storing context information, such as only at the team leader device, is not suitable for emergency situations. In such situations, teams at sites are equipped with devices which are not always highly capable and the network is normally not reliable. Therefore, rather than relying on centralized managers, we believe that context information should be provided by and exchanged among individuals in a dynamic fashion. Context information in emergency situation must be widely shared among different teams at different places and be stored for post-situation studies. Thus, context information exchange should not be limited in a single team/place.

Addressing two different storage mechanisms, distributed storage information in mobile devices and in high-end systems is a challenging issue. Our context management services support any kind of context information modeled in XML. At each CIMS, context information sent by sensors to the service is managed in records. Each record r is represented as a triple (individualID, timestamp, contextDataURI) where individualID is the unique identifier of the team member, timestamp is the time at which the context information was collected, and contextDataURI is the URI specifying the location from which detailed context information can be retrieved. The detailed context information can be stored into files or XML databases, depending on the capability of the hosting device. As mobile devices have a limited resource and context information will partially be stored in the back-end services, not all context information will be kept in a device. Instead, we employ a round robin mechanism to maintain existing context information stored in each CIMS. Depending on the device capability, the round robin database mechanism may be relaxed.

As mentioned before, context information will be collected at CIMSs which hold the context information at the individual level. Aggregation of context information with in a team will be conducted by the team leader. Context information will be available from different places, and the information has to be stored over the time. At a time t, the newest instance of context information collected is called a *snapshot* of context information. We consider the management of context information at five levels: within a device managed by an individual, within a team, within a site, within a response, and within the whole situation. Consequently, we have five different types of snapshots in the scopes of the above-mentioned levels. Spanning the timeline of situation responses, various teams are deployed and context information associated with the situation is collected by the teams. Since response time is a critical issue in the management of emergency situation and the devices used have limited capabilities, we employ simple mechanisms to manage context information. Let $ctx_s(level, time)$ where $level \in \{individual, team, site, response, situation\}$ denote a context snapshot within a *level* at a given *time*. Within a team, each member monitors and collects context information which will be stored and updated locally. A snapshot of context information stored in a device is associated with a timestamp t and is denoted as $ctx_s(individual, t)$. Depending on the capabilities of the device, the number of snapshots kept in a device could be limited to a pre-defined value n.

The context information collected by a team will be stored temporarily at the team leader device or pushed back to the back-end service periodically by CIMS of the team leader. At a given time t, $ctx_s(team, t)$ will be a fusion of $\{ctx_s(individual, t)\}$ for all $individual$ belonging to the team. Similarly, $ctx_s(site, t)$ is a fusion of $\{ctx_s(team, t)\}$ for all teams within the site, $ctx_s(response, t)$ is a fusion of $\{ctx_s(site, t)\}$ for all sites involved in the response. The snapshot of the situation, $ctx_s(situation, t)$, is defined as a fusion of $\{ctx_s(response, t)\}$ for all responses conducted in the situation. While context information at $individual$ and $team$ levels is available at the front-end, the information of the other levels is available at the back-end system only.

5.3 Provenance of Context Information

All context information gathered could be tracked through our provenance support. We design a generic XML schema based on that provenance information of context information can be described. Figure 3 describes main elements of the schema used to describe provenance information. This representation allows us to specify detailed information about the five levels of knowledge by using elements situation, site, response, team and individual. The detailed content of provenance information about these levels can be described in XML and encoded by using <![CDATA [" "]]> section. The element collectedAt indicates the time at which the context information is collected while element contextDataURI specifies the URI through which context information can be retrieved.

The framework will automatically store provenance information into the back-end system whenever context information is pushed back to the back-end system. Provenance information is important because it allows us to correlate all gathered context information to its sources and creators, providing tracing capability and improving the understanding of actions performed within emergency situations.

```
<xsd:complexType name="ContexProvenance">
   <xsd:sequence>
      <xsd:element name="provenanceEntry" type="tns:ContextProvenanceEntry"/>
   </xsd:sequence>
</xsd:complexType>
<xsd:complexType name="ContextProvenanceEntry">
   <xsd:sequence>
      <xsd:element name="situation" type="xsd:string"/>
      <xsd:element name="site" type="xsd:string"/>
      <xsd:element name="response" type="xsd:string"/>
      <xsd:element name="team" type="xsd:string"/>
      <xsd:element name="individual" type="xsd:string"/>
      <xsd:element name="collectedAt" type="xsd:dateTime"/>
      <xsd:element name="contextURI" type="xsd:anyURI"/>
   </xsd:sequence>
</xsd:complexType>
```

Fig. 3. Schema (simplified) for describing provenance of context information

6   Implementation

To implement the architecture mentioned in Section 4, we employ various libraries for handling Web services and XML on mobile devices such as kSOAP2 [19] and CDC-based Xerces[6]. XQuery/XUpdate supports are based on eXist [12] and MXQuery [23]. Our current prototype is implemented in Java ME CDC for PDA and can be customized with Java SE-based libraries for normal laptops to exploit advanced Web services and XML capabilities.

In our implementation, *Service Discovery* and *Team Management* are implemented on top of jSLP [18] which is a lightweight Java implementation of SLP for mobile devices. We use the *SOAP server* in Sliver BPEL [14] and implement our service-based components on top of that. Sliver supports both TCP socket-based and Jetty HTTP-based communications. Within a team, CIMSs can communicate with each other using SOAP by selecting one of those communications.

The back-end context information service is currently implemented based on eXist XML database [12]. CIMSs push data to back-end services by using the REST (Representational State Transfer) interface provided by eXist database.

7   Experiments

7.1   Experimental Application: Supporting Disaster Responses

One of the main motivations for developing this framework is to use it in supporting disaster responses in the WORKPAD project [29]. Being able to collect and provide context data relevant to disaster responses is the key issue to the WORKPAD adaptive process management systems used by team leaders to plan response activities. Furthermore, context information is required by the disaster management support based on geographic information systems (GIS). To this end, we have developed a novel context information model for disaster managements. Figure 4 describes main concepts of the

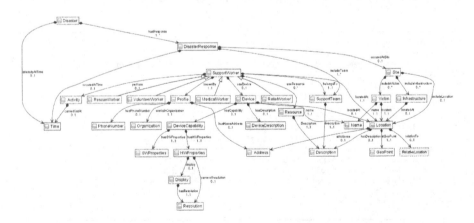

Fig. 4. WORKPAD context information model in disaster responses

first version of the WORKPAD context information model which can describe various context information inherent in a disaster response.

In the case of disaster responses, the five levels of knowledge about context information are *support worker, support team, disaster site, disaster response, disaster* which are mapped to *individual, team, site, response, situation*, respectively. In order to test the current prototype of our framework, we use many simulated sensors whose functionalities are exactly the same as that of real sensors, except that the context information is automatically generated from simulation configuration parameters.

7.2 Testbed

Figure 5 describes our current testbed. We setup a testbed which includes 3 iPAQ 6915 PDAs (Intel PXA 270 416 MHz, 64 MB RAM, Windows CE 5.0, 2GB external MiniSD, IBM J9 WebSphere Everyplace Micro Environment 6.1), a Dell XPS M1210 (Intel Centrino Duo Core 1.83 GHz, 2GB RAM, Windows XP) notebook, and a Dell D620 (Intel Core 2 Duo 2GHz, 2 GB RAM, Debian Linux). Devices in the testbed are connected through a mobile ad-hoc network based on 802.11b. A CIMS is deployed on each device and the Dell D620 laptop is designated as the gateway to the back-end. In our setting, the mobile ad-hoc network bandwidth is limited to 220 Kbits/s but we observed that the average bandwidth is around 150 Kbits/s. The back-end system is based on a Dell Blade (2 Xeon 3.2 GHz CPUs with Hyperthreading, 2GB RAM, and Ubuntu Linux). We use simulated sensors to produce context information according to the WORKPAD context information model. In our experiments, we focus on presenting some preliminary analyses of data transfers and examples of accessing context information using XQuery.

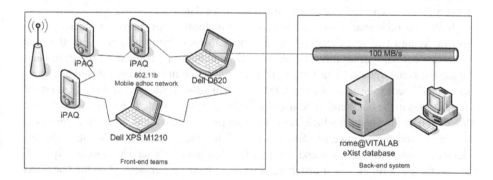

Fig. 5. Testbed deployment

7.3 Performance Analysis

In the first scenario, we consider a team including five members. Three members use PDAs and two members use laptops in our testbed. The Dell D620 is designated as the device of the team leader. Figure 6 shows Jetty HTTP-based and TCP socket-based data transfers between a CIMS deployed in a member and a CIMS deployed the team

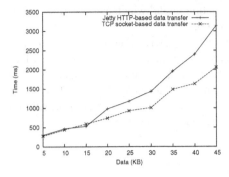 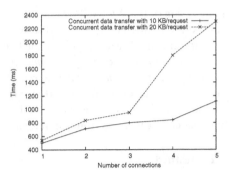

Fig. 6. Data transfer between a CIMS member (PDA) and CIMS team leader (laptop)

Fig. 7. Concurrent data transfer between a CIMS (PDA) and its client (laptop)

leader. Overall when the data size is smaller than 15KB the performance is almost the same. However, TCP socket-based data transfer outperforms Jetty HTTP-based when the transferred size is increased. We also found that the TCP socket-based communication in Sliver is not quite stable. Therefore, in the following experiments we relied on Jetty HTTP-based communication. Figure 7 presented the concurrent data transfer tests between a CIMS in a PDA and a client in the Dell D620. When doubling the data size from 10KB/request to 20KB/request, with the number of concurrent connections is smaller than 3, the transfer time increases but not substantial. However, with more than 3 connections, the transfer time increases significantly. This suggested that we should not use multiple concurrent connections to transfer a large data size to PDAs. Moreover, PDAs might not be used as team leader device in case the number of team member is large and there is a need to transfer large amount of data.

In the second scenario, we modeled a system including four teams. Two team leaders use PDAs and two team leaders use laptops. CIMSs running on devices of the four team leaders will gather information of teams and send the context information back to the back-end system. Non-leader members of a team are simulated through sensors that send data to the team leader. CIMS of each team leader made three concurrent connections to the back-end system and sent totally approximate 17KB every 5 seconds. We conducted the tests in which from one team to four teams send data simultaneously, and measured the average execution time in 5 minutes run. Still in this scenario, all devices connect to the back-end through the designated router Dell D620. Figure 8 shows the performance of transferring data from team leader devices to the back-end system. Overall, we observed that the performance of CIMSs in PDAs is different. These behaviors need to be examined in more detail by analyzing how the back-end system handles requests from teams. The transfer time also increased during the test because the eXist database had to handle more XML documents in the same collection.

All performance data presented are average values determined from various runs. We observed there is a high variation between different runs in PDAs. For example, when measuring parallel data transfer with 3 connections with 20 KB, the fastest transfer time is 190 ms whereas the slowest transfer time is 5170 ms. We also observed that storing the whole big XML document in SCIMS is much faster than updating small XML data

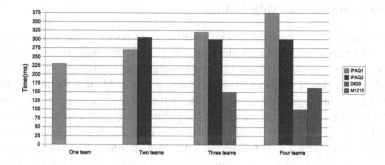

Fig. 8. Performance of transferring data to the back-end system

fragment into an existing XML document using XUpdate when the framework stored context information into separate XML documents and all provenance information into a single XML document. It means that it would be better to store multiple small XML documents than merge them into a big one. Since the framework focuses on bringing data to the back-end system quickly we changed the way we stored provenance information. Instead of storing all provenance information into a single document, we store provenance information in separate files.

7.4 Search for Relevant Context Information

As context information is described in XML, relevant context information can be searched by using XQuery. We outline few examples in the following.

Team A wants to reach to a place "P". Let's check unusable roads lead to "P": this can be done by the team leader when deciding to move the team to a new place. Another team might approach "P" before and notice unusable roads leading to be "P". The following query can be used to find out roads which are unusable.

```
for $infrastructure in collection('/db/contextinformation')//includeInfrastructure
where $infrastructure/category="ROAD" and $infrastructure/condition="UNUSABLE"
return $infrastructure
```

Let's send one worker to place "P" to take a photo: This can be decided by the team leader when she/he needs to send some support workers immediately to one area, e.g., for taking a photo needed for a further analysis. The team leader, for example, can obtain the location information of the support workers, and depending on this information, the leader can assign the task to those people who are near to that place. Similarly the team leader can also look for the activities that the support workers are performing and assign the new task to those who are doing activities with less priority. We can retrieve this information from team level context by using the following query:

```
for $worker in collection('db/contextinformation')//SupportWorker
where $worker//hasCamera and $worker/belongsTo/description="Team 1"
and $worker//Activity/status="LOW"
return $worker
```

8 Discussion of Existing Issues

One of the main issues about context information management is the quality aspects of context information such as incompleteness, duplication and inconsistency. For example, considering the case of context information aggregation. Since the team leader will pull and fuse the latest snapshot from members, it is possible that the team snapshots will miss some information when more information sent to a member than retrieved from the member. However, if we reduce the polling interval, the team snapshots may include duplicate context information. This issue is well known and many methods have been proposed to address it. However, context information is collected by using mobile devices which do not have enough capabilities to conduct these methods. Therefore, in our framework, checking quality of context information could be implemented as a plug-in for the CIMS.

Another issue is the data aggregation at CIMS. Each context management service retrieves various types of context information from different sensors and clients. The information may follow the same model (e.g., in the example of the WORKPAD project) but in practice, different sensors provide different information fragments at different times. Therefore, not only the context management service has to manage multiple fragments of information but also we cannot put the information fragments into a coherent view even they follow the same model. When context information does not follow the same model, CIMS cannot merge information fragments. However, in case context information follows the same model, we can merge information fragments into a single one. We can merge data fragments received in a predefined windows time into a single one, for example based on approximate XML joins [13] or using XUpdate/XSLT[32]. In doing so, we could define plug-ins for CIMSs. However, this might be applicable only to high-end mobile devices as we observed performance issues in updating XML documents in Section 7.3.

Since context information is gathered from various places by various teams, it is important to automatically process context information. We could define rules based on which context information can be evaluated and corresponding actions can be performed based on the evaluation of context information, e.g., inform relevant parties about new emerging issues. To this end, we can apply rule-based systems, event condition action (ECA) and complex event processing (CEP) concepts to the CIMS and the SCIMS.

9 Conclusion and Future Work

Adapting tasks in emergency situations is an important and required feature because both human teams and processes involved in disaster scenarios are established in a dynamic manner and are changed on-demand, depending on specific situations. In this paper, we have presented a novel, generic framework for managing and providing context information in emergency situations, such as natural disasters. We have developed a P2P context management framework that is able to support multiple levels of context information such as individual, team, site, response and situation. We have presented how context information management services in ESCAPE help managing and providing context information to teams at both front-end and back-end sites of the emergency scenarios. We also presented performance analysis of the system and outlined ESCAPE functions for the disaster response management in the EU WORKPAD project. The ESCAPE context management services are based on a SOA model and support XML-based context information. Thus they can be easy used and integrated into emergency situation support systems.

We have not presented different application scenarios. Currently, applications utilizing our framework, such as adaptive process management and GIS-based disaster management systems, are being developed. Our future work is to fully achieve the prototype implementation by addressing issues mentioned in Section 8 and focusing on adaptive aspects within the framework. Moreover, we are working on utilizing context information for adapting processes in disaster responses.

References

1. AWARENESS - Context AWARE mobile NEtworks and ServiceS,
   http://www.freeband.nl/project.cfm?id=494&language=en
2. EU OASIS project - Open Advanced System for dISaster and emergency management,
   http://www.oasis-fp6.org
3. EU ORCHESTRA project, http://www.eu-orchestra.org/
4. JCAF - The Java Context-Awareness Framework,
   http://www.daimi.au.dk/~bardram/jcaf/
5. RCSM Middleware Research Project, http://dpse.asu.edu/rcsm
6. Xerces CDC, http://imbert.matthieu.free.fr/jgroups-cdc/files/
   xerces-cdc.jar
7. Apache Xerces, http://xerces.apache.org/
8. Baldauf, M., Dustdar, S., Rosenberg, F.: A survey on context-aware systems. International Journal of Ad-Hoc and Ubiquitous Computing (January 2006)
9. Bardram, J.E.: Applications of context-aware computing in hospital work: examples and design principles. In: SAC 2004, pp. 1574–1579. ACM Press, New York (2004)
10. Bardram, J.E.: The java context awareness framework (jcaf) - a service infrastructure and programming framework for context-aware applications. In: Gellersen, H.-W., Want, R., Schmidt, A. (eds.) PERVASIVE 2005. LNCS, vol. 3468, pp. 98–115. Springer, Heidelberg (2005)
11. Bellavista, P., Corradi, A., Montanari, R., Stefanelli, C.: Context-aware middleware for resource management in the wireless internet. IEEE Trans. Software Eng. 29(12), 1086–1099 (2003)
12. eXist XML database, http://exist.sourceforge.net/
13. Guha, S., Jagadish, H.V., Koudas, N., Srivastava, D., Yu, T.: Approximate xml joins. In: SIGMOD 2002. Proceedings of the 2002 ACM SIGMOD international conference on Management of data, pp. 287–298. ACM Press, New York (2002)
14. Hackmann, G., Haitjema, M., Gill, C.D., Roman, G.-C.: Sliver: A bpel workflow process execution engine for mobile devices. In: Dan, A., Lamersdorf, W. (eds.) ICSOC 2006. LNCS, vol. 4294, pp. 503–508. Springer, Heidelberg (2006)
15. Henricksen, K., Indulska, J., McFadden, T., Balasubramaniam, S.: Middleware for distributed context-aware systems. In: Meersman, R., Tari, Z., Hacid, M.-S., Mylopoulos, J., Pernici, B., Babaoglu, Ö., Jacobsen, H.-A., Loyall, J.P., Kifer, M., Spaccapietra, S. (eds.) On the Move to Meaningful Internet Systems 2005: CoopIS, DOA, and ODBASE. LNCS, vol. 3760, pp. 846–863. Springer, Heidelberg (2005)

16. Hess, C.K., Román, M., Campbell, R.H.: Building applications for ubiquitous computing environments. In: Mattern, F., Naghshineh, M. (eds.) Pervasive Computing. LNCS, vol. 2414, pp. 16–29. Springer, Heidelberg (2002)
17. Hingne, V., Joshi, A., Finin, T., Kargupta, H., Houstis, E.: Towards a pervasive grid. ipdps, 00:207b (2003)
18. jSLP, http://jslp.sourceforge.net/
19. kSOAP2, http://ksoap2.sourceforge.net/
20. Mantoro, T., Johnson, C.: Location history in a low-cost context awareness environment. In: ACSW Frontiers 2003: Proceedings of the Australasian information security workshop conference on ACSW frontiers 2003, Darlinghurst, Australia. Australian Computer Society, Inc., pp. 153–158 (2003)
21. Meissen, U., Pfennigschmidt, S., Voisard, A., Wahnfried, T.: Context- and situation-awareness in information logistics. In: Lindner, W., Mesiti, M., Türker, C., Tzitzikas, Y., Vakali, A.I. (eds.) EDBT 2004. LNCS, vol. 3268, pp. 335–344. Springer, Heidelberg (2004)
22. Miller, N., Judd, G., Hengartner, U., Gandon, F., Steenkiste, P., Meng, I-H., Feng, M.-W., Sadeh, N.: Context-aware computing using a shared contextual information service. In: Ferscha, A., Mattern, F. (eds.) PERVASIVE 2004. LNCS, vol. 3001, Springer, Heidelberg (2004)
23. MXQuery, http://www.dbis.ethz.ch/research/current_projects/MXQuery
24. Naguib, H., Coulouris, G., Mitchell, S.: Middleware support for context-aware multimedia applications. In: Zielinski, K., Geihs, K., Laurentowski, A. (eds.) DAIS. IFIP Conference Proceedings, vol. 198, pp. 9–22. Kluwer, Dordrecht (2001)
25. Ranganathan, A., Campbell, R.H., Ravi, A., Mahajan, A.: Conchat: A context-aware chat program. IEEE Pervasive Computing 1(3), 51–57 (2002)
26. Román, M., Hess, C.K., Cerqueira, R., Ranganathan, A., Campbell, R.H., Nahrstedt, K.: Gaia: A middleware infrastructure to enable active spaces. In: Mattern, F., Naghshineh, M. (eds.) Pervasive Computing. LNCS, vol. 2414, pp. 74–83. Springer, Heidelberg (2002)
27. Service Location Protocol (SLP), http://tools.ietf.org/html/rfc2608
28. Solarski, M., Strick, L., Motonaga, K., Noda, C., Kellerer, W.: Flexible middleware support for future mobile services and their context-aware adaptation. In: Aagesen, F.A., Anutariya, C., Wuwongse, V. (eds.) INTELLCOMM 2004. LNCS, vol. 3283, pp. 281–292. Springer, Heidelberg (2004)
29. The EU WORKPAD Project, http://www.workpad-project.eu
30. Voida, S., Mynatt, E.D., MacIntyre, B., Corso, G.M.: Integrating virtual and physical context to support knowledge workers. IEEE Pervasive Computing 1(3), 73–79 (2002)
31. XQuery, http://www.w3.org/TR/xquery/
32. XSL Transformations (XSLT), http://www.w3.org/TR/xslt
33. XUpdate, http://xmldb-org.sourceforge.net/xupdate/
34. Yan, H., Selker, T.: Context-aware office assistant. In: IUI 2000: Proceedings of the 5th international conference on Intelligent user interfaces, pp. 276–279. ACM Press, New York (2000)
35. Yau, S.S., Karim, F.: A context-sensitive middleware for dynamic integration of mobile devices with network infrastructures. J. Parallel Distrib. Comput. 64(2), 301–317 (2004)

Capturing Context Requirements[*]

Tom Broens, Dick Quartel, and Marten van Sinderen

Centre for Telematics and Information Technology, ASNA group, University of Twente,
P.O. Box 217, 7500 AE Enschede, The Netherlands
{t.h.f.broens, d.a.c.quartel, m.j.vansinderen}@utwente.nl
http://asna.ewi.utwente.nl

Abstract. Context-aware applications require context information to adapt their behaviour to the current situation. When developing context-aware applications, application developers need to transform specific application context requirements into application logic to discover, select and bind to suitable sources of context information. To facilitate the development of context-aware applications, we propose a Context Binding Transparency that simplifies the process of retrieving context information. A major element of this transparency is the declarative approach to capturing context requirements. This enables application developers to specify their context requirements at a high level of abstraction rather than in programming code, and thus to separate the transformation of context requirements into context binding logic from the development of the actual application logic. In this way, we try to decrease the development effort and facilitate maintenance and evolution of context-aware applications. This paper discusses the design of this binding transparency; especially focusing on the language we developed to capture context requirements.

Keywords: Context-Aware applications, Context Requirements, Context Binding Transparency, Context Binding Description Language (CBDL).

1 Introduction

Ubiquitous computing envisions a situation in which users are surrounded by computing devices that offer unobtrusive services. Unobtrusiveness is defined by Merriam-Webster's dictionary as not being undesirably prominent. In relation to ubiquitous computing this means that, amongst others, offered services should take the current situation of the user into account to tailor the service behaviour to that situation. For example, when a user receives a telephone call but his situation is such that disturbance by audible signals would be inappropriate, his phone should vibrate rather than ring.

A way to enable unobtrusive services is context-aware computing. Context-aware applications use, besides explicit user inputs, context information to adapt the

---

[*] This work is part of the Freeband AWARENESS Project. Freeband is sponsored by the Dutch government under contract BSIK 03025. (http://awareness.freeband.nl).

G. Kortuem et al. (Eds.): EuroSSC 2007, LNCS 4793, pp. 223–238, 2007.

application behaviour to the situation at hand. Context is defined as any information that characterizes the situation of an entity [1] (e.g. user location, availability, weather conditions).

Context information is provided by so-called context sources. These context sources are software entities distributed in the user's environment that acquire context information and make it available to context-aware applications. For a context-aware application to use context information, it has to associate with a suitable context source that can provide the required context information. The association between a context-aware application and a context source that can provide the required context information is called a context binding.

Context sources exhibit certain characteristics that make developing context bindings complex: (i) context information can be offered by a multitude of physically distributed context sources. Problems that arise are how to discover relevant context sources and how to retrieve context information from these (remote) context sources, (ii) (similar) context sources can be provided by different context providers using different data models for storing and accessing context information. Problems that arise are how-to interoperate between context sources and their discovery mechanisms and (iii) context sources are dynamic. Firstly, they can appear and disappear at arbitrary moments (i.e. dynamic availability). Secondly, their quality, which is called Quality of Context (QoC) [2], can vary in time or it can be different from other context sources.

To facilitate the development of context-aware applications, we propose the *Context Binding Transparency* that facilitates the process of developing context-aware applications by simplifying the creation and maintenance of context bindings. Our Context Binding Transparency includes the following three main elements:

1. A *context binding description language* that enables developers to specify their context requirement at an abstract level rather then directly programming them.
2. A *context binding mechanism* that, based on a context requirement specification, creates and maintains context bindings, thereby hiding the distribution, heterogeneity and especially the dynamicity of context producers for the application developer.
3. A *context discovery interoperability mechanism* which hides the heterogeneity and dynamic availability of context discovery mechanisms.

In this paper, we focus on the first element of our proposed transparency: the *Context Binding Description Language* (CBDL). This language enables application developers to specify their context requirements at a high level of abstraction rather than in programming code, and thus to separate the transformation of context requirements into context binding logic from the development of the actual application logic. In this way, we try to decrease the development effort and facilitate maintenance and evolution of context-aware applications. Furthermore, we briefly discuss the second element (i.e. context binding mechanism), however, for details the reader is referred to [3, 4]. For more information on the third element of the proposed transparency (i.e. context discovery interoperability) see [5]. For more information on the overall AWARENESS project see [6].

The remainder of this paper is structured as follows: Section 2 gives a high-level overview of our proposed Context Binding Transparency. Section 3 identifies the requirements of the context binding description language (CBDL). Section 4 presents the design of CBDL. Section 5 discusses the usage of CBDL in the development process of context-aware applications, and it discusses the integration of CBDL with our context binding mechanism. Section 6 gives an example how to apply CBDL in developing a context-aware application and presents a generic reflection on the usability of CBDL. Section 7 discusses related work. Finally, in Section 8, we present a summary and future work.

2   Overview of the Context Binding Transparency

The transparency concept was introduced in the context of distributed system in the Open Distributed Processing (ODP) reference model [7]. Transparencies are offered by mechanisms that hide certain complexities for the application developer to simplify the development of the application at hand. For example, location transparency [7] hides the problems of locating distributed objects by enabling them to be found using logical names rather than physical addresses.

Our Context Binding Transparency hides certain complexities of developing a context binding. A context binding exists between a context consumer and a context producer (see Figure 1). A context consumer is typically a context-aware application, which consumes context information to be able to adapt its behaviour. A context producer is typically a context source, which acquires (produces) context information and makes it available to its environment. We propose to shift the recurring problem of creating and maintaining a context binding from the application to (context) middleware that offers a Context Binding Transparency. This transparency offers a context retrieval and publishing service used for easy exchange of context information. By using these services, the application developer of a context-aware application (context consumer) is unaware of the context producer with which a binding is created, how this binding is created and how this binding is maintained to overcome the dynamicity of context producers.

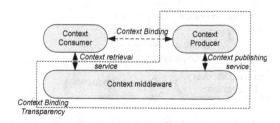

Fig. 1. Context Binding Transparency

Key features of our binding mechanisms offering the proposed Context Binding Transparency are:

- *Initialization:* based on the context requirement specification (expressed in CBDL) the context binding mechanism tries to resolve a context binding by discovering (using available underlying discovery mechanisms), selecting and binding to one ore more suitable context sources.
- *Maintenance:* based on specified criteria (e.g. QoC) the binding mechanism maintains the binding by:
  - o Re-binding at run-time to other suitable context sources when already bound context sources disappear.
  - o Re-bind at run-time to other suitable context sources when the QoC that is provided by the already bound context source may fall below a specified level.
  - o Re-bind to context sources with a higher QoC when they become available.
- *Releasing:* when the application no longer needs context information, the established bindings are released.

For a more elaborate discussion on the Context Binding Transparency, see [8].

3   Context Requirement Analysis

In this section, we discuss the requirements for our context binding description language (CBDL). The context requirement specifications, expressed in CBDL, are used by the binding mechanism to create and maintain context bindings. Thereby, the context binding mechanism has to bridge the gap between the requirements specified by the developers of context-aware applications and the (heterogeneous) context delivery capabilities of underlying context discovery mechanisms capable of discovering available context producers (see Figure 2).

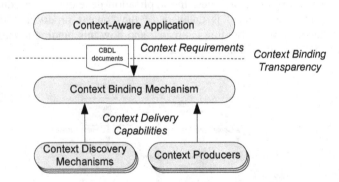

Fig. 2. Bridging the gap between context requirements and context delivery capabilities

We consider the following generic non-functional requirement in the design of CBDL:

- *Generality:* specification of context requirements in CBDL should not be restricted to specific application domains. CBDL should apply to a broad range of context-aware applications.
- *Usability:* specification of context requirements in CBDL should be easy and should not require a steep learning curve.

To capture the functional requirements of CBDL, we take a two-step approach. First, we analyse the capabilities of current context (discovery) middleware mechanisms to identify common capabilities currently offered (Section 3.1). Secondly, we analyse use-cases (from which we present two) to complement our requirements (Section 3.2). Together, these lead to requirements on what should be possible to express in CBDL to be able to capture context requirements used for creation and maintenance of context bindings by our underlying context binding mechanism (Section 3.3).

3.1 Analysis of Context Discovery Middleware

Currently several context middleware mechanisms are developed to facilitate the development of context-aware applications [9]. These mechanisms solve recurring development problems, such as dealing with privacy issues when exchanging context information, creating new context information by reasoning on existing context information, and discovery of distributed context sources. In this section, we analyse current context discovery mechanisms. First, because they implement solutions that fulfil context requirements application developers have and secondly, because our proposed Context Binding Transparency builds on top of these solutions.

We analyse nine different context discovery mechanisms. The first four originate from the Freeband AWARENESS project (CMF, CCS, CDF and Jexci) [10]. These mechanisms are developed for different domains (e.g. telecommunication operator domain, different administrative domains, ad-hoc situations) [10]. Secondly, we review the context discovery mechanism originating from the IST Amigo project (CMS) [11]. Thirdly, we complete our analysis with four external context middleware mechanisms (Context Toolkit [12], PACE [9], Solar [13], and JCAF[14]).

The analysis consisted of reviewing the following aspects of the different discovery mechanisms:

- *Interaction mechanism:* What interaction mechanism do the analyzed discovery mechanisms support?
- *Interaction data:* what type of information is expressed in the context discovery request and response?

The result of our analysis is presented in table 1. From the analysis, we distinguish the following common aspects provided by current context discovery mechanisms:

- All mechanisms support the common request-response and subscribe-notify interaction mechanism to retrieve context information.
- All mechanisms require information on the type of context and the entity to which the context relates, to discovery context sources.
- The majority of the mechanisms introduce the notion of quality of context in the request for context information.

- Some mechanisms require a form of security token (i.e. identity information on the entity that is requesting context) to be able to discover context sources.

Table 1. Comparing context discovery mechanisms

	Interaction mechansism		Interaction data				
Frameworks	Req-Resp	Sub-Not	Entity	Type	QoC	Sec. info	Format
CMF	v	v	v	v	v	v	RDF
CCS	v	v	v	v	v	v	SQL/PIDF
CDF	v	v	v	v	v	-	RDF/PIDF
Jexci	v	v	v	v	v	v	*Negotiable (PIDF/java objects)*
CMS	v	v	v	v	v	-	RDF
Context Toolkit	v	v	v	v	-	v	XML
Pace	v	v	v	v	v	-	Context Modelling Language
Solar	v	v	v	v	-	-	N/A
JCAF	v	v	v	v	-	-	Java objects

3.2 Analysis of Use-Cases

We complement the previous results by analysing use cases. Here we present two uses cases, which we consider representative for a broad range of context-aware application.

Healthcare Use-Case: Epilepsy Safety System (ESS)
The ESS monitors vital signs of epilepsy patients and determines upcoming epileptic seizures. When a likely seizure is detected, the system notifies nearby and available caregivers with instructions on the location (e.g. in lat/long context format) of the patient and route information to the patient. The application uses context information on the location of the patient and the caregiver and context information on availability of the caregivers to provide this functionality. The quality of the location data of the patient should have a minimal precision of 5m (i.e. the specified location of the patient may differ 5m from the actual location) to be able to dispatch caregivers to the right location. The location data of caregivers only has to be minimally 100m precise to be able to determine which one is nearby.

Additionally, the vital signs of the patient are transferred to the healthcare centre where care professionals monitor the patient's state and stays in contact with the dispatched caregiver. Context information on the available bandwidth (e.g. in kb/s) of the patient's device is used to tailor the granularity of transferred vital signs (e.g. increase or decrease sample frequency) and the amount of vital signs (e.g. decrease the number of send channels) to ensure transfer of vital signs to the healthcare centre.

Office Use-Case: My Idea Recorder (MIR)
During meetings, users can use their camera phones to take high-resolution pictures of whiteboard sketches to capture their ideas for future use. The MIR system distributes copies of these pictures to meeting participants. The phone automatically determines the persons that are currently in the meeting based on meeting information (e.g. in Boolean context format) from user's calendars and nearby Bluetooth devices. When the meeting information is not at least 75% correct (i.e. probability of 75% that the participant is actually in/out a meeting), the application asks the participant if he is in

the meeting. The system delays the data transfer until an adequate network becomes available (i.e. GPRS, UMTS, WLAN or Bluetooth) taking into account the cost and bandwidth characteristics of each network type and the battery status of her phone.

Discussion
We analysed multiple use-cases, from which we consider the previously discussed two, representative for a broad range of context-aware applications. From these use-cases, we derive the following characteristics of context and context-aware applications:

- Context is defined by its *context type* (e.g. location, availability, bandwidth, meeting status).
- Context is always related to a *context entity* (e.g. patient, doctor, voluntary care giver, meeting participant).
- Context information can be offered in different *context formats* (e.g. lat/long, xyz, nmea, Boolean).
- Relevancy of context information for applications can depend on different QoC criteria (e.g. precision, probability of correctness). See also [2, 15].
- Context transfer might occur during the whole life-span of the application or during a limited period (e.g. during a seizure).
- Delivery costs resulting from using context (e.g. use of a certain communication mechanism, commercial value of context) might pose criteria for the suitability of context bindings.

3.3 Overall Conclusions and Identification of CBDL Requirements

Based on the analysis of current discovery mechanisms and use cases, we identify the following requirements for CBDL:

- *Basic context elements:* Context type, entity and format are basic elements needed in CBDL to describe context requirements.
- *QoC criteria:* Application have QoC requirements and may react differently when these QoC are not met. Therefore, CBDL should enable application developers to specify quality levels on the required context information.
- *Costs:* Additionally to QoC, context delivery costs pose additional criteria on the suitability of a context binding. Application developers should be able to specify in CBDL cost criteria related to QoC criteria.
- *Binding characteristics:* Transfer of context information can be continuous during the life span of the application or can be limited to a certain period in the life span of the application. Context bindings are therefore not always required. An application developer should be able to specify in CBDL the characteristics of the required binding. This includes re-binding strategy (in case of losing a bound context source) and scope of the discovery. Furthermore, they should be able to specify if re-binding is necessary in case a QoC level cannot be maintained or better quality context sources may appear.

- *Notification:* Although our transparency strives for continuous availability of high quality context information, this might not always be possible. Application developers have to be able to specify in CBDL a notification strategy in case a lost binding cannot be recovered or QoC level cannot maintained, such that the context-aware application can adapt its behaviour to these situations.

4   Design of the Context Binding Description Language

We distinguish three types of information in a CBDL document:

- *Context specification:* basic information on what context information the context-aware application requires.
- *Quality criteria:* information on the quality levels which are acceptable for the context-aware application to function.
- *Binding options:* configuration information required to control the discovery, selection, binding, and maintenance process of a context binding.

These categories are represented in the UML meta-model of the CBDL language as depicted in Figure 3.

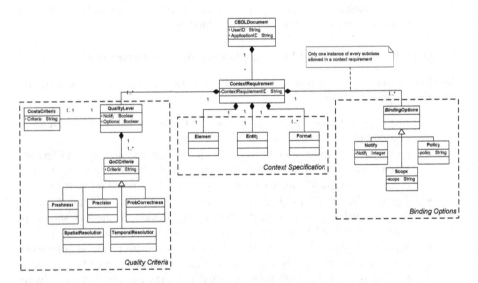

Fig. 3. CBDL language meta-model

The root of the CBDL language is the *CBDLDocument* element, which specifies which user is requesting a context binding (*UserID*) and to which application this binding belongs (*ApplicationID*). This information can be used as security information (e.g. identity to retrieve a security token) to be able to invoke underlying context discovery mechanisms. Furthermore, a CBDL document (*CBDLDocument*) enables application developers to specify multiple context requirements (*ContextRequirement*).

These requirements have to be uniquely identified by an ID (*ContextRequirementID*). This ID can be used to retrieve a handle on the established binding, to enable the context-aware application to retrieve context information.

Every context requirement (*ContextRequirement*) consists of mandatory context specification information. This information specifies: (i) a single type of context information that the application requires (*Element*), (ii) one entity to which the required context is related (*Entity*) and (iii) zero or more data formats the required context may have (*Format*).

Optionally, an application developer can specify multiple quality levels (*QualityLevel*). These quality levels consist of one or more quality criteria coupled with an optional cost criterion. We distinguish five possible types of QoC criteria based on [2, 15]. These are: (i) *Precision*: "granularity with which context information describes a real world situation", (ii) *Freshness:* "the time that elapses between the determination of context information and its delivery to a requester", (iii) *Temporal Resolution*: "the period of time to which a single instance of context information is applicable", (iv) *Spatial Resolution*: "the precision with which the physical area, to which an instance of context information is applicable, is expressed" and (v) *Probability of Correctness*: "the probability that an instance of context accurately represents the corresponding real world situation, as assessed by the context source, at the time it was determined" [15].

Additionally, the application developer may specify if the application needs to be notified when the QoC/Costs of the delivered context information comes into the range of the specified level or falls out of the range (*Notify*, default= true). Furthermore, the application developer specifies if the re-binding mechanisms needs to be triggered when the QoC of the delivered context information falls below the specified QoC level (*Optional*, default=false).

Furthermore, an application developer can optionally specify binding options (*BindingOptions*) to control the binding process of the context binding mechanisms. The following options can be specified:

- *Notify*: the application developer can specify the level of notification he wants to receive on the binding process. The following cumulative levels are identified:
  - *0*: no notifications.
  - *1*: notification when a binding is established.
  - 2: notification when a binding is established and broken.
  - *3*: notification when a binding is (re-)establishing and broken (default).
- *Policy:* the application developer can specify what binding policy should be taken:
  - *Static:* when a binding is broken, no re-binding is necessary.
  - *Dynamic:* when a binding is broken re-binding is necessary (default).
- *Scope:* the application developer can specify if context sources should be searched only inside the scope of the local infrastructure (i.e. producers deployed inside the local application container) or also in external context discovery mechanisms (default = global).

5 Using CBDL and the Context Binding Mechanism

First, we present a general discussion on how to use CBDL in the development of context-aware applications (Section 5.1). Secondly, we describe how CBDL is integrated with our context binding mechanism (Section 5.2).

5.1 Using CBDL for the Development of Context-Aware Applications

Figure 4 presents the development trajectory of a CBDL based context-aware application using our underlying context binding mechanism, called Context-Aware Component Infrastructure (CACI). CACI is the implementation of the context binding mechanism exposing the proposed Context Binding Transparency.

On design-time, the application developer creates the application logic of the context-aware application. Furthermore, he specifies the context requirements relevant for its application in a CBDL document. During the design of the application logic, the application developer has to take the following aspects in mind:

- Create application logic that is able to retrieve context using the interfaces offered by the context binding mechanisms and the context requirement identifiers (*ContextRequirementID*) specified in the CBDL document.
- Create application logic that can receive notification by the underlying binding mechanisms of changes in QoC and binding status based on the notify flags (*notify*) specified in the CBDL document.
- Create application logic that can adapt to unavailability of context or availability of context with too low quality.

Both the application logic and the CBDL document are bundled into a context-aware application component, which can be deployed in the Context-Aware Component Infrastructure (CACI).

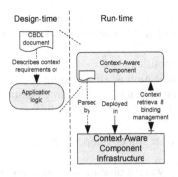

Fig. 4. Development trajectory of CBDL-based context-aware applications

5.2 Integration of CBDL and CACI

Figure 5 presents a functional decomposition of the binding mechanism deployed in the CACI infrastructure. After deployment of the context-aware application component, binding requests are extracted from the CBDL document by the parser. These request are transformed in a discovery request forwarded to available context discovery mechanisms (see [5]). The discovery results are analyzed and a context producer that can fulfil the context requirement (i.e. binding request) is selected. The selected context producer is bound to an internally created context producer proxy (see [3, 8]) from which the context-aware application component can retrieve context information. This proxy is monitored for disappearing of its bound physical context producer. In case of a lost binding to a context source, this triggers a re-binding processes starting from discovery of suitable context producers for a new binding. Furthermore, some context discovery mechanisms offer active discovery, which enable the binding mechanisms to subscribe to discovery changes (e.g. new producers become available), this triggers a new selection process to determine if the new producer is more suitable for the context-aware application component. Status information on the binding can be notified to the application component based on the flags in the CBDL document.

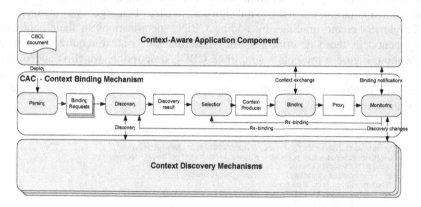

Fig. 5. Functional decomposition of the CACI Context Binding Mechanisms

We chose to represent the CBDL language using XML, as it is currently the de-facto standard for structured data. Consequently, tool support for creating CBDL document is widely available. Furthermore, usage of XML enables easy parsing and validation of the correctness of CBDL documents using for example XML Schema. Therefore, we derived a XML Schema of the CBDL language meta-model. The proof-of-concept of the CACI infrastructure is implemented using java and the OSGi component framework (see [4]). Context-aware components are OSGi components, which have in their component descriptor a pointer to the XML-based CBDL document.

6 Case Example and Reflection

Here we present an example on how to use CBDL for describing context requirements for the Epilepsy Safety System (Section 6.1). Furthermore, we present a general reflection on the usability of CBDL (Section 6.2).

6.1 Case Example: ESS

Let's reconsider the Epilepsy Safety System discussed in section 3.2. The ESS deploys a sensor system on the patient's body (called a Body Area Network (BAN)) which collects and transfers vital signs when a seizure is detected. This data is stored and analyzed in healthcare centres for diagnosis, first aid and treatment. In case a seizure is detected, caregivers dispatched by the healthcare centres may offer help to the patient in this life-threatening situation.

Amongst others, possible beneficial context types in the ESS are: patient and caregiver location, caregiver availability and patient BAN bandwidth usage. Location information helps to decrease travelling time to the patient in case of emergencies. First, because the precise location of the patient (destination) is known and second because a nearby caregiver can be dispatched to the patient. Availability information of caregivers helps to decrease false dispatches of unavailable caregivers. Bandwidth usage information assists to tailor the transferred vital sign data to decrease costs in case a of non-emergency situation, while this information also assists to prevent congestion and failing transfer of vital sign data in case of emergency situations.

Developers create bindings by adding CBDL specifications to their application components. In these descriptions, they describe the context requirements of the application. Figure 6 presents a simplified CBDL description, which is added to the ESS component at the health-care centre to create the binding to location and availability producers. For the other components at the patient and caregiver side, the descriptions are similar.

```xml
<?xml version="1.0" encoding="UTF-8"?>
<CBDLDocument xmlns:xsi="http://www.w3.org/2001/XMLSchema-instance"
xsi:noNamespaceSchemaLocation="CBDL-schema.xsd" UserID="Healthcarecentre"
ApplicationID="ESS_Healthcarecentre">
    <ContextRequirement BindingID="patient_location">
        <Element>Location</Element>
        <Entity>Patient.Tim</Entity>
        <Format>lat/long</Format>
        <QualityLevel>
            <QoCCriteria>
                <Precision>5m</Precision>
            </QoCCriteria>
        </QualityLevel>
    </ContextRequirement>
    <ContextRequirement BindingID="patient_bandwidth">
        <Element>Bandwidth</Element>
        <Entity>Patient.Tim</Entity>
        <Format>kb/s</Format>
    </ContextRequirement>
    <ContextRequirement BindingID="caregiver_location">
        <Element>Location</Element>
        <Entity>Caregiver.John</Entity>
        <Format>lat/long</Format>
        <QualityLevel>
            <QoCCriteria>
                <Precision>100m</Precision>
            </QoCCriteria>
        </QualityLevel>
    </ContextRequirement>
    <ContextRequirement BindingID="caregiver_availability">
        <Element>Availability</Element>
        <Entity>Caregiver.John</Entity>
        <Format>boolean</Format>
    </ContextRequirement>
</CBDLDocument>
```

Fig. 6. Example of CBDL document specifying context requirement of part of the ESS

The CBDL documents are handled by the CACI binding mechanism. The application developer only needs to retrieve the bound producer (see figure 7) by subscribing a call-back to the IContextProducerManager service (i.e. the local services mechanism is provided by OSGi) offered by CACI. This notification strategy is applied to cope with timing differences between the application and the binding performed by CACI. CACI notifies the component when a producer is bound.

```
// standard OSGi code to retreive the CACI service
ServiceReference ref = bc_.getServiceReference(IContextProducerManager.class.getName());
IContextProducerManager manager = (IContextProducerManager)bc_.getService(ref);
// Retreival of the bound context producers (id's correspond with the CBDL document from
fig6)
IContextProducerCallback cb = new Callback(this);
try{
    manager.subscribe("patient_location", cb);
    manager.subscribe("patient_bandwidth", cb);
    manager.subscribe("caregiver_location ", cb);
    manager.subscribe("caregiver_availability", cb);
} catch(ConsumerSubscribeException e){
    System.out.println("Wrong binding ID.");
}
```

Fig. 7. Retrieval of context bindings

6.2 Reflection

Usability of CBDL depends on three major factors (see Figure 8). The first is expressiveness; are context-aware application developer capable of expressing context requirements suitable for their applications. Secondly, learning curve; how difficult is it for the context-aware application developer to learn the CBDL language. Finally, performance; how does the introduced layer of indirection (i.e. transformation of CBDL specification to context bindings) perform.

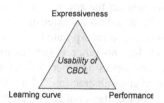

Fig. 8. CBDL usability triangle

- *Expressiveness:* By performing an extensive requirement analysis thereby reviewing current context management mechanisms and analyzing use cases, we created a language capable of specifying a broad range of context-aware applications. Furthermore, we added support for QoC criteria levels and binding process control.
- *Learning curve:* The application developers are required to learn how-to specify context requirements using CBDL and how-to use the CACI infrastructure. We do not think this presents a serious drawback, for two reasons. First, CACI provides simple interfaces to use the established bindings Furthermore, CBDL uses XML to express context requirements. XML

schemas are provided to ease this process. Future extensions could include a GUI that can enable developers to graphically generate CBDL descriptions and CACI integration code. Second, CA application development without CBDL and CACI also requires similar learning efforts to cope with the underlying discovery mechanism.

- *Performance:* Another possible drawback of adding a layer of indirection by CACI is its performance penalty. We performed some initial measurements on the time spent deploying a component, parsing the CBDL, discovery of context producers (making sure a discovery match can be made), selection of a context producer and returning the selected producer to the deployed component. This resulted both on the PC and the windows mobile device in insignificant overhead (less than 1ms time spent). Although more performance measurements are needed, our preliminary conclusion is that the delay introduced by CACI is considerably less or can be neglected compared to the delay for the (remote) discovery of context producers (i.e. communication and processing delay).

7 Related Work

In this paper, we propose a language to specify context requirements, which can be interpreted by our CACI infrastructure [3-5], to create and maintain context bindings in dynamic environments. The importance of coping with the dynamicity of (context) bindings in the infrastructure has also been recognized by others, who proposed several mechanisms for this purpose, such as context-sensitive bindings [16], service-oriented network sockets [17] and OSGi (Extended) Service Binder [18, 19]. Compared to CACI, these mechanisms have a similar goal but are not tailored to more advanced context-aware applications. Context producers and consumers have distinct characteristics that have to be incorporated in the binding mechanism to be able to fully support the application developer. For example, context binding mechanisms should be based on an extensible context model and the notion of quality of context (QoC) should be incorporated in the mechanisms.

To the best of our knowledge, no other initiatives exist to develop a language, which enables application developers to specify requirements for bindings with context producers at a high level of abstraction. Although Hong [20] recognizes the need for such a language, coined the context specification language (CSL), this language has not been detailed.

On the other hand, several types of languages have been proposed for other purposes, facilitating the development of context-aware applications in different ways. For example, Chan et al. [21] define a mathematical rule-based context request language. This language, implemented in XML, enables developers to specify context reasoning rules, using predicate calculus, interpreted by an infrastructure inference engine to retrieve required context information. Yau et al. [22] define a Situation-Aware object interface definition language (SA-IDL), which can be used to generate base classes for a situation-aware object. Etter et al. [23] describe a rule-based approach to specify context-aware behaviour in the ECA-DL language and to delegate the execution of this behaviour to the infrastructure using Event-Condition-Action rules. Robinson et al. [24] describe the Context Modelling Language (CML) which can be used to capture context information requirements to be used in the design of context-aware applications. Chen [25] discusses a context ontology (SOUPA) that can be used to exchange context among entities in a uniform manner.

8 Summary and Future Work

In this paper, we discuss the Context Binding Description Language (CBDL). This language enables application developers of context-aware applications to specify their context requirements at a high level of abstraction rather than at the programming code level. CBDL thus enables a separation between the development of the application logic and the development of context bindings. The responsibility for creating and maintaining context bindings is shifted to our Context-Aware Component Infrastructure (CACI), which can interpret context requirements and use these to drive its discovery and binding mechanisms. In this way, we try to decrease the development effort and facilitate maintenance and evolution of context-aware applications.

The requirements for CBDL are derived from an analysis of current context management systems and future use scenarios. Elements incorporated in the CBDL language support (i) specification of context, (ii) specification of quality criteria, and (iii) specification of binding control information. We implemented the language using XML and integrated it with our CACI infrastructure. We believe that the CACI infrastructure offers a useful new transparency, which we call the Context Binding Transparency.

We plan the following research activities to further improve the CBDL concept and CACI prototype:

- Extending the CBDL language to support the development of applications with context producer capabilities or both context consumer and producer capabilities.
- Further evaluation of usability of the Context Binding Transparency featuring the CBDL language for development of context-aware applications.

References

1. Dey, A.: Providing Architectural Support for Context-Aware applications, Georgia Institute of Technology (2000)
2. Buchholz, T., Kupper, A., Schiffers, M.: Quality of Context: What it is and why we need it. In: 10th Workshop of the HP OpenView University Association (HPOVUA 2003), Geneva, Switzerland (2003)
3. Broens, T., Halteren, A., Sinderen, M.v.: Infrastructural Support for Dynamic Context Bindings. In: Havinga, P., Lijding, M., Meratnia, N., Wegdam, M. (eds.) EuroSSC 2006. LNCS, vol. 4272, Springer, Heidelberg (2006)
4. Broens, T., et al.: Dynamic Context Bindings in Pervasive Middleware. In: Middleware Support for Pervasive Computing Workshop (PerWare 2007) White Plains, USA (2007)
5. Broens, T., Poortinga, R., Aarts, J.: Interoperating Context Discovery Mechanisms. In: 1st Workshop on Architectures, Concepts and Technologies for Service Oriented Computing (ACT4SOC 2007), Barcelona, Spain (2007)
6. Sinderen, M.v., et al.: Supporting Context-aware Mobile Applications: an Infrastructure Approach. IEEE Communications Magazine 44(9), 96–104 (2006)

7. Blair, G., Stefani, J.: Open Distributed Processing and Multimedia. Addison-Wesley, Reading (1998)
8. Broens, T., Quartel, D., Sinderen., M.v.: Towards a Context Binding Transparency. In: Broens, T. (ed.) 13th EUNICE Open European Summer School, Enschede, the Netherlands. LNCS, vol. 4606, Springer, Heidelberg (2007)
9. Henricksen, K., et al.: Middleware for Distributed Context-Aware Systems. In: DOA 2005, Agia Napa, Cyprus, Springer, Heidelberg (2005)
10. Benz, H., et al.: Context Discovery and Exchange. In: Pawar, P., Brok, J. (eds.) Freeband AWARENESS Dn2.1, Freeband AWARENESS Dn2.1 (2006)
11. Ramparany, F., et al.: An Open Context Management Information Management Infrastructure. In: Intelligent Environments (IE 2007) Ulm, Germany (2007)
12. Dey, A.: The Context Toolkit: Aiding the Development of Context-Aware Applications. In: Workshop on Software Engineering for Wearable and Pervasive Computing, Limerick, Ireland (2000)
13. Chen, G., Kotz, D.: Solar: An open platform for context-aware mobile applications. In: International Conference on Pervasive Computing, Zurich, Zwitserland (2002)
14. Bardram, J.: The Java Context Awareness Framework (JCAF) - A Service Infrastructure and Programming Framework for Context-Aware Applications. In: Pervasive Computing, Munchen, Germany (2005)
15. Sheikh, K., Wegdam, M., Sinderen, M.v.: Middleware Support for Quality of Context in Pervasive Context-Aware Systems. In: PerWare 2007. IEEE International Workshop on Middleware Support for Pervasive Computing, New York, USA (2007)
16. Sen, R., Roman, G.: Context-Sensitive Binding, Flexible Programming Using Transparant Context Maintenance, in Technical Report WUCSE-2003-72. Technical Report WUCSE-2003-72, Washington University (2003)
17. Saif, U., Palusak, M.: Service-oriented Network Sockets. In: MobiSys 2003. International conference on mobile systems, applications and services, San Francisco, USA (2003)
18. Cervantas, H., Hall, R.: Autonomous Adaptation to Dynamic Availability Using a Service-Oriented Component Model. In: 26st International Conference on Software Engineering, Edinburgh, Scotland (2004)
19. Bottaro, A., Gerodolle, A.: Extended Service Binder: Dynamic Service Availability Management in Ambient Intelligence. In: FRCSS 2006 International Workshop on Future Research Challenges for Software and Services, Vienna, Austria (2006)
20. Hong, J.: The Context Fabric: An Infrastructure for Context-Aware Computing. In: CHI 2002. Doctoral Workshop, Human Factors in Computing Systems Minneapolis, USA (2002)
21. Chan, A., Wong, P., Chuang, S.N.: CRL: A Context-Aware Request Language for Mobile Computing. In: Cao, J., Yang, L.T., Guo, M., Lau, F. (eds.) ISPA 2004. LNCS, vol. 3358, Springer, Heidelberg (2004)
22. Yua, S., Wang, Y., Karim, F.: Development of Situation-Aware Application Software for Ubiquitous Computing Environments. In: COMPSAC 2002. International Software and Applications Conference, Oxford, England (2002)
23. Etter, R., Dockhorn Costa, P., Broens, T.: A Rule-Based Approach Towards Context-Aware User Notification Services. In: ICPS 2006. International Conference on Pervasive Services, Lyon, France (2006)
24. Robinson, R., Henricksen, K.: XCML: A runtime representation for the Context Modelling Language In: PerCom 2007. Pervasive Computing White Plains, USA (2007)
25. Chen, H., Finin, T., Joshi, A.: The SOUPA Ontology for Pervasive Computing. Ontologies for Agents: Theory and Experiences (2005)

Deployment Experience Toward Core Abstractions for Context Aware Applications

Matthias Finke[1], Michael Blackstock[2], and Rodger Lea[1]

[1] Media and Graphics Interdisciplinary Centre, University of British Columbia
FSC 3640 - 2424 Main Mall, Vancouver, B.C., Canada
[2] Department of Computer Science, University of British Columbia
201-2366 Main Mall, Vancouver, B.C., Canada
{martinf@ece, michael@cs, rodgerl@ece}.ubc.ca

Abstract. Despite progress in the development of context aware applications and supporting systems, there is still significant diversity in the models and abstractions they expose. This work describes an effort to gain a better understanding of the situation and develop a core set of abstractions by deploying several context aware applications, using a rapid prototyping platform. From this experience we propose and demonstrate a set of abstractions shown to be useful for a range of context aware applications. Combined with a survey and analysis reported elsewhere [1] we then provide an analysis toward providing a core set of abstractions that we argue can be used as the basis for modeling many context aware systems, including not only context, but other aspects such as entities, their relationships and associated events, services and content. We then provide several practical lessons learned from the use of our model and abstractions during analysis and our iterative platform development process.

1 Introduction

Despite significant experimentation and deployment of context aware platforms and applications over the last 15 years, there is surprisingly little agreement on core abstractions and models for such systems. Individual research groups have developed abstractions suited to their application or research target [2-5] and often built bespoke systems to implement these [6-12]. While there is some overlap in the models and abstractions they have developed, there is also significant diversity. In an attempt to understand this situation, and in particular to try and develop a core set of common abstractions for context aware applications, we have, over the last two years, taken a dual research approach. First, we have surveyed and analyzed a set of key ubicomp systems with a goal of identifying abstractions and models in support of context aware applications. Secondly, and in parallel, we have implemented and deployed a set of context aware applications using a rapid prototyping platform with a goal of using practical experience to design, experiment with and validate core abstractions and models suitable for a range of context aware services. We have reported on the survey and analysis elsewhere [1]. This paper reports on our experiences developing

G. Kortuem et al. (Eds.): EuroSSC 2007, LNCS 4793, pp. 239–254, 2007.

and deploying four context aware applications and the underlying evolution of our platform as we improve our abstractions and model.

Our work has been carried out within the framework of the Mobile Multimedia urban shared experience (MUSE) project, a multi-disciplinary research project focused on exploring mobile multi-media services suitable for an urban environment. A key aspect of this project is its use of context aware services and its focus on real-world deployments [13]. To date Mobile MUSE has explored a variety of services deployed using traditional carrier networks as well as experimental WiFi based infrastructure and WiFi enabled cell phones. These include location aware games [14, 15] context aware tourist guides [16], tagging and folksonomy applications [17] as well as local event support services such as location based film festival services. Although Mobile MUSE has a strong technology and deployment focus it also includes significant research on business and sociological aspects of context aware services [18]. Within Mobile MUSE, the work of the MAGIC lab at the University of British Columbia (UBC) has primarily been to explore advanced services that exploit broadband wireless networks. As a basis for this research, we have developed the MUSE context aware platform (MUSEcap) and deployed a variety of services across the UBC WiFi network, one of the largest campus WiFi networks in North America with over 1700 access points.

1.1 Background and Motivation

While there have been many research applications and systems developed for places such as tourist destinations [19], campuses [20], meeting rooms [21], homes [9], and hospitals [22], there has been little consensus on the high level abstractions exposed to context aware applications from supporting platforms. With the wide variety of research and commercial systems available, using the same system for all context aware application domains is not realistic. One environment may differ significantly from another in terms of the entities (people, places and things) and the capabilities such as context, and services available. Researchers have justifiably proposed and built systems deemed important for different context aware application types and paradigms.

That said, when the same systems are used in different situations, practitioners have shown it is possible to seamlessly move user tasks between domains allowing them to make use of resources there [11]. However, given the variety of systems available and their specialization for different domains, using the same system in all places is not realistic. To address this, practitioners have demonstrated that data and control level interoperability can be achieved using various techniques. For example, the use of an intermediary such as the Patch Panel [23] to transform control messages as they flow through the iROS Event Heap [3] has been shown to be useful in addressing control flow interoperability. Component oriented systems like Obje/SpeakEasy [5] and the Equip Component Toolkit [24] have shown that the use of a small, standard set of component interfaces with mobile code or the use of component properties can allow users to configure components to interoperate. Friday et al [25] demonstrated the provision of an abstraction layer on top of heterogeneous service architectures from within the infrastructure, while the ReMMoC system shows that it is possible to provide a generic service abstraction in device-side middleware [26] to address service interoperability. Henrickson et al. designed a model for context [4] as the basis for a context aware application supporting infrastructure [10].

Based the experience outlined in this paper and previous analysis [1] we believe that it is possible to express the run time environment of *any* context aware application or supporting system using a set of common abstractions. We can use these abstractions for analysis and design, and to provide an abstraction layer for not only heterogeneous context producers or service infrastructures, but on the environment's computing resources as a whole. Our immediate aim is to develop a common model that we can use as a base set of abstractions for our MUSEcap platform for a variety of context aware applications.

This paper provides 3 key contributions: (1) it proposes a common model and abstractions we believe are suitable for a variety of context aware applications and services, (2) it validates this model and abstractions using deployed context aware applications and (3) it offers practical lessons learned from real world deployment and several iterations of the underlying systems platform.

This paper is organized as follows: In section 2 we discuss two initial context aware applications/services we have developed and explore the abstractions they needed. In particular, we explain the evolution of the abstractions to support increased application functionality and our experiences balancing abstractions against domain specific services. In addition, we briefly outline the implementation of our architecture and how we supported our core abstractions. In section 3 we present two further prototype deployments we used to validate our initial model and explore its ability to support a range of context aware services. In section 4 we combine the results of our practical experience with our parallel survey and analysis of existing context aware platforms to propose a more generic common model for context aware services. We then relate this common model to our practical experiences and discuss the mapping between our initial abstractions and those supported by the common model. In section 5 we discuss some of the lessons we have learned during this practical investigation with a particular focus on two issues; the tension between core system abstractions and domain specific services and secondly on the drawback of a purely practical approach to exploring and developing common abstractions and underlying system models. Finally in section 6 we conclude and discuss future work.

2 Context Aware Model and Abstractions: Evolution

We began our practical experimentation in late 2005 with the development of a simple location aware game played by teams on campus. Our goal was to develop an initial platform for context aware services driven out of the application needs. During early 2006 we deployed the service and used our experiences to refine our model and evolve our implementation. We deployed our second context aware service in summer of 2006 again using our experiences to refine the model and improve the implementation. We adopted a standard web services architecture for our underlying platform to ensure rapid prototyping and ease of development. Below we discuss the two deployments and the platform development.

2.1 CatchBob! to the Fugitive

The Fugitive [27] is a mobile multi-user location based game that extends the functionality of CatchBob! [28]. The game is played on a university campus by a team of three using Tablet PCs. The team attempts to locate a virtual character (the Fugitive) that is initially hidden on a digital map displayed on each participant's Tablet PC. The playing field (digital map) on the Tablet PC shows every player's present position while providing visual cues to signal one's proximity to the Fugitive.

Fig. 1. The Fugitive user interface

The goal of the game involves two parts, a *catch* phase and a *chase* phase. In the catch phase, players physically move around the campus with their position being updated accordingly on their digital map. The objective is to trap the Fugitive by physically forming a triangle with the team members. When the triangle area has been reduced to a certain size by the team, the Fugitive becomes visible and starts to move to other locations on campus: the chase phase begins. In the chase phase, participants re-position themselves on the digital map to chase and trap the now visible, moving Fugitive by again forming an even smaller physical triangle than before. Map and ink messaging are available to enable communication in the game. Communications are augmented by auditory beeps to alert players of incoming messages from other teammates. Figure 1 shows a screenshot of the application installed on a Tablet PC.

2.1.1 Discussion

The Fugitive has a very simple set of system abstractions as shown in Figure 2. The core abstractions include the notion of users (with the Fugitive itself a special case) and context - primarily location information. A containment abstraction, the environment, was added to provide a framework for the overall application. Initially we considered associating communications, that is, ink messages with a particular user, but eventually placed this as a service associated with the environment since map annotations and ink messages are broadcast to all users in the environment.

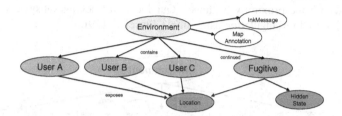

Fig. 2. The Fugitive environment model

2.2 PlaceMedia

PlaceMedia is a more ambitious system designed to explore how users can create and share media via the notion of location (or place). It allows users to define and maintain their personal profiles, to create and manage contact list of friends, to communicate with friends using instant messaging and to share media with friends. The physical location of all friends is visualized on a digital map as part of the user interface along with their profile and present status (i.e. online, busy, offline, etc.) as shown in Figure 3. Instant messaging enables friends to see each other's presence and to chat with each other while moving around campus.

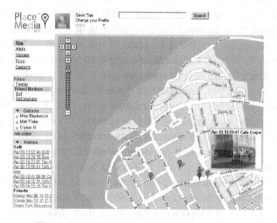

Fig. 3. PlaceMedia Tablet PC user interface

Users can create multimedia artifacts that are combined with context data such as location, creation time etc. This content can then be uploaded on a server to share with friends. When media is shared, icons representing the content on the digital map indicate the place where artifacts were created. One key feature of the PlaceMedia application is support for *Context Sensitive Alerts*. These alerts are triggered by a context variable, e.g. person or place, and can be used to build dynamic context aware applications. For example, with our prototype users can create an alert containing a multimedia message combined with a specific place (e.g. a coffee bar) and a proximity variable. Other users will get the alert when they approach the predefined location (i.e. within the proximity distance). Our location tracking subsystem supported either GPS or WiFi triangulation using Intel's PlaceLab [29].

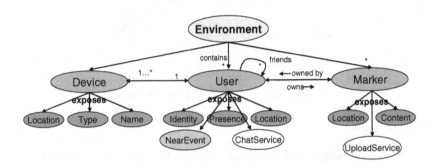

Fig. 4. PlaceMedia environment model

The abstractions exposed by the PlaceMedia prototype extended the simple Fugitive model as shown in Figure 4. The Fugitive game was designed for a specific device, the Tablet PC, and so there was an implicit mapping between the device and the user. In the more generic PlaceMedia system, we aimed to support multiple devices for a given user, and so we introduced *Devices* as a core abstraction in our model and provided support for a one-to-many relationship between *User* and *Device* entities. *Device*s now expose their location context rather than the end user. Where the Fugitive game structure defined a fixed number of users, in PlaceMedia we support an unlimited number of users. The PlaceMedia application required several new types of context such as device location, type and name, user identity and presence. In addition to these simple context types, each user in PlaceMedia has a friends list or a *roster* to relate friends to one another. Communications in PlaceMedia is now supported through a *Chat* service associated with a *User* entity. Messages are directed to a single user, rather than broadcast to all users in the environment as in the Fugitive.

The notion of *place* was another important abstraction introduced into PlaceMedia so that content like photos, videos and text could be left by users at particular places. To support these content-enhanced places, we introduced a *Marker* abstraction that not only exposes its location context but also allows users to associate multimedia artifacts such as pictures or video recordings with these Markers. To support context-sensitive alerts, an event abstraction called a *NearEvent* was added to the model. This event, associated with a user, supports the subscription of events that are fired when a user is within a specified range of another user, or a Marker. Finally, we added two services to facilitate direct user communications and to upload marker content. To reflect that Chat is between two users, the Chat services are associated with users involved in communications. A user can *retrieve* messages exchanged with another user, and *send* messages to another user. To send messages to many users, they must be sent individually in contrast to the broadcast model in the Fugitive.

2.2.1 Discussion

As can be seen from Figure 4, the core abstractions of PlaceMedia have been expanded and are more generic than the original Fugitive model. The types of entities and our support for context has also been expanded to include device, user and marker, position, identity, type, and content. The Marker entity is interesting in that it elevates location to a first class object. Typically in context aware systems, location is a key context attribute associated with users or other objects. However, the PlaceMedia application forced us to rethink location – in some cases it is simply a context attribute like a longitude and latitude, but in others it constitutes a first class object in its own right. We found that the PlaceMedia application required friendship and ownership relationships between registered users and markers to more easily find other relevant entities in the system for display. Another key abstraction that PlaceMedia required was the notion of context sensitive alerts which we generalized through an event system.

2.3 MUSE Context Aware Platform Implementation (MUSEcap)

MUSEcap was developed using the JBoss [30] Java 2 Platform Enterprise Edition (J2EE) [31] application server following a classic three tier system architecture; a simplified system diagram is shown in Figure 5. As is typical in three tier architectures, the top tier is for presentation and user interface, the middle tier for functional processing logic, and the bottom tier for data access. This architecture allows the user interface, application logic and database to change independently without affecting the other tiers.

To map the PlaceMedia system model (see section 2.2) to this architecture, we used standard entity-relationship modeling techniques[1] for the persistent data in our system. This included database tables for *Markers, Users, Devices*, and the relationships between them such as *friendship* and *ownership* (see Data Tier in Figure 5). We then refactored the single application-specific PlaceMedia interface into several, more general purpose interfaces in the logic tier, with each corresponding to an entity in the system such as Users, Device and Markers for easier reuse in subsequent applications. These interfaces provide functions to *find, add* and *remove* the entities they handle (e.g. Users and Devices). Methods are also provided to *get* and *set* context values such as location and presence, to call the associated service methods such as sending and retrieving messages, and to subscribe to events. The SessionBean logic accesses objects in the data tier wrapped using J2EE EntityBean interfaces. In the presentation tier we provided servlets, HTML and Java Server Pages (JSP) and a J2ME based application for access using a Tablet PCs, handheld PCs, and mobile phones.

---

[1] http://en.wikipedia.org/wiki/Entity-relationship_diagram

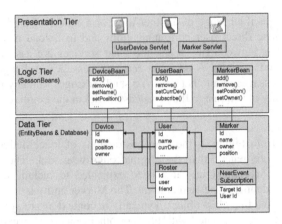

Fig. 5. Simplified MUSEcap architecture

3   Validating the Model Through Additional Prototypes

To evaluate MUSEcap and its associated abstractions and model we developed two additional applications. The first application developed and deployed in late 2006 focused on the notion of a *Tour* while the second application called *MoCoVo* supports mobile social networking and was deployed in spring of 2007.

3.1   Tour Prototype

The Tour prototype supports the notion of tours: a set of connected locations, organized around a theme and employing media and context data to guide people from location to location. This prototype was created as a logical extension of the Place-Media application and as another instantiation of the canonical 'tour guide' context aware application [19].

We created a mobile phone application using Java Micro Edition (J2ME) [32] that automatically records location and content captured by the end user over time. The idea is to allow friends to create tours for each other using mobile phones. Users can see these tours on the screen, download them to their mobile phones, and then "play back" these downloaded tours when walking around later. This could be extended to support hikers, cyclists or tourists who carry a mobile device. The application tracks their location and allows them to annotate places with media such as photographs, video clips and audio commentaries.

When the trip is captured, the user can upload the tour data to the platform. A digital map visualization is created on a web page that shows the locations and content recorded during the tour. Icons are placed on the map along the tour path to indicate where multimedia artifacts were created. A simple web-based editing tool allows tour creators to edit, add locations and media after the basic location data has been captured. Finally, any tour can be published and can be downloaded to a mobile device and followed by others. Friends can visualize tours, and download them to their mobile phone to follow a tour. During tour playback on a mobile phone, location is again tracked to trigger play back of media when a user is within a defined distance of a tour location. In addition users can add media to an existing tour thus building up a richer tour narrative over time.

Fig. 6. Mobile phone tour prototype user interface. The tour capture interface is shown along the top row, and the tour playback along the bottom row.

3.1.1 Discussion

The purpose of our Tour prototype is to attempt to reuse the core abstractions developed for the Fugitive and PlaceMedia and implemented in MUSEcap in an effort to assess how useful these abstractions were in building additional applications. Despite the tour prototype's additional requirements, we found that the abstractions provided by PlaceMedia did facilitate reuse. The tour application made use of users, markers, and associated context including location and content provided by the MUSEcap platform, requiring no changes to the underlying model or abstractions. However, the Tour application did require new application-specific code for grouping Markers into tours, adding support for paths between markers, and the association of content with these paths, and services for uploading and downloading tours.

3.2 Mobile Comments and Voting (MoCoVo) Prototype

MoCoVo is a social networking application that allows groups of users to share media via their phones and to network based on the media. One of the major objectives of the application idea is the creation and sharing of new multimedia artifacts paired with group-based communication support. Users can easily create a new group and invite friends to join it. Such a group has a more of dynamic character than our friends roster in PlaceMedia and is designed to support dynamic ad-hoc groupings such as special interest groups organized around events or locations. For instance, a group of people visiting "New York" could create a group that shares pictures (including location context) and comments about the city – obviously the grouping is dynamic and time limited to the visit.

Fig. 7. Mobile phone interface for the MoCoVo application. The first row illustrates taking and commenting on a picture, the second row browsing, and voting on a picture.

Once part of a group, users are able to create media, such as pictures, using their mobile phone and write comments or tag them before uploading to the server for sharing within the group. Once shared on the server any group member can access the media and download it back on their phones. While browsing pictures or other media, group members have the ability to comment and vote on each media clip. Current vote results are presented with media while browsing. Furthermore, each member of the group can provide comments/tags that will be associated with the pictures and can be accessed by others.

3.2.1 Discussion

Despite the different application domain, we were able to reuse several facilities of MUSEcap - specifically user management, and content uploading, in the development of MoCoVo. We decided to refactor the platform to support Content as a separate entity in itself, associated with either Users or Markers, recognizing that content may be associated with other entities in the environment, not just Markers. Group entities were also added to the application to contain multiple users. We added support for ranking, comments and tags context types associated with Content. While there were some changes made to the platform to support MoCoVo, most of the development work of this prototype was spent on the client application and only limited work on the server side implementation was needed.

While the Tour and MoCoVo application highlighted the fact that our core model and abstractions were adaptable to a range of context aware applications we realized that to go beyond this class of applications, we would need to extend and generalize our model further. We must be able to easily add new reusable entity types such as social groups, and new capabilities such as a "ranking" context, voting services and text comments that can be reused by new classes of social networking applications. To accomplish this we found the need to generalize our context aware application models further. We describe our analysis toward these general abstractions in the following section.

4 Analysis

During the development of our context aware applications, we found that we could generalize our abstractions further for greater reuse during analysis and in future platform iterations. The Fugitive, for example, can be modeled as a game hosting a set of players and the Fugitive itself. If we generalize a game to an *environment*, and the users and fugitive as *entities*, the Fugitive environment hosts several entities: the players of the game (users), and the Fugitive, a virtual agent. To generalize further, an *environment* hosts *entities*.

The user entities in the Fugitive expose context: their location. The agent entity has two types of context: its hiding state, and current location. In general, entities like users *expose* different forms of *context*. Finally, the Fugitive platform provides several *services*. In the case of the Fugitive, these are most obviously associated with the environment as a whole: a map annotation service, and an ink chat service to facilitate group communications. To generalize, the *environment* entity *exposes* these *services*.

Placemedia extended the capabilities of the Fugitive. Like the Fugitive it included mobile user entities in a campus environment model, but we also introduced Device entities as shown in Figure 4. PlaceMedia users expose not only location, but a presence-state context associated to indicate their instant messaging status (on-line, away, busy), and information about registered users' identity. Marker entities were added to the system to mark places of interest on the campus. Markers have static location context and content associated with them. In PlaceMedia, Devices, and Markers entities are owned by specific users. That is, there is an ownership *entity relationship* between PlaceMedia users, and their Devices and Markers. Similarly, users can be friends with one another, in this case a social entity relationship using the roster functionality. PlaceMedia also introduced support for event notifications that are signaled when one user is near another, or near a Marker on the campus. Event support has been found to be useful in supporting follow-on applications such as the tour to notify the user when a tour marker is nearby. In general events are also a capability associated with an entity.

While the Tour application was able to reuse specific entities and exposed capabilities in MUSEcap, the MoCoVo application required more changes to the platform to satisfy its new requirements. For a platform to support MoCoVo, the Tour application, PlaceMedia and the Fugitive without changes to the system's API, we need to move toward a more general set of abstractions.

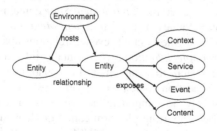

Fig. 8. Core abstractions of the Ubicomp Common Model

Based on the experience outlined in this paper and on an analysis of other ubiquitous computing systems [1] we have identified the abstractions common not only to PlaceMedia and follow on applications, but to a significant number of infrastructure-based context aware applications independent of the environment domain and infrastructure used. These abstractions represent a more generic model than those supported in MUSEcap and can be summarized as follows:

- **Environment Model** that encapsulates the current state of the environment including entities in the environment such as users and devices, the types of context and services components provide, and other aspects of the environment as a whole.
- **Entities** are base-level abstractions such as people, places, computing devices and other things, groups and activities. They can be specialized to environment-specific entities such as game players, living rooms, class rooms, mobile phones and meetings.
- **Entity Relationships** such as location, social, ownership, and activity-related relationships between people, places and things.
- **Context** associated with entities. Context information can include values such as location, status.
- **Services** or functionality associated with entities such as users, places and devices. Ubiquitous systems either provide their own service infrastructure or build on existing middleware systems.
- **Events** that can signal a change of state such as a person entering a room, a light turning on or a presence state change (online to offline).
- **Data or Content** related to an entity such as a place, user, or activity. This could include a user's personal photo, or an audio description of a location.

To summarize, all of the context aware applications outlined in this paper expose an environment model to the user, hosting the entities (people, places and things) that are relevant to the application. These entities have capabilities associated with them, where capabilities are the types of context, relationships, events, content or services they expose to the application. The abstractions and how they relate to one another in an environment model are illustrated in Figure 8.

The deployment of these applications and a survey of existing systems has informed the design of a general model for all ubiquitous computing environments called the Ubicomp Common Model [1]. The UCM not only describes entities and their capabilities as in our analysis, but three related aspects of a context aware environment. The first aspect, called the *Environment State* contains entity instances, their types, and current context values. The *Environment Meta-State* aspect contains entity types and their *capabilities*: the context, events, relationships and services they expose. The *Environment Implementation* aspect contains information about the components of an underlying system that implement the capabilities described in the Meta State. For example, if a location context is supported by User entities, a SessionBean in the PlaceMedia system is the component that will implement this context.

With the UCM and a supporting platform called the Ubicomp Integration Framework (UIF) we have begun to integrate the MUSEcap capabilities described here with other context aware systems [6, 24, 33] into a single environment model in an effort to evaluate our model's suitability for addressing interoperability and portability across smart spaces. The UIF has also been designed to support native context aware development, where the environment model and implementation components are hosted by the UIF itself, rather than an integrated system. Due to space limitations, for more information please refer to [1].

5 Lessons Learned

Based on our experience developing our applications and platform, and our subsequent analysis, we have derived some useful guidelines for context aware application and systems designers related to the use of our abstractions, and platform deployment experience summarized here.

Abstractions: Model applications and systems as environments. We have found that it is possible and useful to model all of our systems as an environment that hosts entities and their capabilities. We have found this technique to be valuable in determining how the requirements of applications supporting systems are similar or different since the model highlights the new entities and capabilities required by a context aware application. We have also found it to be valuable in designing an API for context aware applications that is independent of the underlying sensor and service implementations [1].

Abstractions: Associate capabilities with entities. We have found that it is possible to not only associate context with an entity as suggested by Dey et al. [34], but also the other capabilities of an underlying system, such as the services, events, and content. While the location of a user is obviously related to the user entity, the chat service and "near event" should also be associated with the users they serve. When a service is not obviously associated with one entity in particular, such as a user, place or device, it can be associated with the environment entity as a whole, as is in the case for the map annotation and ink chat facilities of the Fugitive.

Abstractions and Platform: Clearly differentiate application-specific services from reusable platform abstractions. During our iterative development of MUSE-cap we initially implemented a requirement as part of the PlaceMedia application, only later realizing that it was more appropriate as a reusable MUSEcap platform service. For example, the Marker entities were developed as part of the PlaceMedia abstraction, and then were found to be a useful abstraction for the Tour application. We recognized that the supporting platform, such as the enterprise application platform we used, should provide a general purpose and extensible facility for providing a range of reusable capabilities. This way, application-specific services can be migrated into the general purpose platform when they have been found to be reusable by more than one application.

Platform: Traditional three tier enterprise application environments are excellent rapid prototype platforms. There is a tendency in the context aware and ubiquitous computing (ubicomp) systems community to develop bespoke platforms in support of experimental context aware applications. Our experience has shown us that by using existing web services and enterprise application infrastructures (e.g. J2EE and JBoss) and standard 3 tier architectures, we are able to rapidly develop and more importantly, evolve and change, our underlying platform, MUSEcap. While we understand the eventual need for more sophisticated middleware systems, we were surprised by how far a more standard systems environment could carry us. Given their ease of development, reliability, sophisticated tool chains and rapid deployment support we believe that such development environments have a stronger role to play in the systems and middleware community.

Platform: Development of architectural models through practice needs to be augmented with survey and analysis. Our exploration of a context aware platform and common abstractions in support of context aware applications has reinforced the lesson that prototyping and real world deployment is necessary but not sufficient to develop generic system abstractions. The breadth and scope of the ubicomp space dictates that any group will struggle to prototype context aware applications which cover more than a small subset of the possible application domains. It is crucial therefore that when considering common models and abstractions, a comprehensive survey and analysis of existing systems and applications is carried out. While this lesson may be perhaps obvious it is surprising how many ubicomp systems are developed that provide little evidence of wide applicability.

6 Conclusions

From the deployment experience described here we have found that several context aware applications and their supporting infrastructure can be described by an environment model consisting of related entities (people, places, things), and their associated capabilities as described in our analysis. Furthermore, we have implemented this model using a standard three tier architecture. We have described the development of an evolving platform to deploy four context aware applications, and how the features of these systems can be categorized into one of seven high level abstractions: an environment model, entities, entity relationships, context, events, services and content.

With this work in parallel with an analysis and survey of other ubicomp systems [1] we have designed the Ubicomp Common Model (UCM), a comprehensive model for context aware computing environments. We have provided some guidelines into how to use this model for context aware systems analysis, and lessons learned based on our deployment experience. In future work we aim to evaluate the model and integration system further by integrating several existing ubicomp systems into a composite environment model.

References

1. Blackstock, M., Lea, R., Krasic, C.: Toward Wide Area Interaction with Ubiquitous Computing Environments. In: Havinga, P., Lijding, M., Meratnia, N., Wegdam, M. (eds.) EuroSSC 2006. LNCS, vol. 4272, Springer, Heidelberg (2006)
2. Dey, A.K.: Understanding and Using Context. Personal Ubiquitous Comput. 5, 4–7 (2001)
3. Johanson, B., Fox, A.: The Event Heap: A Coordination Infrastructure for Interactive Workspaces. In: Proceedings of the Fourth IEEE Workshop on Mobile Computing Systems and Applications, IEEE Computer Society, Los Alamitos (2002)
4. Henricksen, K., Indulska, J., Rakotonirainy, A.: Modeling Context Information in Pervasive Computing Systems. In: Proceedings of the First International Conference on Pervasive Computing, Springer, Heidelberg (2002)

5. Newman, M.W., Sedivy, J.Z., Neuwirth, C.M., Edwards, W.K., Hong, J.I., Izadi, S., Marcelo, K., Smith, T.F.: Challenge: Recombinant Computing and the Speakeasy Approach. In: Proceedings of Mobicom 2002 (2002)
6. Ponnekantia, S.R., Johanson, B., Kiciman, E., Fox, A.: Portability, extensibility and robustness in iROS. In: Proceedings of IEEE International Conference on Pervasive Computing and Communications, Dallas-Fort Wirth (2003)
7. Dey, A.K.: Providing Architectural Support for Building Context-Aware Applications. PhD Thesis. College of Computing, Georgia Institute of Technology (2000)
8. Román, M., Hess, C.K., Cerqueira, R., Ranganathan, A., Campbell, R.H., Nahrstedt, K.: Gaia: A Middleware Infrastructure to Enable Active Spaces. IEEE Pervasive Computing, 74–83 (2002)
9. Brumitt, B., Meyers, B., Krumm, J., Kern, A., Shafer, S.A.: EasyLiving: Technologies for Intelligent Environments. In: Proceedings of the 2nd international symposium on Handheld and Ubiquitous Computing, Springer, Bristol (2000)
10. Henricksen, K., Indulska, J.: A Software Engineering Framework for Context-Aware Pervasive Computing. In: PerCom 2004. Proceedings of the Second IEEE International Conference on Pervasive Computing and Communications, IEEE Computer Society, Los Alamitos (2004)
11. Sousa, J.P., Garlan, D.: Aura: an Architectural Framework for User Mobility in Ubiquitous Computing Environments. In: Proceedings of the 3rd IEEE/IFIP Conference on Software Architecture, Kluwer, B.V (2002)
12. Bardram, J.E.: The Java Context Awareness Framework (JCAF) - A Service Infrastructure and Programming Framework for Context-Aware Applications. In: Pervasive Computing: Third International Conference, Springer, Berlin (2005)
13. Mobile MUSE, http://www.mobilemuse.ca/
14. The Digital Dragon Boat Race (DDBR). Mobile MUSE (2005), http://www.mobilemuse.ca/projects/digital-dragon-boat-race
15. Jeffrey, P., Blackstock, M., Deutscher, M., Lea, R.: Creating Shared Experiences and Cultural Engagement through Location-Based Play. In: Computer Games and CSCW workshop at ECSCW 2005, Paris, France (2005)
16. The Re:call Project. Mobile MUSE (2005), http://www.mobilemuse.ca/projects/re-call-project
17. Metrocode. Mobile MUSE (2007), http://www.mobilemuse.ca/projects/metrocode
18. Smith, R.: Cell in the city: Is cellular phone use eroding the distinction between public and private space? In: Greenberg, J., Elliott, C. (eds.) Communications in question: Canadian perspectives on controversial issues in communication studies, Thomson-Nelson, Toronto, Canada (2007)
19. Cheverst, K., Davies, N., Friday, A., Mitchell, K.: Experiences of Developing and Deploying a Context-Aware Tourist Guide: The Lancaster GUIDE Project. In: Mobicom 2000, Boston, USA (2000)
20. Griswold, W.G., Shanahan, P., Brown, S.W., Boyer, R., Ratto, M., Shapiro, R.B., Truong, T.M.: ActiveCampus: Experiments in Community-Oriented Ubiquitous Computing, vol. 37. IEEE Computer Society Press, Los Alamitos (2004)
21. Johanson, B., Fox, A., Winograd, T.: The Interactive Workspaces Project: Experiences with Ubiquitous Computing Rooms. IEEE Pervasive Computing 1, 67–74 (2002)
22. Bardram, J.E., Hansen, T.R., Mogensen, M., Soegaard, M.: Experiences from Real-World Deployment of Context-Aware Technologies in a Hospital Environment. In: Dourish, P., Friday, A. (eds.) UbiComp 2006. LNCS, vol. 4206, pp. 369–386. Springer, Heidelberg (2006)

23. Ballagas, R., Szybalski, A., Fox, A.: Patch Panel: Enabling Control-Flow Interoperability in Ubicomp Environments. In: PerCom 2004 Second IEEE International Conference on Pervasive Computing and Communications, Orlando, Florida, USA (2004)

24. Greenhalgh, C., Izadi, S., Mathrick, J., Humble, J., Taylor, I.: ECT: a toolkit to support rapid construction of ubicomp environments. In: Davies, N., Mynatt, E.D., Siio, I. (eds.) UbiComp 2004. LNCS, vol. 3205, Springer, Heidelberg (2004)

25. Friday, A., Davies, N., Wallbank, N., Catterall, E., Pink, S.: Supporting service discovery, querying and interaction in ubiquitous computing environments. Wirel. Netw. 10, 631–641 (2004)

26. Grace, P., Blair, G.S., Samuel, S.: A reflective framework for discovery and interaction in heterogeneous mobile environments. SIGMOBILE Mob. Comput. Commun. Rev. 9, 2–14 (2005)

27. Jeffrey, P., Blackstock, M., Finke, M., Tang, T., Lea, R., Deutscher, M., Miyaoku, K.: Chasing the Fugitive on Campus: Designing a Location-based Game for Collaborative Play. Loading..Journal 1(1). Special Issue from Canadian Games Studies Association (CGSA) Workshop, vol. 1 (2006)

28. Nova, N., Girardin, F., Dillenbourg, P.: 'Location is not enough!': an Empirical Study of Location-Awareness in Mobile Collaboration. In: IEEE International Workshop on Wireless and Mobile Technologies in Education, Tokushima, Japan (2005)

29. Place Lab: A Privacy-observant location system. Intel Research Seattle, http://www.placelab.org/

30. JBoss Home Page (2006), http://www.jboss.com/

31. Java 2 Platform, Enterprise Edition (J2EE) Overview. Sun Microsystems, http://java.sun.com/j2ee/overview.html

32. Java Platform Micro Edition at a Glance, http://java.sun.com/javame/index.jsp

33. Salber, D., Dey, A.K., Abowd, G.D.: The context toolkit: aiding the development of context-enabled applications. In: Proceedings of the SIGCHI conference on Human factors in computing systems, ACM Press, Pittsburgh, Pennsylvania (1999)

34. Dey, A.K., Abowd, G.D.: Toward a Better Understanding of Context and Context-Awareness. Georgia Institute of Technology, College of Computing (1999)

Ambient Energy Scavenging for Sensor-Equipped RFID Tags in the Cold Chain

Christian Metzger, Florian Michahelles, and Elgar Fleisch

Information Management
ETH Zurich, Switzerland
{cmetzger, fmichahelles, efleisch}@ethz.ch

Abstract. The recent introduction of passive RFID tags into leading retailers' supply chains has had a tremendous impact on their ability to manage the flow of goods. Tags on cases and pallets increase the supply chain's visibility and allow for accurate tracking and tracing. Currently, additional efforts are being made to use RFID to actively monitor a product's shipping condition (e.g. temperature, shock-vibration, etc.). This is of substantial interest for perishable goods and pharmaceuticals, whose shipping conditions are regulated. For continuous tracking of environmental data during the process of the movement of goods through the supply chain, RFID tags need to be equipped with sensors. Such sensors require continuous power, which is usually supplied by a battery on the tag. However, a battery not only significantly increases the cost of the hardware and makes it heavy and bulky, but also limits a tag's lifetime. We propose ambient energy scavenging as a method to power sensors on battery-free RFID tags for continuous temperature monitoring and we show its applicability to the cold chain. Through detailed analysis of typical transport conditions we have identified ambient power sources which allow us to specify chip requirements and to make informed decisions about tag placement and total cost. We conclude that efficient monitoring ability is available at significantly lower cost than comparable implementations with active tags. Due to reduced costs, we predict high market penetration, which will result in more detailed information about multi-echelon supply chains. The fine-grained measurements will reveal failures and inefficiencies in the cold chain at a level of detail that would be hard to achieve with active tags. The elimination of the short-comings in the cold chain will result in reduced shrinkage, better quality and freshness of goods, and an overall reduction of losses of revenue.

Keywords: Shipping transportation, energy scavenging, RFID, sensors, monitoring.

1 Introduction

Roughly 20 percent of all perishable goods become unusable before they reach consumers. This amounts to a global loss of an estimated $20 billion per year [1], [2]. Perishable goods such as meat, poultry, dairy products, fish, vegetables, pharmaceuticals, etc. are commonly shipped in an environment-controlled supply

G. Kortuem et al. (Eds.): EuroSSC 2007, LNCS 4793, pp. 255–269, 2007.

chain – the cold chain. Legislation for various perishable products regulates the strict enforcement of shipping temperature and storage conditions [3], [4]. Monitoring equipment is not ubiquitous and the majority of key tasks in food production is still carried out manually. Additionally, producers often have their goods delivered by distributors and therefore cede control over product handling. The complexity of today's global supply networks, consisting of a number of logistics providers and several echelons along a supply chain, call for good control mechanisms and monitoring systems – both to certify a product's sound quality and to detect and locate failures in the supply chain. The localization of inefficiencies in the cold chain results in reduced temperature-related shrinkage and quality-related rejections. This has a direct impact on the loss of revenue caused by shrinkage.

According to European statistics on transportation, 72% of shipping is carried out by trucks [5]. Cargo trucks that are designed to transport cooled and frozen goods are already partly equipped with temperature monitoring devices. However, the data collected during transportation is not passed to the next echelon of the supply chain, along with the products. Additionally, the devices' capability to capture harmful situations is limited to the cargo area of the truck and excludes the processes of loading and unloading, or truck idling time. Therefore, it is desirable that each product monitors its transportation conditions and keeps a record of its product history.

Retailers seek to increase the visibility of their supply chains to allow tracking and tracing of all products in order to better manage the flow of goods and to certify the origin of a product. Recent advances in radio-frequency identification (RFID) technology have lead to considerable roll-outs of RFID in the retail industry, headed by retailers such as Wal-Mart, Metro, Target, and Tesco [6]. Since 2005, Wal-Mart has demanded that its top suppliers tag all of their products. Further, they are planning on making this a requirement for all suppliers. However, visibility through simple track-and-trace is insufficient to guarantee that a product's quality remains unaltered. It is necessary to monitor the shipping conditions in order to detect discrepancies from the prescribed environmental influences. Having located a deviation, one can intervene and modify the particular process. Monitoring and simple collection of temperature and time data can be accomplished by sensor-equipped RFID tags. Additionally, wireless data transmission does not require unloading of pallets at the delivery gate for data collection or retrieval, and therefore the flow of supply chain movements is not interrupted.

Various companies (KSW Microtech, Savi Technology, Intelliflex, etc.) have contributed to the problem of capturing data from the ambient environment with dedicated solutions; many of them involve active RFID tags that are equipped with monitoring sensors (e.g. temperature sensor, shock sensors, etc.). While sensor-equipped RFID tags allow for continuous data collection along with remote data reading, these tags require a battery to provide energy to the chip and the sensors. However, a battery not only limits the lifetime of a tag to the capacity of the battery but also significantly increases hardware and assembly costs. Additionally, a battery's capability to provide energy is further reduced in an environment with low air temperature such as that in the cold chain.

Even though expensive active tags may be feasible within a closed-loop system where tags on containers are reused repeatedly, it is cheap disposable tags that are required for an open-loop system. This is due to the fact that the ownership of a tag in an open-loop system changes while it moves through the supply chain, and in most cases, it is not cost-effective to return the tags.

In the following, we propose an energy harvesting system as a cheap replacement for batteries to power RFID chips and sensors that monitor the shipping environment. We present a detailed analysis of ambient energy sources in truck transportation and the design requirements for a monitoring system in the cold chain.

2 State of the Art

The broad range of RFID tags can be divided into three types with respect to their power supply: passive, semi-active, and active tags. Passive tags derive their energy from the electro-magnetic field of a reader while semi-active tags contain a battery that powers their logic parts. Semi-active tags can operate even if they are outside the reader's range, due to their battery. Active tags make use of their battery not only to power their logic parts but also to actively broadcast. This results in long reading distances [7].

The cost of passive tags has significantly decreased over the past years down to about 15-40 cents per tag [8] due to large scale integration into supply chains (estimated revenues of $1.54 billion in 2006 [9]). However, seamless monitoring of environmental conditions requires semi-active or active tags, but equipping a tag with a battery significantly increases the cost of the tag. Additionally, proper operation of a tag is limited to the battery's lifetime and disposal may be affected by governmental regulations [10]. A separation of the electronic parts from the battery may be required for disposal. Furthermore, a battery's form factor may exceed the unobtrusive attachment constraints for tag placement on pallets and boxes. Therefore, a power source is necessary that can be integrated directly into the chip.

We propose energy scavenging, a technique also referred to as energy harvesting or energy harnessing, as an alternative means of providing continuous power to tags. Energy scavenging allows the collection of ambient energy and its transformation into usable electrical energy to power sensors for periods when RF power is unavailable. Energy scavenging offers high integration into current RFID chips and provides energy from inexhaustible sources. Light (photo cells), motion (kinetic), pressure (from sound waves), deflection (piezoelectric), and temperature differences (Carnot effect) may serve as such ambient power sources. Preliminary research in this area has mainly focused on applications in wearable computing [11]-[13] and construction monitoring, but little has been done to analyze and specify the requirements for logistics transportation.

The power consumption of RFID chips has substantially decreased over the past years. Currently, chip manufacturers offer passive UHF RFID chips that require as little as $25\mu W$ in programming mode [14]. These chips contain about 1kBit of memory (EEPROM), which retains data beyond periods of power starvation.

Therefore, a continuous power supply is not needed. Energy only needs to be provided to the chip for short periods of time so that measurements can be taken and stored in the chip's memory.

The electrical characteristics of a chip are fixed, but the power consumption of a sensor substantially depends on the number of measurements taken. A simple temperature sensor, which is designed to monitor the temperature in a cold chain [15], requires about $2\mu W$. The power requirement of the chip and the sensor results in a total power consumption of $27\mu W$. The length of the measurement cycle is chosen according to the resolution required for the application. However, there is a tradeoff between increases in the level of detail (more measurements) and reduction of power consumption.

3 Ambient Energy Sources in Transportation

To replace batteries, ambient energy sources need to be made accessible. There are several different forms of energy - mechanical, thermal, magnetic and bond energy. These forms of energy can all serve as sources of potential electrical energy. In the following, the most promising sources of ambient energy are reviewed.

3.1 Possible Energy Sources

Solar cells offer high power density in direct sunlight, and the transformation of solar energy into electrical energy is an evolved technique [16]. However, the presence of sufficient light in storage houses and cargo trucks is rather unlikely; and it is extremely difficult to estimate the time the goods in transit are exposed to light during loading and unloading, and the intensity of light exposure.

Another inexhaustible source of ambient energy in urban environments is noise. Noise creates an acoustic pressure at a microphone's membrane, which leads to a deflection of the membrane. This deflection can be transformed into electrical energy either electro statically (capacitive) or electro dynamically (magnetic induction). To calculate the energy gain from noise, specific knowledge about the noise distribution in the cargo area is required. Unfortunately, there is no adequate data available. Rough calculations with data from statistics of noise pollution along highways [17] show that membrane diameters of about 2.5cm are required. This size, however, exceeds the restrictions for a chip's form factor.

In order to generate energy from temperature drops (Carnot effect) a significant temperature gradient is required (a drop of several degrees of Celsius over the chip's physical dimensions of less than $1cm^3$). However, the temperature difference between frozen goods and a cooled cargo truck (the temperature of the chip) may be minimal. The temperature gradient is further reduced when the sensing element is not directly attached to the frozen goods but to the container the goods are shipped in. Thus, energy harvesting would not start before a significant change in the shipping environment occurs. This results in a significant delay in the first measurement.

The most promising method for power generation is vibration energy scavenging because vibrations are abundantly available in transportation. The three main options for converting vibrations into electrical energy include variable capacitor generators, electromagnetic generators, and piezoelectric generators. Piezoelectric materials have the greatest potential power density [16] and therefore piezoelectric generators have been chosen for further analysis.

3.2 Vibration Levels in Shipping Transportation

Ambient vibrations are available in many environments and are very distinctive during transportation. The U.S. Department of Agriculture [18] analyzed the common carrier shipping environment in great detail and collected vibration data from various studies. Extensive measurements with regard to road vehicles have been conducted by Sharpe and Kusza [19]. They recorded vibrations on scheduled common motor carrier semi-trailers by analyzing 3 different trucks with cargo weight of 26,500 pounds for a 45-foot trailer, 6,000 pounds for a 40-foot trailer, and 9,000 pounds for a 40-foot trailer. Measurements were taken of the rear vertical accelerations for all runs, including rough roads, smooth highways, and several loading conditions at different speeds. The overall root-mean-square (rms) acceleration with respect to the according frequency is presented as power spectral density (g^2/Hz) in Figure 1. Lateral measurements are smaller by a factor of two or more, and therefore, they are not further explored. Results of other studies [20], [21] that present analyses of vibration levels in truck transportation in conjunction with the damage of perishable goods show similar results.

The magnitude of the vibrations as well as the exact frequency of excitation depends on many factors such as suspension system, load, speed, road conditions, and location of cargo. Sharpe et. al attribute frequency peaks between 0Hz and 5Hz to the natural frequencies of the suspension system. Intermediate peaks ranging from 10 to 20Hz occur due to the unsprung mass, which consists of the wheels, axles, and suspension systems. The third significant peak appears between 50Hz and 100Hz and is related to the tyres' natural frequency. Higher frequency peaks are caused by structural elements of the truck vibrating at their natural frequencies.

Sharpe also reports on an investigation to determine the in-service vibration environment of a Hy-Cube, 70-foot, 100-ton boxcar (train). However, these vibration deflections are smaller by two orders of magnitude compared to trucks (Figure 1).

For aircraft, frequency excitations are highly dependent on the type of aircraft (turbojet, turboprop, reciprocating engine, or helicopter) while the amplitudes depend on the flight mode (take-off, climb, cruise, and landing). Figure 1 shows the frequency spectrum for a Being 707 containing different modes; ground run-up, taxi, takeoff, and cruise. The vibration amplitude peaks occur with moderate accelerations at significantly higher frequencies than for trucks and railcars. Even though aircraft show the highest available accelerations, no further analysis has been undertaken because the distribution of the vibrations of acceleration over the different operation modes and type of aircrafts is very inhomogeneous and therefore, it is difficult to classify the results.

The truck is the predominant means of transportation for shipping goods within Europe [5]. Additionally, the analysis of a truck's power spectral density shows substantially higher acceleration amplitudes than that of trains. This results in higher potential energy because the vibrations are proportional to the vehicle's acceleration amplitudes (according to Newton (F = m·a) a higher acceleration amplitude results in higher potential energy due to maximized applied force). Therefore, the truck has been chosen as the means of transportation for the following design of an energy scavenging system that powers monitoring devices.

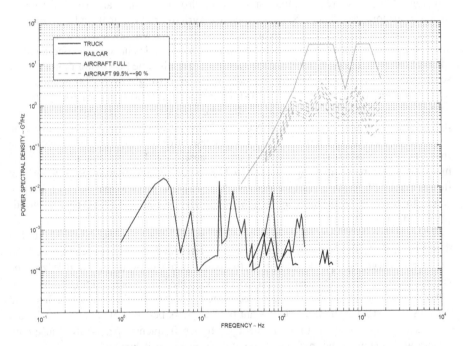

Fig. 1. Power density spectrum for trucks, railcars, and aircraft (adapted from [19])

3.3   Design of an Energy Scavenging System

Vibration energy is collected through a piezoelectric bending element. The deflection of piezoelectric material leads to an electrical charge (in contrast to capacitive systems that require a charge to trigger the transformation), and therefore provides a transduction from mechanical to electrical energy. This transducer can be mounted in many ways. Roundy [22] suggests a cantilever beam configuration with a mass placed on the free end because this setup results in the lowest stiffness for a given size and in a relatively high average strain for a given force input, which maximizes the magnitude of the oscillation (Figure 2).

Short beam length is a prerequisite for small form factors. However, low frequencies require long beam lengths and therefore these frequencies cannot be utilized as a source of energy for our design. But the beam length can be reduced through the deposition of material on the tip of the beam.

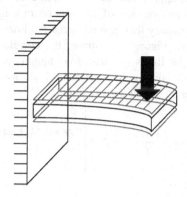

Fig. 2. Cantilever beam design with deposited piezoelectric material on the top and bottom sides. Force applied at the tip leads to a deflection.

Silicon is the common material for high density packaging in large scale integration (chip design). However, silicon is unsuitable for the design of an energy harvesting system because the high stiffness of the material requires long beam lengths. The polymer SU-8 is commonly used for applications in MEMS (Micro-Electrical-Mechanical Systems) due to its low stiffness and the fact that it can be processed at low cost. A layer of piezoelectric material is deposited on both sides of the SU-8 beam to make it piezoelectric (Figure 2). The oscillation of the cantilever beam leads to a deflection of the piezoelectric layer, which results in an electrical charge.

There are many variables in this design for which the system can be optimized. However, in shipping transportation the emphasis is on minimal form factor and maximal power output – two opposite variables.

The physical dimensions of the cantilever beam are described by its length l, width b and height h. Information about the stiffness of the beam is derived from the deflection s, which is affected by the mass load on the tip of the beam. The most massive mass load is chosen to give the lowest stiffness. In this application, lead is deposited on the tip of the beam. The mass load is described by its weight m and the area l_a x b, which is covered by the deposited material. Equations (1), (2), (3), and (4) reveal the correlations of these parameters (these formulas are partly derived from [23]). A heavy mass load minimizes beam length (Eq. (1)) and maximizes power output (Eq. (4)) for a given frequency. However, the maximum mass load is limited by the stiffness of the beam material (Eq. (2)) and by the height of the deposited lead (Eq. (3)). Additionally, a small beam width requires only a short beam length (Eq. (1)) but increases the height of the mass load (Eq. (3)).

Power output is inversely proportional to the resonant frequency (Eq. (4)) and by maximizing the ratio A^2/ω, the greatest potential for power generation is achieved.

Because lower frequencies have greater power output potential but also require longer beam lengths, the lowest resonant frequency is chosen that just meets the constraints for total beam length. The appropriate size of a cantilever beam is highly application dependent. Figure 3 gives an overview of different beam lengths at given frequencies. 75Hz seems to be the frequency that best balances minimal size and maximal power output for this application. Figure 4 illustrates the correlation of beam length and mass load and shows upper limits for mass load heights at 75Hz. Additionally, the upper bound for the deflection s is illustrated. It is assumed that there are no implications with total mass load for a given length.

$$m_A[Kg] = l \cdot b \cdot \alpha \left[\frac{3 \cdot \hat{E} \cdot \frac{1}{12} h^3}{(2\pi \cdot f)^2} \cdot \left(\frac{1.875}{L} \right)^4 - \rho \cdot h \right], \quad \alpha = 0.24, \text{ E=4GPa for SU-8} \tag{1}$$

$$s = -4 \cdot \frac{l^3}{\hat{E} \cdot h^3 \cdot b} \cdot F, \qquad F = m_A \cdot g \tag{2}$$

$$h_A = \frac{m_A}{\rho \cdot S}, \qquad S = l_a \cdot b \qquad \rho = 12000 \text{kg/m}^3 \text{ for lead (Pb)} \tag{3}$$

$$P = \frac{(m_a \cdot A^2)}{16 \cdot 2 \cdot \pi \cdot f \cdot \zeta_e}, \zeta_e = 0.01 \tag{4}$$

m_a: mass load, A: acceleration magnitude, f: resonant frequency,
ζ_e: electrical damping constant

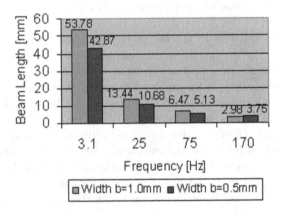

Fig. 3. Resonant frequencies in correlation to beam length (m_a = 85mg, S = 1.5mm x 1mm^2)

If the height of the mass load is restricted to 5mm at a given area of 1.5mm x 1mm (to fit within 1cm^3), the minimum beam length and maximum mass load is derived from Figure 4 at the intersection of the mass load and the beam length, given the constraints for h=5mm (l_a=1.5mm, b=1mm). It shows a beam length of 6.47mm and a mass load of 85mg. The power output for a mass load of 85mg and an acceleration magnitude of 0.082g (Figure 5) at 75Hz results in 0.7295μW (Equation (4), Figure 6). The same calculations are repeated for the next frequency peak at 170Hz with an acceleration amplitude of 0.0447g. This results in a power gain of only 0.00956μW, which is 8.5 times smaller than the power output at 75Hz (Figure 6). Therefore, a design for 75Hz shows the best performance at sufficiently short beam lengths.

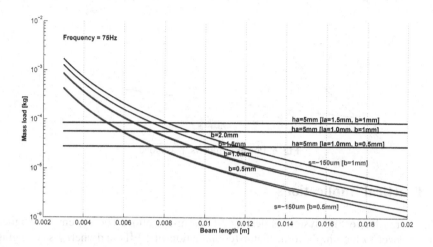

Fig. 4. Mass load vs. beam length at 75Hz

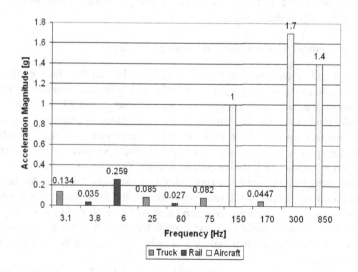

Fig. 5. Frequency peaks and acceleration amplitudes for trucks, railcars, and aircraft

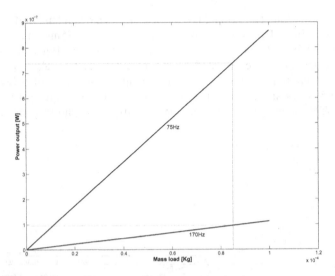

Fig. 6. Power output vs. mass load

4 Practical Implications

The energy harvesting system is designed for transportation. The system is designed with respect to the chip's form factor (combination of RFID and energy scavenging system), which has to meet transportation and handling requirements. The chip's restricted physical dimensions limit maximal power output and therefore has constraints regarding minimal cycle time, energy storage, and mount. The next sections will address these issues.

4.1 Measurement Cycle Times and Energy Storage

Previous calculations stated that most passive UHF RFID chips equipped with a temperature sensor currently require as little as 27µW. Calculations in the previous section show that adequate energy scavenging generates 0.73µW. This is orders of magnitude smaller than what is required for continuous operation of the chip and its sensor. However, usually there is no need for seamless monitoring. Goods in a cold chain do not spoil because of occasional spikes in temperature but due to temperatures that exceed a threshold for a long period of time. In order to capture these events and store their temperature values along with time stamps for post-delivery analysis, periodic recording seems sufficient. Not only can the cycle time be reduced, but also the period of time that is required to read a sensor's value and to store it in the chip's memory. One measurement can be completed within 100ms, and therefore the system's power requirement is reduced to only 2.7µJ, which is a tenth of its original power requirement.

If the generated power is accumulated and stored, 3.7s are required until enough energy is available to take one single measurement. This allows for periodic measurements with a rate of 16.2 samples per minute. A capacitor serves as an energy reservoir and by using Equation (5) the capacitor's capacity is determined to be 937.5nF. The capacitor is charged until it reaches a threshold voltage of 2.4V. Then, a switch is triggered and the capacitor starts providing energy.

$$E = 0.5CV^2, V = 2.4V \tag{5}$$

Even though a resolution of more than 16 samples per minute is possible, cycle times of less than 5 to 10min do not appear appropriate for two reasons: (1) fine-grained measurements are not needed in an environment of slow temperature change (2) the amount of data that has to be stored exceeds the memory's capacity on a regular passive UHF tag. What is important is to store additional energy for periods of time during which transportation vibrations are minimal and only a very limited amount of energy is generated. These may be situations when a truck is stuck in a traffic jam or when the driver has to stop to meet her/his rest requirements.

4.2 Mount

The design considerations described above suggest best performance with a cantilever beam of 6mm length and a lead mass load of 5mm in height. Additional encapsulation for protection against environmental influences leads to a chip volume of 1cm^3. Rough calculations of the die's weight result in 1.2g (density of SU-8 is 1200kg/m^3).

Currently, 80μm thick paper serves as mount for passive RFID tags. Deploying a 1.2g die on a paper mount would lead to significant deflection of the paper. This is a result of the low stiffness of the paper – vibration forces applied to the die would directly pass to the paper. Hence, the die would oscillate itself instead of providing a fixed frame within which the beam can oscillate. A firmer mount is necessary and plastics may replace paper with the drawback of increased cost and thickness.

5 Discussion

Currently, there are solutions to capture environmental condition deviations from required shipping conditions by means of special devices added to the truckload. However, these devices only give a very limited view of the shipping process. We argue for an object-centric solution in combination with RFID, where each product holds information about its origin and its shipping record. Given this information, grocery store personnel can make an informed decision on whether to move the products onto the shop floor or to reject them. In addition, a continuous shipping record allows detection of failures in the cold chain, even if products are re-sorted numerous times along a global supply chain. Furthermore, if produce is detected to be perishing, trucks can be dynamically redirected to a closer retailer.

In order to monitor individual shipping conditions, each product's RFID tag must be equipped with monitoring sensors, which require energy to operate. Even though an RFID reader placed in a truck's cargo area can provide energy to the tags and power their sensors during transportation, this approach is inadequate for situations such as loading and unloading – the system would fail to capture environmental changes, thus compromising seamless monitoring.

5.1 Cost Considerations

An active or semi-active RFID tag's battery costs about $0.23/year. Replacing it with a low cost energy scavenging system will significantly reduce manufacturing costs. The cost of an energy scavenging polymer chip comprises material, processing, and assembly costs. The expenses of the material, assuming a $1cm^3$ SU-8 polymer, are about $0.001 per chip. The cost of chip processing is estimated to be about $0.01 while the assembly cost comes to about $0.1 (based on [24]). This results in a total cost of $0.111, which is about half the price of a battery alone. There is a handling cost for exhausted battery replacement, which increases the price gap. The difference between the prices is further increased by the high volume of tags sold each year. Revenues for cheap passive RFID tags for 2006 are estimated at $1.96 billion and at $1.09 billion for expensive active tags [9].

Deploying an energy scavenging system involves additional costs such as handling and placing costs as well as consequential costs due to insufficient energy accumulation. But it offers an infinite lifetime, which is of significant interest if tags on containers are used several times. Even if tags are discarded after their first use, taking advantage of energy harvesting, rather than a battery, allows the tag costs to be cut by about 25%.

5.2 Periods of Energy Depletion

While energy scavenging offers many advantages over conventional batteries, a solution provider has to consider periods during which the truck is at rest and therefore the system does not generate any energy. Measurements can be taken with long cycle times, and energy for several measurements can be stored in the tag's capacitor in advance. Nevertheless, there remains the uncertainty that a significant change in the shipping environment may not be captured because the system has run out of energy. This situation may only occur during lengthy periods of rest; e.g. when the driver has to stop to meet her/his rest time requirements. The current EU regulation is that the rest time can be taken in two periods – the first one an interrupted 3 hours and the second an uninterrupted 9 hours [25]. The system presented is not concerned with the latter as it is not usually the case that a transport company will have a truck with perishable goods on the road when a driver has to meet the 9-hour rest requirement. Our system is capable of providing 3 hours' worth of energy in advance.

5.3 Special Constraints for Tag Placing

The proposed cantilever beam design absorbs vibrations in one direction only. In truck transportation, the acceleration magnitudes in the vertical direction are predominant. Therefore, special attention needs to be paid to tag placement. To operate most efficiently, the energy scavenging system needs to be exposed to the strongest vibrations. This requires careful handling of the goods to guarantee that they are loaded and stockpiled so that the tag either faces the top or the bottom side. However, only a small adjustment to the containers and boxes would insure that they are placed correctly.

6 Conclusion

Detailed information about the shipping environment is of special interest for perishable goods supply chains. Control and visibility over product handling is limited due to several echelons of the supply chain. Currently, battery-powered devices are used to monitor the shipping environment of goods in a cold chain but the cost of these devices, their bulkiness, and their limited lifetime prevent high market penetration. As a result, only limited information is available about the cold chain, which precludes useful insights as to its efficiency.

In this paper, we have presented the design of a vibration energy scavenging system that powers monitoring devices on products at significantly lower cost than comparable battery-powered approaches. This proposed system, designed for the truck transportation environment, generates 0.73µW of power. This offers a maximum temperature measurement resolution of 16.2 samples per minute. The analysis of shipping environments, along with extensive calculations of physical effects, shows the feasibility of our approach.

The system presented comes at the cost of $0.111. The complete system, including a RFID chip, is about 25% cheaper than comparable battery-powered systems. Additional costs for battery replacement and handling, which would further leverage the price gap, are not included. However, the fact that the vibration energy scavenging system needs to be exposed to the most intensive vibrations (vertical vibrations) results in additional costs for tag placement.

The availability of cheaper monitoring devices with an infinite lifetime will result in increased market penetration, and therefore, will lead to more detailed information about the shipping environment. This will support the analysis and improvement of a cold chain's efficiency and will result in reduced shrinkage and extended shelf life. Additionally, customer satisfaction can be improved through better quality and freshness. Therefore, a minimal investment in sensor-equipped RFID tags based on energy scavenging will result in an overall reduction of revenue losses in the perishable goods industry.

Acknowledgments

The authors would like to thank Svetlena Taneva and Cyril Vancura for their contribution.

References

1. Kader, A.: Increasing Food Availability by Reducing Postharvest Losses of Fresh Produce. In: V International Postharvest Symposium, International Society for Horticulutral Science, Verona, Italy (2005)
2. Houghton, S.: Commonwealth Scientific and Industrial Research Organisation (CSIRO), Innovative cold chain education and training solutions, Reference 2006/45 (2006)
3. EU Directives 92/1 on the Monitoring of Temperatures in the Means of Transport, Warehousing and Storage of Quick-Frozen Foodstuffs intended for Human Consumption
4. Regulation (EC) No 178/2002 of the European Parliament and the Council, Official Journal of the European Communities L31 (2002)
5. Eurostat, Europe in figures - Eurostat yearbook 2005, Eurostat (2005)
6. Lee, Y., Cheng, F., Leung, Y.: Exploring the Impact of RFID on Supply Chain Dynamics. In: Proceedings of the 2004 Winter Simulation Conference (2004)
7. Want, R.: An Introduction to RFID Technology, IEEE Pervasive Computing 5(1) (2006)
8. Frost & Sullivan, World RFID-based Application Markets, A686-11 (2004)
9. Swan, R.: FDA Counterfeit Drug Task Force, Interim Report, Docket Number 2003N-0361 (2003)
10. Directive 2002/96/EC of the European Parliament and of the Council on waste electrical and electronic equipment (WEEE), Official Journal of the European Union (January 2003)
11. Starner, T.: Human-powered wearable computing, IBM Systems journal 35(3&4) (1996)
12. Paradiso, J., Starner, T.: Energy Scavenging for Mobile and Wireless Electronics, IEEE Pervasive Computing 4(1) (2005)
13. Want, R., Farkas, K., Narayanaswami, C.: Energy Harvesting and Conservation, IEEE Pervasive Computing 4(1) (2005)
14. Atmel, Document 4843A-RFID, ATA5590 Tag Antenna Matching, Application Note, p. 5 (February 2005)
15. Analog Devices, Document C01316, AD7816, Rev. C (September 2004)
16. Starner, T., Paradiso, J.: Human-Generated Power for Mobile Electronics. In: Piguet, C. (ed.) Low-Power Electronics Design, ch. 45, pp. 1–35. CRC Press, Boca Raton (2004)
17. Utah Department of Transportation, Final Report, Roadway Pavement Grinding Noise Study (2003)
18. Ostrem, F., Godshall, W.D.: An Assessment of the Common Carrier Shipping Environment, General Technical Report FPL 22, Forest Products Laboratory, U.S. Department of Agriculture, U.S. Government Printing Office, 1979-651-089
19. Sharpe, W.N., Kusza, T.: Preliminary measurement and analysis of the vibration environment of common carrier shipping damage prevention, Michigan State University, School of Packaging, Tech. Rep. (1972)
20. Jarimopas, B., Singh, S., Saengnil, W.: Measurement and Analysis of Truck Transport Vibration Levels and Damage to Packaged Tangerines during Transit. Packaging Technology and Science 18, 179–188 (2005)
21. Singh, S., Antle, J., Burgess, G.: Comparison Between Lateral, Longitudinal, and Vertical Vibration Levels in Commercial Truck Shipments. Packaging Technology and Science 5, 71–75 (1992)
22. Roundy, S.: Toward Self-Tuning Adaptive Vibration Based Micro-Generators. In: Proceedings of SPIE, vol. 5649, p. 373. SPIE Press, San Jose (2005)

23. Nalwa, H.S. (ed.): Encyclopedia of Nanoscience and Nanotechnology, American Scientific Publishers, vol. 1, pp. 499–516 (2004)
24. Polytronic Systems Department, Frauenhofer Institute for Reliability and Microintegration, Munich, Germany (2006)
25. Roca, A., Markov, H.: Report on the joint text approved by the Conciliation Committee for a directive of the European Parliament and of the Council on minimum conditions for the Implementation of Council Regulations (EEC) Nos 3820/85 and 3821/85 concerning social legislation relating to road transport activities and repealing Directive 88/599/EEC, European Parliament, A6-0005/2006, PE365.140v02-00 (January 2006)

Escalation: Complex Event Detection in Wireless Sensor Networks

Michael Zoumboulakis and George Roussos

School of Computer Science and Information Systems,
Birkbeck College, University of London
Malet Street, London WC1E 7HX, UK
{mz,gr}@dcs.bbk.ac.uk

Abstract. We present a new approach for the detection of complex events in Wireless Sensor Networks. Complex events are sets of data points that correspond to interesting or unusual patterns in the underlying phenomenon that the network monitors. Our approach is inspired from time-series data mining techniques and transforms a stream of real-valued sensor readings into a symbolic representation. Complex event detection is then performed using distance metrics, allowing us to detect events that are difficult or even impossible to describe using traditional declarative SQL-like languages and thresholds. We have tested our approach with four distinct data sets and the experimental results were encouraging in all cases. We have implemented our approach for the TinyOS and Contiki Operating Systems, for the Sky mote platform.

Keywords: Event Detection, Complex Events, Parameter-free Detection, Data Compression, Network Control, Reactive Sensor Networks.

1   Introduction

Our approach is aimed at the extremely resource-constrained devices known as *motes*; motes combine a radio component, a microcontroller, an array of sensors and an energy source (e.g. battery). They are designed for operating unattended for long periods of time (typically for as long as their energy source lasts). Our target platform for which we are developing code is the TMote Sky motes [1].

The event-driven model in sensor networks broadly prescribes that individual or groups of sensor nodes should react programmatically to events, very much in the same manner a traditional DBMS reacts to violating transactions. The database abstraction for sensor networks offers the capability of querying with the use of a declarative SQL-like language and specifying reactive functionality with the use of triggers and predicates. The problem arises from the fact that the use of predicates and thresholding with relational and logic operators, are not expressive enough to capture most of the complex events that occur in sensor networks. Complex events are sets of data points that constitute a pattern; this pattern shows that something interesting or unusual is occurring in the underlying process monitored. While viewing the sensor network as a database

G. Kortuem et al. (Eds.): EuroSSC 2007, LNCS 4793, pp. 270–285, 2007.

is useful for running queries over streaming data, it offers little help in detecting interesting patterns of complex events.

A separate problem arises when users do not know in advance the event type they are looking for. A benefit of our approach is that it also allows non-parametric event detection. In this sense, motes become context-aware and they are capable of detecting relative change. Practically, this is implemented with a concept borrowed from Machine Learning: motes go through a learning phase that is known to be normal. During this phase, they continuously compare strings (that correspond to temporally adjacent sets of readings) and compute distances among them. Once the learning phase is complete, these distances are used for non-parametric detection — the distances effectively constitute the normal context; a distance never seen before represents some unseen and new change.

Another mode of operation offered by our approach is approximate event detection: users can search for events that are *like* other events by submitting an event template. Practically, the template (that is a set of real-valued readings) gets translated into a string representation and submitted to the relevant motes. Streaming sensor readings are then transformed to strings and comparisons are performed against the template. A distance (between the template and the newly-generated string representations of streaming readings) that approaches zero signifies that the reporting mote is experiencing a very similar event to the one submitted. Of course a zero distance denotes that the exact same event has occurred.

Perhaps the strongest point of our approach is that it makes motes context-aware: since they can sense relative change they can adjust the sampling frequency dynamically and prolong their lifetime by conserving power resources. Furthermore, optimisation of sleep scheduling can ensure that in periods of low activity the majority of the nodes are asleep. If distances start to increase, then explicit wake-up messages can be sent to sleeping nodes tasking them with sensing, processing or sending.

In the sections to follow, we will discuss in great detail our approach and we will provide a high-level description of our algorithm. We will present results from our experiments with sensor data sets, and we will conclude by outlining future plans (including an indoor test deployment).

1.1 Target Applications

Our approach can be beneficial to a fairly large selection of applications, namely:

- *Pervasive Health Monitoring.* One aspect of this has to do with monitoring the activity of human organs over time (such as ECG, EEG, etc.) and projects such as [2]. An example of a Complex Event is a heart attack — if there is any pattern in the signal just before the heart attack. then a detected event can help save lives. The other aspect has to do with Pervasive Healthcare Monitoring [3,4]: monitoring patients at their sensor-equipped homes rather than hospitals. Any "abnormal" behaviour can flag an event — for instance an elderly person not getting up at their normal time may indicate illness. The system can react to this by informing a social worker to visit or call. We are also testing our approach to a patient classification problem using the same distance metric for similarity / dissimilarity. The data comes from patients suffering from Cystic Fibrosis and healthy controls. Data has been collected using the Cyranose 320 electronic nose — a device with 32 sensors that can be used for bacterial classification [5].

- *Environmental (Outdoor or Indoor) Monitoring.* Normally, this involves monitoring physical attributes of the environment such as temperature, humidity, light, etc. Examples include: meteorological stations, wildfire monitoring (and projects such as the Firebug [6,7], indoor monitoring (projects such as the Intel Lab Deployment [8] of which data we have used for our experiments).
- *Athlete Performance Monitoring.* Approaches like the SESAME project [9] use motion sensors (such as accelerometers) in order to monitor attributes related to the performace of elite athletes. A Complex Event in this case can be a pattern that denotes a change in performance — for instance a sub-optimal start to a sprint.

To generalise the above, our approach can benefit any system or process that exhibits long periods of normality and relatively much smaller periods of rare events.

2   Conversion of Streaming Sensor Data to a Symbolic Representation

For the conversion to string we use the Symbolic Aggregate Approximation (SAX) algorithm [10,11] which is a very mature and robust solution for mining time-series data. SAX creates an approximation of the original data by reducing its original size while keeping the essential features — this fact makes it a good choice for the sensor network setting because it offers the advantage of data *numerosity reduction* which is a very desirable property since it can contribute to saving power (sensors transmit a string which is a reduced approximation of the original data set).

The distance between SAX-representations of data has the desirable property of lower-bounding the Euclidean distance. Lower-bounding means that for two time-series sets Q and C there exist symbolic representations $\hat{Q}$ and $\hat{C}$ respectively such that $D_{LB}(\hat{Q}, \hat{C}) \leq D(Q, C)$ where $D(Q, C)$ is the Euclidean distance.

Once the time series has been converted to a symbolic representation, other distance metrics & algorithms can be applied for measuring similarity as well as for pattern-matching. We have tested a variety of metrics such as the Hamming Distance, Levenshtein Distance, Local & Global Alignment, Kullback-Leibler Divergence and Kolmogorov Complexity (as described in [12]).

3   *Escalation*: Applying SAX to Streaming Sensor Data

We have combined SAX and the distance metric used by SAX with an algorithm that detects complex events without the need of the user supplying an explicit threshold. The algorithm learns what constitutes an interesting change, very much in the machine learning sense. The main beenfits of introducing such an approach to sensor networks are:

- *Dynamic Sampling Frequency Management.* This refers to the ability to autonomously make a decision on whether to increase or decrease the sampling frequency. The decision is based upon the rate of change.
- *Parameter Free Detection.* This refers to event-detection without supplying thresholds or generally an event description.
- *Successful Detection of Complex Events.* This refers to the ability to detect complex events that are impossible to describe or capture using traditional techniques such as triggers and thresholding.

Firstly, longevity in a sensor network is of crucial importance, so by monitoring how quickly attribute values change a sensor node can make an autonomous decision to increase or decrease the sampling frequency accordingly. Our algorithm essentially compares a symbolic representation of a current set of values to a previous one and calculates the distance. If the distance is zero or a very small number then the sampling frequency can be reduced. Conversely, sampling frequency can be doubled as soon as the distance starts to increase and moreover local broadcasts can be made to neighbouring nodes in order to issue wake-up calls or increase sampling frequency commands. This way the sleep schedule of nodes can be naturally adapted to the rate of the physical change. This advantage is important because it can significantly prolong the useful lifetime of the sensor network. In addition it means that sleep scheduling can be optimised and the biggest percentage of the network can be asleep for the majority of the time. If, for example, the network is split into regions, then only two nodes per region need to be awake at any given time. The second node acts as a verifier or a fail-over for the first node.

Secondly, specifying a threshold based on user experience or expectancy can be an arbitrary process. A small threshold means that the nodes are overworking and similarly a high threshold may mean that interesting changes are lost. Moreover, threshold are susceptible to outliers. Thresholding is also fairly fixed in nature; the threshold is either compiled into a binary image at pre-deployment or it is injected as a VM command at runtime. Either way it involves human intervention to the system. Our algorithm offers complex detection by allowing the user to either specify the pattern they are looking for and then search for proximity to that pattern (in terms of approximate string matching) or having the node make an autonomous decision on what is interesting based on monitoring some normal data. This second advantage allows us to successfully detect a whole class of events with no programming effort — all the user will have to do is to press start (or to be precise compile the binary image into the node) and the node will autonomously do the rest. There is no need for specifying what events to look for by using cumbersome and error-prone code. The node will take care of learning normality and then deducing what is not normal.

Lastly, we have mentioned earlier that events in sensor networks are conceptually different than events in conventional DBMS: in sensor networks we deal with signals or streams of continuous real-time data. We do not deal with transactions. So a complex event in that sense is a set of data points that satisfies some condition e.g. this condition could be a state change. Or to use an alternative description, a complex event is a pattern that its frequency of occurrence is greatly different from that which we expected, given previous experience or observations [11]. By using SAX with the sensor network data, we are able to detect such complex patterns with good accuracy.

Algorithm 1. Complex Event Detection Algorithm

1: **variables** counter, bufferSize, distanceBufferSize, saxWindow, stillLearning;
2: **call** initialise (set values to the above, set stillLearning to true)
3: **while** (stillLearning is true) **do**
4:    get data from sensors;
5:    **if** counter $\leq$ bufferSize **then**
6:       fill buffer with data; break;
7:    **else**
8:       update buffer with new reading and shift elements;
9:       extract two subsections from buffer (say, a and b);
10:       call SAX passing a and b and update distances (if dist(a,b) $\neq$ 0);
11:       **if** distance buffer is fully populated **then**
12:          stillLearning = false;
13:          select maximum distance observed during learning phase;
14:       **end if**
15:    **end if**
16: **end while**
17: **while** true **do**
18:    get data from sensors;
19:    update buffer with new reading and shift elements;
20:    extract two subsections from buffer (say, a and b);
21:    call SAX passing a and b and update distances (if dist(a,b) $\neq$ 0);
22:    **if** dist(a,b) $\geq$ maximum distance set in learning phase (line 13) **then**
23:       report distance and counter (representing timestamp) or do some other action
24:    **end if**
25: **end while**

3.1 High-Level Description

Let us start by describing the algorithm is pseudocode. We have developed 2-3 variants of this, but the basic idea is the same: namely, monitor distances during learning phase, choose minimum, maximum or average distance, and then compare adjacent string representations. If their distance is greater than the maximum distance observed during training, then it is highly likely that an event is occurring.

There is a further variation of the algorithm that offers the escalation functionality, in that it calls a third optional phase to investigate a reported complex event even further. This is implemented by a function call on line 23 (not shown).

The whole of the readings set is passed to this function, and the subsections are expanded in order to distinguish true from false positives. This way, the sensitivity (high true positive rate) and selectivity (low false positive rate) of the algorithm are maximised. Empirically, from running our experiments we notice that an escalation phase is not always necessary as in most cases the algorithm will detect the exact point of the event with great accuracy.

Summarising, there are two main phases to the algorithm, with an optional third;

- *Learning Phase*, where the algorithm keeps track (learns) of the distances it has seen.
- *Initial Detection Phase* during which SAX is called to convert adjacent sets of numeric readings to strings and calculate their distance. An event is flagged if the distance between two strings is greater that the maximum distance observed during learning.
- *Escalation Phase* is a temporally deferred operation that varies the window size over the numeric sets and produces progressively larger strings. It is used to increase the accuracy of the algorithm (by pruning false positives).

Essentially, the algorithm puts the sensor in an infinite loop that forces them to convert readings to strings, compute their distance and test it against a distance learned during the initial learning phase. Each phase is further discussed below.

Learning Phase. During this phase the algorithm keeps track of distances between strings. The set of observed distances during the normal period is then used to select the maximum distance that will be used for future event detection. Zero distances are dropped automatically — from experience with our experiments, most of the time distances are 0 or 0.5; 0.5 represents a 1-character difference in the string; by default adjacent characters have a distance of zero, eg. a and b have a 0 distance, but a and c have a distance of 0.5. The algorithm can be modified to drop 0.5 distances and choose not to store them as they represent minor changes.

When the learning phase completes — after a given time period elapses — the maximum distance observed is selected. The learning phase passes through $\frac{1}{4}$ of the total data in our experiments, and in the sensor-mote implementation this is not known in advance, so either the user of the system has to specify a fixed learning phase duration or otherwise the algorithm can set an arbitrary learning limit and test it against a counter incremented every time the timer fires (after some arbitrary time period has elapsed, stop the learning phase). The learning phase will largely depend on the data type monitored: for instance if we monitor light readings, then a reasonable choice for a learning phase would be twenty-four hours since in a day-night cycle we would record most of light variations and their relative distances.

Once the maximum distance is set in the learning phase it doesn't change. This is the way the algorithm works at the moment, and it is also a potential limitation. We intend to investigate ways that introduce reinforcement learning in the near future. Once we have the maximum (learned) distance we can enter the second phase of the algorithm which is continuous (it is a function called for every new reading entering the set).

Initial Detection Phase. SAX is called to convert adjacent regions of the readings set and calculate the distance of the two strings. The readings set is essentially a sliding window of the streaming sensor data. It is stored in a data structure of a fixed size; every single new reading that enters the structure at the front, causes the oldest reading at the back to be dropped, working very much like a FIFO queue. In addition, this phase is optimised by not calculating distances for identical strings; these are dropped in silence. Generally there are two alternatives for this stage: to either call the conversion and distance comparison function at every timer tick (as shown on the algorithm) or to call it every t timer ticks. In the latter case, t has to be equal to the size of the readings set to ensure no events are undetected and a detection latency is introduced which is at worst t timer ticks. At the time of this writing we are investigating the trade-offs between the two alternatives.

Escalation Phase. The Escalation Phase is an optional, temporally deferred phase that searches through the whole readings set in order to determine if the underlying process continues to change in the future — the detection phase indicates if something has changed and the point of the change while the Escalation phase essentially determines whether the point of change was a true or false positive and whether the physical process continues to change. Usually, a defer period of 5 data points (or timer fires) is enough to distinguish a true from a false positive, but this largely depends on the sampling frequency.

During the Initial Detection Phase we pass two temporally adjacent subsections of the buffer to the conversion and comparison function — we do not pass the entire readings set (unless we use the *every t timer ticks* variant described in the previous phase). The reason we pass subsections to the function and not the entire set is because we aim to minimise the computation overhead; since we choose to call the function at every timer tick, we must ensure that the function completes before the timer expires. We also need to cater for other components such as the radio and ADC competing for CPU time.

If we choose to call the Escalation Phase, we post a `task` that is added to the queue. It takes the entire readings set and searches through it to determine whether the event was a true positive and if the process continued to change or if it stopped. The searching is done by varying the SAX window size, starting with a small value and progressively increasing it until the size of the set is reached. The compression ratio is also, optionally, reduced. For instance if the previous phase was using a 4 : 1 compression, this phase can reduce compression to 2 : 1 to produce more accurate results. Both of these actions (increasing the SAX window and reducing the compression) will expectedly produce higher distances. If these distances are sufficiently higher than the initial detection distance, then we can report that the process monitored either continues to change and/or that the initial detection was a true positive. Sufficiently higher is quantified by multiplying the initial detection distance by a factor empirically learned and dependent upon the process monitored and the data type.

The performance of this phase will very much depend upon the type of the event and process monitored. If an underlying process changes gradually from a normal state a to an abnormal state b, and then it remains in this abnormal state which is stable in terms of difference between data points, the escalation call will able to confirm the change, provided the change from state a to b is in the readings set. The size of the readings set is limited by the sensor's RAM and therefore if the defer period from the occurrence of the event (change of states from a to b) is large, and all the readings in the set are now stable the escalation phase will flag the report as a false positive as distance will be smaller than the initial reported distance. This indicates that the escalation phase can not be a general-purpose call but it has to be tailor-made to the underlying process monitored and the the event type.

4   Experimental Results

All the experiments in this section were ran in MATLAB using real sensor data that was fed to the algorithm in a streaming fashion with the use of a `Timer` object to simulate sensor network data acquisition.

4.1   Data from an Indoor Deployment

The majority of the experiments were ran on the Intel [8] data set, which is a collection of approximately 2.3 million readings from 54 sensor nodes collecting temperature, relative humidity, light and voltage. On our experiments we have used all the readings from all the nodes except voltage readings. Voltage is quite well-behaved in a sense that voltage readings do not fluctuate greatly. In contrast, light is the most unpredictable attribute and it tends to fluctuate a lot, partly due to human factors (e.g. a person switching a light on or off).

Due to space limitations, we will not discuss the outcome of every single experiment; instead we will present a few selected cases.

In the example shown on Figure 1 (a), we are searching for a specific pattern — we are looking in a week's segment of data for a pattern that is not included in that week but the most similar pattern is found. The search for a pattern works in the following manner: we take a set of data points which we convert to a string. This string represents the pattern we are searching for. Searching is accomplished by comparing this string to a string representation of the entire dataset (this can be done by using alignment algorithms instead of standard distance used in most of our trials); if distance becomes zero then it means that the exact same pattern has repeated (e.g. the event has occurred). If distance approaches zero then it means that a very similar pattern is observed. In the second example (Figure 1 - (b)) we did not supply a pattern but instead we were looking for the most interesting change. The most interesting change in this case is the spike shown on the figure — which was successfully detected. Searching for the most interesting change is the non-parametric version of the algorithm that trains on a normal set and then uses distances learnt for future detection (described in more detail in section 3.1).

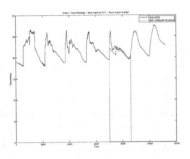   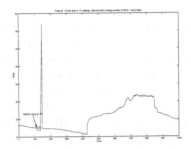

(a) Temperature Readings at node 4   (b) Temperature Readings at node 27

Fig. 1. Temperature Readings at nodes 4 and 27 — (a) approximate detection — in this case we are searching for a pattern (the readings of one day) in the readings for one week; the closest match is shown delimited by the two vertical lines and (b) non-parametric detection of a synthetic event; the spike in the graph represents an abrupt change in values — we successfully detect the exact point of change

Similarly for humidity (Figure 2 (a)), we were searching for interesting changes (in this instance, a synthetic event corresponding to the spike shown in the figure) without supplying a pattern or a threshold. One can see in the diagram that the change is correctly identified (shown by the small bounding box and arrow).

The Light attribute is somewhat different in a sense that it changes frequently and the changes are aperiodic — for example they occur whenever a human switches a light on or off. Having said that, our algorithm using SAX still performed well enough on the Light attribute. In some cases there were a few false positives, but no false negatives. This is shown in Figure 2 (b) that contains a false (first bounding box) and a true positive (second bounding box).

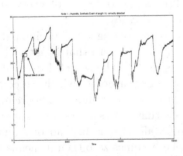   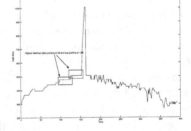

(a) Humidity Readings at node 1   (b) Light Readings at node 1

Fig. 2. Humidity and Light Readings at node 1 — (a) non-parametric detection of a synthetic event correctly identified, shown by the arrow and (b) non-parametric detection - the first match is a false positive (shown by the first bounding box) and the second match is the true positive. The false positive was discarded in the escalation phase of the algorithm, as the second match was reported with a much higher distance than the initial detection.

Overall, we were satisfied that the algorithm can successfully detect complex events of different sizes on different attributes. We had no false positive reports for Temperature and Humidity attributes, and real and synthetic events were detected accurately (accuracy in this sense refers to reporting the event very near at the beginning of its occurrence, typically within 4–8 data points). On the light attribute we had approximately 13 per cent false positives and 2 per cent false negatives. This is not entirely surprising especially if we take into account the arbitrary value changes that are largely unpredictable. Outdoor environmental monitoring data is usually more well behaved and changes can be predicted more accurately.

Finally it is worth mentioning that we have experimented with both synthetic and real events (we have used synthetic events in the parameter-free scenario, while we have used real events in the pattern-matching scenario).

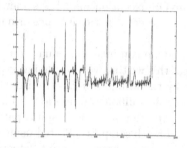
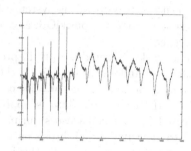

(a) ECG — normal turning to Super Ventricular

(b) ECG — normal turning to Malignant Ventricular

Fig. 3. ECG — Two different ECG datasets that start normal but change at point 512. We successfully detected the point of change in both cases.

4.2   ECG Data

We have also tested our approach on ECG data obtained from [13]. We have tested two cases (Figure 3), the first one being a normal ECG that turns to Super Ventricular and the second case was normal ECG turning to Malignant Ventricular. The detection was accurate in both cases and our algorithm using SAX managed to detect the exact points of change.

5   Future Work

We have implemented a simple first version of our algorithm for the TMote Sky platform [1]. The current implementation runs on both the TinyOS2 [14] and the Contiki OS [15,16,17]. The current implementation runs solely on motes, so we are not assuming a tiered architecture but rather aiming autonomous operation. However, we aim to investigate the performance cost in comparison with a tiered architecture where are all the motes report readings to a base-station that is not resource-constrained. The base-station will then perform the symbolic conversion and distance comparison and will make workload decisions and task the motes accordingly. One simplified execution scenario may involve tasking motes in areas of the network with high activity (e.g. reporting values that denote eventful underlying processes) to sample at higher frequencies while tasking motes in areas of the network with low activity to sample readings at low frequencies. Sleep scheduling decisions can be made based on distance information too; for example if distances over reported readings are more or less constant then a percentage of motes can be asleep or at a Low-Power Listen state — conversely when something interesting occurs (denoted by increasing distances over reported readings) explicit wake-up calls can be issued.

Our aim is to explore the trade-offs between the autonomous operation and tiered architecture. In addition we will also investigate the performance difference between the two operating systems; the TinyOS execution model is somewhat different to that of ContikiOS; as a result we expect a noticeable performance difference.

The work presented is work-in-progress and we continue to evaluate our approach with different data sets. At the time of this writing we are running experiments on motion sensor data [18] collected by the Mitsubishi Electric Research Labs (MERL) — this data is approximately 30 million raw motion readings collected by over 200 sensors during a year. We are also using hospital patient data to identify Cystic Fibrosis patients from breath samples collected using the Cyranose 320 [19] (essentially this is a classification rather than event-detection, but nevertheless we intend to evaluate the usefulness of our approach for different application settings).

Our plans for future work are further discussed in the subsections below.

Dynamic Sampling Frequency Management. As mentioned in section 3, this refers to the desired ability for individual nodes to make autonomous decisions about the Sampling Frequency, based on the distances monitored over adjacent symbolic representations of readings. A high relative distance means that the underlying physical process monitored is undergoing some change. We mention the word relative because distance is very much relative to the attribute being monitored and other application-specific factors. Please note that not every application will be able to benefit from this. For example monitoring an ECG requires more-or-less static sampling frequency and the minimum is bound by the time it takes for a single heartbeat. Other applications such as environmental monitoring can benefit greatly from adjustments in the sampling frequency. From our experience with the experiments on the Intel Lab data set, we notice that distance for great periods of time is zero or less than one (for example the light attribute during the night is almost constant — having the motes to sample at a fixed interval during nighttime is wasteful). During those times, the sampling frequency can be decreased until change starts to occur again. Once distance begins to increase, sampling frequency can be increased accordingly. And this brings us to the next point.

Local Group Coordination. So far in our discussion we have assumed that our algorithm is ran autonomously and in isolation by single nodes. And this is currently the case, but in the near future we aim to explore local coordination in such a way that the sensor field is split into regions where only two nodes are adequate to cover a specific region. The number two may seem arbitrary, but we choose two nodes so one can act as a fail-over (other choices are also valid). This can a huge energy-saving impact, especially on dense deployments. In simple terms, it means that the majority of the nodes in the network can be asleep for most of the time. The two awake nodes per region effectively monitor distances and can issue wake up calls to the rest of the nodes if they sense a big change. "Big" is defined by observing a normal period and learning what constitutes normal change (in terms of distances), as described in section 3.1. This form of sleep-scheduling can be performed in-the-network without any human interaction at all.

Reinforced Learning. We also aim to explore the possibility for reinforced learning in the machine learning sense. At present, our algorithm trains for some time and then, when it has learned an appropriate value for distance, it stops. We are planning to test our hypothesis for reinforcement learning, where learning is an ever-lasting process and new readings can affect the distance used for event detection.

Plan an Indoor Deployment. We plan to deploy a number of motes running our algorithm to monitor changes in temperature (and other attributes) in one of the university buildings.

Evaluate the Potential Use of Markov Models. We have already implemented an approach based on Markov Models in MATLAB. However, maintaining the two matrices (transition and emission) needed by such a model is a computationally-intensive process. We plan to investigate if an implementation for the sensor motes is feasible. Event-detection using probabilies is an alternative to the algorithm we currently use (which is distance-based rather than probability-based).

Our first priority is to implement and evaluate dynamic sampling frequency management and group coordination. The latter is a by-product of the former, and in our opinion the two combined can bring significant energy savings to networked sensors by making them context-aware and responsive to change.

6   Related Work

The approach by [20] is the one that come conceptually closer to our framework: in that work the authors classify gestures by comparing them to strings and performing string matching. The main difference with our approach is that we primarily aim for complex event detection and in addition we do generalise our approach for many application scenarios. In addition we offer non-parametric event detection. In the remainder of this section we will discuss other work that is related or has affected our own research.

One of the first papers that presented an Event-Based Approach was Directed Diffusion [21]. In that approach a node would request data by sending interests which are conceptually similar to subscriptions in a publish/subscribe system. Data found to match those interests is then sent towards that node. A different framework that is based on event classification, is the Online State Tracking [22] approach. This technique consists of two phases: the first phase is the learning process where new sensor readings are classified to states and the second phase is the online status monitoring phase during which nodes are collaborating to update the overall status of the network. This is interesting work in a sense that it moves away from individual nodes' readings and views the whole network as a state machine.

Another event-based technique based on thresholding is Approximate Caching [23] whereby nodes only report readings if they satisfy a condition. A more recent paper [24] suggests a mixture of hardware and software as a solution for detecting rare and random events. The event types they consider are tracking and detecting events using the eXtreme Scale Platform (XSM) mote equipped with infrared, magnetic and acoustic sensors. Central to their architecture is the concept of passive vigilance, which is inspired from sleep states of humans where the slightest noise can wake us up when we are asleep. This is implemented with Duty Cycling and recoverable retasking. A similar approach [25] proposes a sleep-scheduling algorithm that minimises the surveillance delay (event detection delay) while it maximises energy conservation. Sleep scheduling is coordinated locally in a fair manner, so all nodes get their fair share of sleep. A minimal subset that ensures coverage of the sensing field is always awake in order to be able to capture rare events. Sleep scheduling is related to our approach since we aim to introduce Dynamic Sampling Frequency Management and base upon it a sleep coordination mechanism. This was discusses in more detail in section 5.

A first paper that addresses the need for complex event detection is the one by Girod et al [26]. In their work the authors suggest a system that would treat a sequence of samples (a signal segment) as a basic data type and would offer a language (WaveScript) to express signal processing programs as declarative queries over streams of data. The language would be able to execute both on PCs and distributed sensors. The data stream management system (called WaveScope) combines event-stream and data management operations. Unfortunately, the paper describing the system is a position paper so there is only a high-level description of the architecture (and no current implementation or mention of plans).

A paper that falls under both the Event-Based and the Query-Based subcategory, is the one describing REED [27]; REED is an improvement on TinyDB [28]. Basically it extends TinyDB with its ability to support joins between sensor data and static tables built outside the network. The tables outside the network describe events in terms of complex predicates. These external tables are joined with the sensor readings table, and tuples that satisfy the predicate indicate readings of interest (eg. where an event has occurred).

A somewhat different method that supports geographic groupping of sensor nodes was presented in Abstract Regions [29,30]. Abstract Regions is essentially a family of spatial operators for TinyOS that allows nodes to form groups with the objective of data sharing and reduction within the groups by applying aggregate operators such as `min, max, sum`, etc. The work by [31] extended the types of aggregates supported by introducing (approximate) quantiles such as the median, the consensus, the histogram and range queries. Support for Spatial Aggregation was also suggested by [32] where sensor node would be grouped and aggregates would be computed using Voronoi diagrams. Another approach [33] models the sensor network as a distributed deductive declarative database system. Their suggested method allows for composite event detection, and the declarative language used (SNLog) is a variant of Datalog.

7   Conclusions

We have presented our approach for detecting complex events in sensor networks. Complex events are sets of data points hiding interesting patterns. These patterns are difficult or even impossible to capture using traditional techniques such as thresholds. Moreover, often the users of the system may not know what they are looking for, a priori.

We use SAX to convert raw real-valued sensor readings to symbolic representations and then we use a distance metric for string comparison. Our framework can either perform exact-matching or approximate-matching of some user-supplied pattern. Moreover, it can perform non-parametric event detection; in this case the user need not specify anything. The sensor trains for some time and over some data that is known to be normal and it learns appropriate values for distance. Then, once training is finished, the learned distances are used for the detection phase.

Our implementation currently runs on the Tmote Sky platform on TinyOS2, but we have an implementation for ContikiOS for comparison purposes. Our approach appears to match very well the unique resource constraints of sensor networks.

Acknowledgements

The authors would like to thank Dr Eamonn Keogh who patiently answered our questions regarding SAX and provided relevant code and test data.

References

1. MoteIV Corporation: TMote Sky,
   http://www.moteiv.com/products/tmotesky.php
2. Katsiri, E., Ho, M., Wang, L., Lo, B.: Embedded real-time heart variability analysis. In: Proceedings of 4th International Workshop on Wearable and Implantable Body Sensor Networks (2007)

3. MIT House_n Project: http://architecture.mit.edu/house_n/
4. Biosensornet: Autonomic Biosensor Networks for Pervasive Healthcare,
   http://www.doc.ic.ac.uk/~mss/Biosensornet.htm
5. Dutta, R., Hines, E., Gardner, J., Boilot, P.: Bacteria classification using Cyranose
   320 electronic nose. In: BioMedical Engineering Online (2002)
6. Doolin, D., Sitar, N.: Wireless sensors for wildfire monitoring. In: Proceedings of
   SPIE Symposium on Smart Structures and Materials (2005)
7. Firebug: Design and Construction of a Wildfire Instrumentation System Using
   Networked Sensors, http://firebug.sourceforge.net/
8. Madden, S.: Intel Lab Data (2004),
   http://berkeley.intel-research.net/labdata/
9. Hailes, S., Coulouris, G., Hopper, A., Wilson, A., Kerwin, D., Lasenby, J., Kalra,
   D.: SESAME: SEnsing for Sport And Managed Exercise,
   http://www.sesame.ucl.ac.uk/
10. Lin, J., Keogh, E., Lonardi, S., Chiu, B.: A symbolic representation of time series,
    with implications for streaming algorithms. In: Proceedings of ACM SIGMOD
    Workshop on Research Issues in Data Mining and Knowledge Discovery, ACM
    Press, New York (2003)
11. Keogh, E., Lin, J., Fu, A.: HOT SAX: Efficiently Finding the Most Unusual Time
    Series Subsequence. In: Proceedings of IEEE International Conference on Data
    Mining, IEEE Computer Society Press, Los Alamitos (2005)
12. Keogh, E., Lonardi, S., Ratanamahatana, C.A.: Towards parameter-free data
    mining. In: Proceedings of the tenth ACM SIGKDD international conference on
    Knowledge discovery and data mining, ACM Press, New York (2004)
13. Keogh, E.: The Time Series Data Mining Archive,
    http://www.cs.ucr.edu/~eamonn/TSDMA/index.html
14. TinyOS: An open source OS for the networked sensor regime,
    http://www.tinyos.net/tinyos-2.x/doc/html/overview.html
15. Dunkels, A., Finne, N., Eriksson, J., Voigt, T.: Run-time dynamic linking for
    reprogramming wireless sensor networks. In: SenSys 2006: Proceedings of the 4th
    international conference on Embedded networked sensor systems (2006)
16. Dunkels, A.: The Contiki OS, http://www.sics.se/contiki/
17. Dunkels, A., Schmidt, O., Voigt, T., Ali, M.: Protothreads: simplifying event-
    driven programming of memory-constrained embedded systems. In: SenSys 2006:
    Proceedings of the 4th international conference on Embedded networked sensor
    systems (2006)
18. Workshop on Massive Datasets: MERL Data, http://www.merl.com/wmd/
19. Smiths Detection: Cyranose 320,
    http://www.smithsdetection.com/eng/1383.php
20. Stiefmeier, T., Roggen, D., Trster, G.: Gestures are strings: Efficient online ges-
    ture spotting and classification using string matching. In: Proceedings of 2nd
    International Conference on Body Area Networks (BodyNets) (2007)
21. Govindan, C.I.R., Estrin, D.: Directed diffusion: A scalable and robust communi-
    cation paradigm for sensor networks. In: Proceedings of the Sixth Annual Inter-
    national Conference on Mobile Computing and Networking (2000)
22. Halkidi, M., Kalogeraki, V., Gunopulos, D., Papadopoulos, D., Zeinalipour-Yazti,
    D., Vlachos, M.: Efficient online state tracking using sensor networks. In: MDM
    2006: Proceedings of the 7th International Conference on Mobile Data Manage-
    ment (2006)
23. Olston, C., Loo, B.T., Widom, J.: Adaptive precision setting for cached approxi-
    mate values. In: Proceedings of SIGMOD Conference (2001)

24. Dutta, P., Grimmer, M., Arora, A., Bibyk, S., Culler, D.: Design of a wireless sensor network platform for detecting rare, random, and ephemeral events. In: IPSN 2005: Proceedings of the 4th international symposium on Information processing in sensor networks (2005)
25. Cao, Q., Abdelzaher, T., He, T., Stankovic, J.: Towards optimal sleep scheduling in sensor networks for rare-event detection. In: IPSN 2005: Proceedings of the 4th international symposium on Information processing in sensor networks (2005)
26. Girod, L., Mei, Y., Newton, R., Rost, S., Thiagarajan, A., Balakrishnan, H., Madden, S.: The Case for a Signal-Oriented Data Stream Management System. In: CIDR 2007 - 3rd Biennial Conference on Innovative Data Systems Research (2007)
27. Abadi, D.J., Madden, S., Lindner, W.: Reed: robust, efficient filtering and event detection in sensor networks. In: VLDB 2005: Proceedings of the 31st international conference on Very large data bases (2005)
28. Madden, S.R., Franklin, M.J., Hellerstein, J.M., Hong, W.: Tinydb: an acquisitional query processing system for sensor networks. ACM Trans. Database Syst. 30(1) (2005)
29. Welsh, M.: Exposing resource tradeoffs in region-based communication abstractions for sensor networks. SIGCOMM Comput. Commun. Rev. 34(1) (2004)
30. Welsh, M., Mainland, G.: Programming sensor networks using abstract regions. In: NSDI 2004: Proceedings of the 1st conference on Symposium on Networked Systems Design and Implementation (2004)
31. Shrivastava, N., Buragohain, C., Agrawal, D., Suri, S.: Medians and beyond: new aggregation techniques for sensor networks. In: SenSys 2004: Proceedings of the 2nd international conference on Embedded networked sensor systems (2004)
32. Sharifzadeh, M., Shahabi, C.: Supporting spatial aggregation in sensor network databases. In: GIS 2004: Proceedings of the 12th annual ACM international workshop on Geographic information systems, ACM Press, New York (2004)
33. Chu, D., Tavakoli, A., Popa, L., Hellerstein, J.: Entirely declarative sensor network systems. In: VLDB 2006: Proceedings of the 32nd international conference on Very large data bases (2006)

Multi-sensor Cross Correlation for Alarm Generation in a Deployed Sensor Network

Ian. W. Marshall[1], Mark Price[2], Hai Li[2], N. Boyd[3], and S. Boult[4]

[1] Lancaster Environment Centre, University of Lancaster, Lancaster LA1 4YQ
i.w.marshall@lancaster.ac.uk
[2] Infolab21, Dept. of Computing, University of Lancaster, Lancaster, LA1 4WA, UK
pricemc@comp.lancs.ac.uk
[3] Salamander Group, Williams House Manchester Science Park, Manchester, M15 6SE
n.boyd@salamander-group.co.uk
[4] School of Earth, Atmospheric and Environmental Sciences, Manchester University
Manchester M60 1QD
s.boult@manchester.ac.uk

Abstract. We are developing a sensor network to assess the hydro-dynamics of surface water drainage into Great Crowden Brook in the Peak District (UK). The complete network will observe soil moisture, temperature and rainfall on a number of vertical slope transects. GSM access for remote real time reporting of network status is only available from the hilltops so a multihop communication strategy is being used. To minimise radio usage and maximise battery life we are reporting only those alarms and events that are judged to be of high priority by a simple rule based decision engine, based on spatio-temporal cross-correlation of the available sensor inputs. In this paper we present the data handling strategy, report the findings from the initial technology trial and discuss the implications of the recovered environmental data samples for the design of effective alarm generation rules. It is clear that the measurements can always be interpreted more reliably when richer contextual information is captured, but care must be taken with the choice of observables.

Keywords: Sensor network, experience, cross-correlation, data analysis, rules, comparison, context.

1 Introduction

It has widely [1-3] been proposed that sensor networks are a good solution for environmental monitoring. However, this application presents a number of major challenges for current technology. In particular environmental science involves the study of coupled non-equilibrium dynamic processes that generate time series with non-stationary means and strongly dependent variables and which operate in the presence of large amounts of noise/interference (thermal, chemical and biological) and multiple quasi-periodic forcing factors (diurnal cycles, tides, etc). This typically means that any

G. Kortuem et al. (Eds.): EuroSSC 2007, LNCS 4793, pp. 286–299, 2007.

analysis must be based on large data samples obtained at multiple scales of space and time. In addition the areas of interest are large, relatively inaccessible and typically extremely hostile to electronic instrumentation. Our analysis of these factors has encouraged us to focus on this list of generic requirements;

- Node lifetime (between visits) should be 1 yr or greater .
- Communication range should be ~250m
- Nodes should be portable, unobtrusive, low cost, etc.
- Networks are expected to be sparse since areas of interest are large and budgets are small

However, it should be noted that in our experience to date, the characteristics of each environment, the dominant processes operating in it and the measurements that are of interest are sufficiently different that the design of an appropriate sensor network solution is normally most determined by site specific constraints. Most importantly the opportunities for exploiting contextual correlation to disambiguate observations and improve the maintenance and robustness of a deployed sensor network are always site specific. In this paper we describe the design and initial deployment of a hydrological sensor network designed for operation in an upland setting. We report on our initial experiments with correlating readings for management purposes, and offer some initial hypotheses regarding aspects of this that might be generic based on a comparison with data obtained in an earlier experiment in a marine setting.

2 Network Design

The trial network was designed to test the feasibility of obtaining real time hydrological event alarms from a steep sided upland stream valley (with no GSM availability) via a relay site on nearby hilltop peat moorland where GSM service was available (from operator equipment sited on a prominent television mast on a hilltop around 3 km North East of the site). The site is at Great Crowden Brook just north of Crowden, Derbyshire in the Peak District National Park. It is near the Pennine way and is visited regularly by hikers, but is not subject to intense human interference. The site is used as rough grazing for sheep, and also forms part of the catchment for a local water supply reservoir (the full catchment is approximately 18 square kilometres, a relatively small setting for environmental science). It should be noted that Crowden was chosen as it is an exceptionally richly instrumented site with 5 existing nodes measuring rainfall and water quality. The brook is prone to occasional flash flooding events. The site is about 1hrs walk from the nearest car park (Crowden), which itself is about 1 hrs drive from the office in Manchester. All equipment has to be carried to the site on foot and removed when experiments are complete. To avoid interference by passing hikers it is advisable for equipment to be hard to spot but it is not necessary that it is completely hidden. Malfunction due to various types of interaction with the sheep is difficult to completely avoid. A view of the site is shown in figure 1, where the rough nature of the terrain is apparent. The vertical distance from valley base to Hilltop is just over 100m however the topography is convex so there is no simple line of sight between the point of interest and the nearest available GSM service.

Fig. 1. An overall view of the deployment site. Sensor 1 is below the feet of the photographer, sensor 2 is on the slope to the right, sensor 3 is on the slope of the hill in the distance and sensor 4 is on the peak of the hill in the distance between the two outcrops on the skyline. The brook is just out of sight in a cleft below the rocks at the bottom left of the picture.

The initial trial network is illustrated schematically in Fig 2. It consists of 4 nodes equipped with:

1. A motherboard with a *Texas Instruments* MSP430F1611 microcontroller. This microcontroller runs at 8 MHz, is low power (~3 mA at 3.3 V), has eight 12-bit A/D inputs, two UARTs (Universal Asynchronous Transmitter/Receiver, used for serial communication) and 28 Kbytes of NVRAM. This board also carried an embedded temperature sensor.

2. An "Antilog" serial logger. This is connected to one of the MSP's UARTs and is used to record data and events on a 32 Mbyte MMC card.

3. A Radiometrix 173.5 MHz radio transceiver (chosen to give the desired range). This has a current drain of approx 20mA, a range of 300m (when placed on rough ground) and a 9.6 Khz transmission bandwidth. The radios were equipped with a simple quarter wave whip antenna (since directional antennae were judged to be at risk of being knocked off target by the sheep)

4. A high capacity (19 Ah) 3.6V Lithium Thionyl battery. Provides power for all the components, as there is no power regenerative capability on the node. This battery delivered a measured lifetime of >60 days.

In addition one (source) node was equipped with an EC-5 soil moisture sensor (available from *Labcell Technologies, UK)*, and one (sink) node was equipped with a "Fargo Maestro 20" GSM modem which used SMS to send alarms back to the user (and a 4R25 6 volt lantern battery used to separately power the modem). The other two nodes were deployed as intermediate relays.

In the multihop relay each sensor node passes summary data and any alarms to its neighbour. Its neighbour then reads the data, adds its own summary data and alarms, and passes them on. At the base-station node (which has GSM capability) the alarms and any associated summary data are transmitted (via SMS) back to the users in Manchester. For the trial the internode spacings were all in excess of 100m as shown in figure 2 and illustrated geospatially in figure 3. To achieve this multihop behaviour we implemented a simple messaging service. The radios were configured with a point to point link layer protocol with a Manchester encoding for error checking. Each node was programmed to turn on its transceiver for 3 minutes each hour, listen for any hello messages, acknowledge them, and log any subsequent data messages received. Each node then read its log, calculated any alarms (compare received data with local data) turned on its transceiver a second time and sent the messages it wanted to send (hello-ack-send_data). Each hello message was sent 5 times or until an acknowledgement was received (whichever was sooner). If no acknowledgement was received after the fifth hello, the data was sent anyway. Clock drifts were expected (and observed) to be well within the tolerances of this scheme. Sophisticated dynamic routing protocols are not thought to be appropriate for the sparsely populated networks typical of environmental monitoring, where each node can rarely communicate with more than 3 others even in ideal conditions.

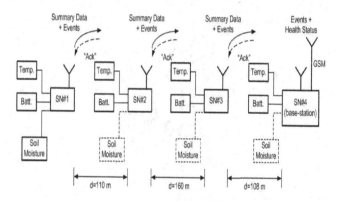

Fig. 2. Schematic of deployed network

For the initial trial various alarm conditions were programmed into each node, in the form of rules. The rules were checked each hour when the logger was read, and if a condition was satisfied an alarm code was generated. This alarm code was then propagated to the base-station then acted upon by the base-station. The alarm code is in a very simple format: "A<node ID><sensor ID><alarm number>". This made it easy to parse, and also easily extensible to include more nodes, sensors and alarm conditions. The rules were as follows;

Soil moisture sensor (sensor ID = 0)
'0' = avg_ec5 > 750 mV (very wet)
'1' = avg_ec5 < 250 mV (very dry)
'2' = avg_ec5 > 1000 mV (sensor error)
Temperature (sensor ID = 1)
'0' = avg_ctemp < -10 (very cold)
'1' = avg_ctemp > 50 (very hot)

'2' = avg_ctemp < -20 or avg_ctemp > 60 (sensor error)
Battery (sensor ID = 2)
'0' = avg_batt < 3.4V   (low)
'1' = avg_batt <= 0   (error)
Radio (sensor ID = 3)
'0' = no message received

So, for example, an "A100" error code would indicate that node #1's soil moisture sensor was registering very wet conditions which would be strongly indicative of a major precipitation event. Similarly, an "A420" error code is a warning that node #4's battery is running low.

Each node would only generate an alarm the first time that an event with that code occurred in any day. In order to reduce the number of SMS messages sent by the base-station, it also only sent one type of alarm code out per node in any 24 hour period. For example, if the battery level was flagged as low, even if subsequent measurements also showed it to be low, only *one* SMS was sent that day).

The base station also generated a message at midnight each day so that the operators were confident that any alarms had not been blocked by a fault at the base station.

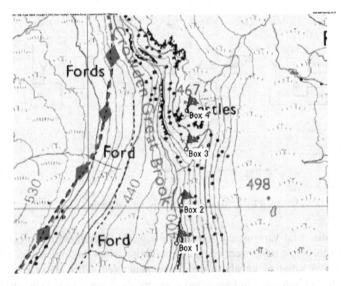

Fig. 3. Map of the deployment. Area shown approx 500x500m Grid lines cross at SE06000 02000, box 1 is at SE 06259 01896 (GPS). The underlying mapping is reproduced by kind permission of the United Kingdom Ordnance Survey (© Crown copyright Ordnance Survey. All rights reserved).

These rules were not expected to be particularly effective, but to serve as a test of the message oriented strategy.

The nodes were configured to read and log temperature, soil moisture, and battery power every 15 minutes (250ms time window). In addition, the number of times that communication was completely unsuccessful between two nodes was also recorded. A more advanced procedure (logging of both RSSI and bad packets when the radio is on) will be implemented in Phase 2. The GSM node also logged the GSM signal strength and every successful message send.

The physical hardware is shown in Figure 4 (traditional boards were used for prototyping to avoid the need for last surface mount soldering of late design changes) and the physical enclosure is shown in figure 5.

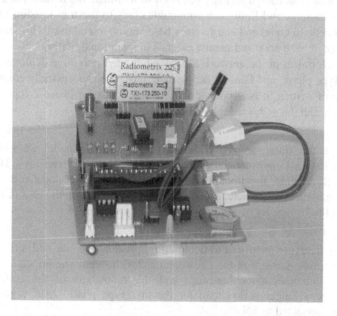

Fig. 4. Photograph of the custom made MSP motherboard (top) connected to the 173.25 MHz radio board (bottom). The combined unit is common to all nodes.

Fig. 5. The physical enclosure (20x10x8cm)

This network currently acts as a technology demonstrator for two EPSRC projects; PROSEN [4] and DIAS [5].

A. Hardware design and development strategy

From the outset our design strategy has been to minimise the duration and frequency of communications, and limit such communications to the transmission of summary data (from node to node) and alarms from base-station to end user. This minimises the power usage of each node and greatly extends it operational lifetime.

The development of the network can be subdivided into three deployment phases. The prototype phase, and Phase 1 deployment have been successfully completed and, at the time of writing, it is envisaged that modifications for Phase 2 deployment will be complete, and the nodes redeployed by mid-August 2007.

B. Prototype phase

Here four nodes (three sensing nodes and a base-station node) were placed in a controlled, external environment with good GSM signal for five days. The purpose was to check the basic functionality of the nodes and the inter-nodal communications prior to deployment at the actual field site. This took place on the roof of the Physics building at the University of Kent, and was entirely successful.

C. Phase 1

Phase 1 of the experiment was to deploy a minimal set of four nodes (three sensing nodes plus a base-station) in the Crowden catchment area. The principal aim was to check the communications hardware and the robustness of the communication algorithm for a longer period than for the prototype phase. Also, a soil moisture sensor was connected to SN #1.

D. Phase 2

Phase 2 of the experiment will be to redeploy the nodes each (including the base-station) equipped with two soil moisture sensors (to give short range variation), and an external thermometer. Two nodes (1 and 4) will also be equipped with a bucket sensor for measuring local rainfall. This arrangement will enable an assessment of the dynamics of flow down the slope and will allow an initial estimation of the relative importance of surface run-off and sub-surface drainage in the local hydrodynamics. This in turn will lead to an improved understanding of the processes dominating the erosion of the peat on the hilltops and the impacts of this on water quality in the stream.

Based upon the data we have acquired from Phase 1, we are optimising our alarm generation strategy. The details will be described elsewhere [6], but simple static thresholds will not be used. We will also modify the onboard software so that we have a more thorough test mode for evaluation of GSM communication from the base-station to the end user, and so that the user can send update commands via SMS.

E. Phase 3

During 2008 we plan to deploy up to 20 nodes monitoring 5 transects within the area shown in fig 3.

3 Results from Phase One

The network was initially deployed on the 13th April 2007, and then recovered on the 10th June 2007, 59 days later.

Nodes #1 and #2: These nodes were still fully functional when retrieved, and showed no signs of damage or interference.

Node #3: The 173 MHz antenna showed signs of having been chewed by passing sheep. This does not seem to have seriously impaired its performance. This could be avoided by use of sheep excrement, but trampling/kicking/nudging would not be avoided this way. This node had a minor leak due to dust compromising the seal during a mid deployment inspection that required the box to be opened. Clearly opening boxes in the field should be avoided in any full trial. The battery in this node was flat (probably due to the water ingress).

Node #4 (base-station): This node was also intact, but both antennae (the GSM and 173 MHz) also showed signs of being chewed. The GSM antenna was also corroded, and a more weather proof version will be required in future tests. The battery in this node was also flat. In this case the battery was expected to have a shorter life as it was powering the serial interface driver for the GSM interface. This was intended to be powered by the lantern battery, but due to time constraints was powered off the lithium battery. This will be fixed in phase 2.

A. Soil Moisture Readings
The moisture sensor was placed at a depth of 15 cm in a sphagnum (moss) swale near the brook. This swale dries out when there is no rain and fills with both direct rainfall and runoff after rain. Figure 6 below shows the soil moisture readings as measured by the sensor on node #1. Also shown is the daily rainfall record from a weather station sited around 3 kilometres away near the television mast supporting the GSM operator's equipment.

Clearly some of the soil moisture events are caused by rainfall that was also observed several kilometres away. Equally clearly the amount of water detected does not correlate well. This is expected – upland rainfall is often quite patchy (spatially bursty) on quite short length scales, and in any case the degree of swale filling is determined by both rain quantity and the degree of saturation of the surrounding slopes. There is one rainfall event that did not result in increased moisture – this is likely to be the result of the rain being in the next valley and illustrates the need for the proximal rainfall measurements (at both the top and the foot of the slope) planned for phase 2. There are 4 moisture events that seem to have no rainfall. These could be times when the swale filled with run-off from the West (the rain bucket is North-East). The preponderance of Rain to the West and not to the East over the inverse is consistent with the prevailing weather conditions (Westerlies) at this site.

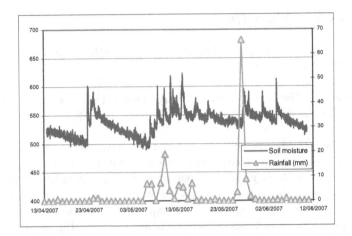

Fig. 6. The soil moisture (left hand axis - mV) and rainfall (right hand axis – mm) records for the two month trial deployment period

The daily sampling at the weather station is clearly insufficient to resolve dynamic process details, but the data captured indicates that a 15 min sampling rate is probably sufficient for us to learn a considerable amount about this environment. In a similar network built in Australia [7] the sampling rate was set very low (to increase lifetime) and increased when rain was detected. In a flat semi-arid setting where soil moisture can only relate to rainfall, and rainfall is normally relatively homogeneous this strategy made sense. However in our case we clearly cannot use detection of rain to guarantee capture of soil moisture increases.

Almost all the moisture events captured show a sharp increase and a slow decrease. Only one (middle of graph) shows a slower rise (that might be expected for drainage only events). To confirm the hypothesis that slower rise time indicates drainage dominated a co-located rain detector is necessary.

The data suggest that an event notification strategy based on detecting changes in moisture level of 15 mV or more, that are sustained for a full hour will flag all true events, and no false events triggered by noise. We also expect occasional transient events relating to sheep urine that can be filtered by correlation of observations between sites. A sustained inconsistency is evidence of sensor malfunction. In this case there are no diurnal cycles. Cross-correlation with radio signal strength (due to range restriction in rainy conditions) is also a useful cross-check that we have used elsewhere [11]

B. Battery Voltage and Temperature

The battery voltage record from node 1 is shown in figure 7. As expected there is no obvious trend (the battery was still working). Lithium batteries of this type do not exhibit a gradual drop in voltage (like lead acid batteries), but a rapid drop at end of life. The early warning of battery failure in the nodes we have designed is a loss of radio range (the radio is powered by a charge pump which is highly sensitive to small impedance changes in the battery, which can be observed by noting the RSSI and the number of messages received with errors, and checking that the errors are not due to rain or temperature related voltage drops at the battery. In other words, if errors are observed consistently and soil is not wet and temp is not low then battery is near end of life and an alarm should be generated to initiate a site visit to change the battery (or replace the node). The temperature sensitivity of the battery is clearly indicated in figure 8 where the battery voltage is clearly synchronised with the diurnal temperature variation and with the weather related changes from day to day.

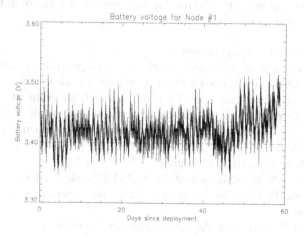

Fig. 7. Battery Voltage vs. time

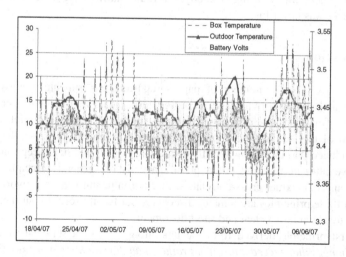

Fig. 8. Temperature (left hand scale - Celsius) at both in box and external sensors. Battery voltage is also shown (right hand scale - V) to illustrate the correlation discussed ion the text.

The main purpose of figure 8 is to illustrate the temperature readings and compare them with the temperature readings taken at the TV mast weather station. The two traces are broadly similar but the weather station seems to record higher temperatures during some periods. The periods of difference do not correlate well with the observed periods of rainfall in figure 6. However they do correlate extremely well with periods of bright sunshine (also recorded by the weather station but not shown here. The weather station is on an open site (necessarily) whilst the sensor is in a site heavily shaded for most of the day by the sides of the valley (running approx. North South). Thus differences between the temperature sensors cannot be used in sensor diagnostics without also cross-correlating with insolation measurements. Rainfall is not sufficient - it does not always rain when it is cloudy.

C. Alarm generation and propagation

Of major interest was the successful generation and propagation of alarm signal along the sensor nodes. Figure 7 shows the battery voltage of Node #1. Shown by the dotted line is the lower alarm threshold (3.4 volts) at which an alarm code is generated. Analysis of the data shows that the battery level dipped below 3.40 volts on 844 separate occasions, corresponding to at least once per day where the battery level breached the lower threshold. Therefore 59 (one per deployment day) alarm codes should have been sent to the base-station.

Analysis of node #4's data showed that 59 "A120" codes were indeed received and parsed. We are thus confident that the alarm and communication mechanism is working correctly.

Figs 6 and 8 reveal that the temperature and moisture thresholds were never breached and no alarms should have been generated. In fact no alarms were generated so the equipment was working correctly. No radio alarms were generated as contact with node #1 was never lost.

4 Discussion

There are now quite a number of papers reporting experimental deployments of environmental sensor network and recording sensor data that show evidence of the variability of natural environments. Examples include [7-12]. We do not propose to discuss all of the existing examples here as space is limited, however they provide strong evidence that the findings of this paper are typical and not exceptional. In this section we attempt to draw out a preliminary list of general principles for data handling and contextual cross-correlation in environmental sensor networks. We do not claim these principles are a major discovery, but felt it useful to put them and the evidence for them into a clear and short document.

The first principle derived from the ubiquitous lack of a stationary mean is to: *Report changes that exceed a threshold rather than deviations from a fixed threshold.*

The traces above illustrate the fact that environmental data is sometimes dominated by forcing cycles and sometimes bursty. This leads directly to a second principle: *Any change threshold must be larger or more sustained than the cycles or noise in the data. If this is not possible a more sophisticated strategy must be used*

A third observation is that the timescales of significant changes are highly variable from milliseconds for landslip and earth tremor, through seconds for wind and atmosphere, to years/decades for climate change. However most phenomena ideal for existing sensor network technology occur on spatial scales of 10-10000m and have characteristic timescales in the minutes to hours window. Our third principle is thus; *Know the timescales that are significant and design your data handling strategy accordingly.*

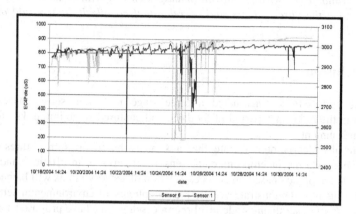

Fig. 9. Conductivity data from the Secoas trial in 2004 [11]

In figure 9 we show a data trace from the experiments in SECOAS (that we assisted with) during 2004. This site was on the seabed (6 m depth) 2 km offshore from Great Yarmouth, a holiday resort on the East coast of the UK, and crucially about 1 km downcurrent in the longshore tidal drift from the main sewage outfall. The plot shows seawater conductivity. Often conductivity is used in oceanography as a crude indicator of sensor site integrity since in the open ocean it does not vary very much. The expected variation can be seen in the small (tidal) ripples and the end to end trend (seasonal). The large bursts of drops in conductivity are not usual. However, they are not indicative of sensor malfunction. The peaks correlate very well with the maxima in the tidal current and correspond to when the water movements are most turbulent. It is precisely these times when mixing of the fresh water from the sewage outfall will be most partial, so it is likely that the events are detecting partially mixed pockets of brackish water. Further evidence confirming this interpretation was obtained by noting that the events were most noticeable at the weekend (e.g. middle of graph) when the population of the town is significantly increased at this time of year. The most noticeable event captured (shown on graph) was at a weekend when there was a major (horse) race meeting in the town and the population was maximal. This illustrates that sometimes the cross-correlations required to achieve an interpretation can be very subtle, and not amenable to automated capture. We propose as a general principle: *Event generation can never be 100% accurate so always report sustained (over two or more readings) major changes in any time series.*

A further example available at [12] illustrates further issues with diurnal cycles. In this data set temperature was monitored in the intertidal zone near Vancouver Island. Not only are the cycles highly variable (tides vary according to positions of sun and moon), but during the summer when the tide is in the temperature cools and during the winter when the tide is in the temperature increases. Any in-situ preprocessing must in this case be aware of astrophysical and seasonal changes aswell as more local changes. Fortunately long term wide area phenomena are relatively well understood and in-situ learning is not required. This leads us to a final general principle: *Always embody available knowledge into the management algorithms as this will signify-cantly reduce the observational and correlational complexity that would otherwise be needed.*

5 Conclusion

We have successfully constructed a three stage multi-hop sensor network that operated in the field for a period of 59 days, and captured some initial data about the target environment.

The results clearly demonstrate that a strategy based on simple thresholds for generation of alarms in environmental sensor networks is insufficient and a more sophisticated approach based on cross correlation of sensory inputs will be necessary.

The results also demonstrate vividly the complexity of environmental settings and the difficulty of generalising node level preprocessing. We have proposed some ideas for generic rules that will be subject to future experimental testing in a range of environmental settings

Analysis of the data, and the lessons learnt during the Phase 1 field trial, give us an excellent knowledge base on which to proceed to a full deployment and long term operation leading to capture of new science quality data.

Acknowledgement

The authors would like to thank the Department of Trade and Industry for the funding to carry out this project. Special thanks also goes to Terry Rockhill and Clive Birch in the mechanical workshop of the University of Kent for their invaluable assistance with fabrication, and to the SECOAS project team for Fig 9.

References

1. Websites for work in USA: http://www-mtl.mit.edu/research/icsystems/uamps/ http://www.isi.edu/scadds/pc104testbed/guideline.html http://webs.cs.berkeley.edu/ http://cens.ucla.edu/ http://dsn.east.isi.edu/ http://www.ices.cmu.edu/sensornets/
2. Websites for work in Europe: http://eyes.eu.org/ http://www.smart-its.org/
3. Websites for work in Australia: http://www.ee.unimelb.edu.au/ISSNIP/ http://www.csse.uwa.edu.au/adhocnets/WSNgroup/ http://www.csiro.au/resources/pfik.html
4. PROSEN: Proactive Condition Monitoring of Sensor Networks, http://www.prosen.org.uk
5. DIAS: Design Implementation and Adaptation of Sensor Networks, http://www.dcs.gla.ac.uk/dias/

6. Li, H., Price, M.C., Stott, J., Marshall, I.W.: The development of a Wireless Sensor Network sensing node utilising adaptive self-diagnostics. LNCS. Springer, Heidelberg (2007)
7. Cardell-Oliver, R., Smettem, K., Kranz, M., Mayer, K.: A Reactive Soil Moisture Sensor Network: Design and Field Evaluation. International Journal of Distributed Sensor Networks 1(2), 149–162 (2005)
8. Turau, V., Witt, M., Weyer, C.: Analysis of a real multi-hop sensor network deployment: The Heathland Experiment. In: INSS 2006. Proceedings of the Third International Conference on Networked Sensing Systems, Chigago, Illinois (June 2006)
9. Xu, Y., Heidemann, J., Estrin, D.: Geography-informed Energy Conservation for Ad Hoc Routing. In: Hughes, D., Greenwood, P., Coulson, G., Blair, G., Pappenberger, F., Smith, P., Beven, K. (eds.) Proc. ACM/IEEE Int. Conf. on Mobile Computing and Networking, July 2001, Rome, Italy. An Experiment with Reflective Middleware to Support Grid-based Flood Monitoring, submitted to the Wiley InterScience Journal on Concurrency and Computation: Practice and Experience, pp. 70–84 (June 2007)
10. Grace, P., Hughes, D., Porter, B., Coulson, G., Blair, G.: Middleware Support for Dynamic Reconfiguration in Sensor Networks. Invited submission to Distributed Systems Online (DS Online) (April 2007)
11. Tateson, J., Roadknight, C., Gonzalez, A., Khan, T., Fitz, S., Henning, I., Boyd, N., Vincent, C., Marshall, I.W.: Real World Issues in Deploying a Wireless Sensor Network. In: REALWSN 2005. Workshop on Real-World Wireless Sensor Networks, Stockholm, Sweden (June 2005)
12. http://faculty.washington.edu/mdethier/Fucus/

Author Index

Amft, Oliver   126

Bargh, M.S.   93
Bigham, John   175
Birchfield, David   46
Blackstock, Michael   239
Boult, Steve   286
Boyd, Nathan   286
Broens, Tom   223

Campbell, Andrew T.   1

David, Pierre   110
de la Hamette, Patrick   29
Dustdar, Schahram   207

Ebben, P.W.G.   93
Eisenman, Shane B.   1
Esfandiyari, Sohail   159

Finke, Matthias   239
Fleisch, Elgar   255

Gluhak, Alexander   159

Havinga, Paul   62
Holzmann, Clemens   77
Hulsebosch, R.J.   93

Iacob, S.M.   93
Idasiak, Vincent   110

Jenny, Reto   29
Juszczyk, Lukasz   207

Kidane, Assegid   46
Kratz, Frédéric   110
Kupschick, Stefan   159

Lane, Nicholas D.   1
Lea, Rodger   239
Lenzini, G.   93
Li, Hai   286

Lijding, Maria   62
Lombriser, Clemens   126
Lu, Kaiyuan   190

Manzoor, Atif   207
Marshall, Ian.W.   286
Meratnia, Nirvana   62
Metzger, Christian   255
Michahelles, Florian   255
Miluzzo, Emiliano   1
Muthukrishnan, Kavitha   62

Naeem, Usman   175
Nussbaum, Doron   190

Presser, Mirko   159
Price, Mark   286

Qian, Gang   46
Quartel, Dick   223

Rajko, Stjepan   46
Rangarajan, Sankar   46
Roggen, Daniel   29
Roussos, George   270

Sack, Jörg-Rüdiger   190
Shimakawa, Hiromitsu   142
Stiefmeier, Thomas   126

Takada, Hideyuki   142
Tröster, Gerhard   29, 126
Truong, Hong-Linh   207

van Sinderen, Marten   223

Wang, Jinfu   175

Yamahara, Hiroyuki   142

Zhu, Ling   159
Zoumboulakis, Michael   270

Author Index

Lecture Notes in Computer Science

Sublibrary 5: Computer Communication Networks and Telecommunications

For information about Vols. 1– 4465
please contact your bookseller or Springer

Vol. 4793: G. Kortuem, J. Finney, R. Lea, V. Sundramoor-thy (Eds.), Smart Sensing and Context. X, 301 pages. 2007.

Vol. 4785: A. Clemm, L.Z. Granville, R. Stadler (Eds.), Managing Virtualization of Networks and Services. XIII, 269 pages. 2007.

Vol. 4773: S. Ata, C.S. Hong (Eds.), Managing Next Gen-eration Networks and Services. XIX, 619 pages. 2007.

Vol. 4745: E. Gaudin, E. Najm, R. Reed (Eds.), SDL 2007: Design for Dependable Systems. XII, 289 pages. 2007.

Vol. 4725: D. Hutchison, R.H. Katz (Eds.), Self-Organizing Systems. XI, 295 pages. 2007.

Vol. 4712: Y. Koucheryavy, J. Harju, A. Sayenko (Eds.), Next Generation Teletraffic and Wired/Wireless Ad-vanced Networking. XV, 482 pages. 2007.

Vol. 4686: E. Kranakis, J. Opatrny (Eds.), Ad-Hoc, Mo-bile, and Wireless Networks. X, 285 pages. 2007.

Vol. 4685: D.J. Veit, J. Altmann (Eds.), Grid Economics and Business Models. XII, 201 pages. 2007.

Vol. 4581: A. Petrenko, M. Veanes, J. Tretmans, W. Grieskamp (Eds.), Testing of Software and Communi-cating Systems. XII, 379 pages. 2007.

Vol. 4572: F. Stajano, C. Meadows, S. Capkun, T. Moore (Eds.), Security and Privacy in Ad-hoc and Sensor Net-works. X, 247 pages. 2007.

Vol. 4549: J. Aspnes, C. Scheideler, A. Arora, S. Madden (Eds.), Distributed Computing in Sensor Systems. XIII, 417 pages. 2007.

Vol. 4543: A.K. Bandara, M. Burgess (Eds.), Inter-Domain Management. XII, 237 pages. 2007.

Vol. 4534: I. Tomkos, F. Neri, J. Solé Pareta, X. Masip Bruin, S. Sánchez Lopez (Eds.), Optical Network Design and Modeling. XI, 460 pages. 2007.

Vol. 4517: F. Boavida, E. Monteiro, S. Mascolo, Y. Koucheryavy (Eds.), Wired/Wireless Internet Commu-nications. XIV, 382 pages. 2007.

Vol. 4516: L.G. Mason, T. Drwiega, J. Yan (Eds.), Managing Traffic Performance in Converged Networks. XXIII, 1191 pages. 2007.

Vol. 4503: E. Airoldi, D.M. Blei, S.E. Fienberg, A. Gold-enberg, E.P. Xing, A.X. Zheng (Eds.), Statistical Net-work Analysis: Models, Issues, and New Directions. VIII, 197 pages. 2007.

Vol. 4479: I.F. Akyildiz, R. Sivakumar, E. Ekici, J.C.d. Oliveira, J. McNair (Eds.), NETWORKING 2007. Ad Hoc and Sensor Networks, Wireless Networks, Next Generation Internet. XXVII, 1252 pages. 2007.

Vol. 4465: T. Chahed, B. Tuffin (Eds.), Network Control and Optimization. XIII, 305 pages. 2007.

Vol. 4427: S. Uhlig, K. Papagiannaki, O. Bonaventure (Eds.), Passive and Active Network Measurement. XI, 274 pages. 2007.

Vol. 4396: J. García-Vidal, L. Cerdà-Alabern (Eds.), Wireless Systems and Mobility in Next Generation In-ternet. IX, 271 pages. 2007.

Vol. 4373: K.G. Langendoen, T. Voigt (Eds.), Wireless Sensor Networks. XIII, 358 pages. 2007.

Vol. 4357: L. Buttyán, V.D. Gligor, D. Westhoff (Eds.), Security and Privacy in Ad-Hoc and Sensor Networks. X, 193 pages. 2006.

Vol. 4347: J. López (Ed.), Critical Information Infras-tructures Security. X, 286 pages. 2006.

Vol. 4325: J. Cao, I. Stojmenovic, X. Jia, S.K. Das (Eds.), Mobile Ad-hoc and Sensor Networks. XIX, 887 pages. 2006.

Vol. 4320: R. Gotzhein, R. Reed (Eds.), System Analysis and Modeling: Language Profiles. X, 229 pages. 2006.

Vol. 4311: K. Cho, P. Jacquet (Eds.), Technologies for Advanced Heterogeneous Networks II. XI, 253 pages. 2006.

Vol. 4272: P. Havinga, M. Lijding, N. Meratnia, M. Weg-dam (Eds.), Smart Sensing and Context. XI, 267 pages. 2006.

Vol. 4269: R. State, S. van der Meer, D. O'Sullivan, T. Pfeifer (Eds.), Large Scale Management of Distributed Systems. XIII, 282 pages. 2006.

Vol. 4268: G. Parr, D. Malone, M. Ó Foghlú (Eds.), Au-tonomic Principles of IP Operations and Management. XIII, 237 pages. 2006.

Vol. 4267: A. Helmy, B. Jennings, L. Murphy, T. Pfeifer (Eds.), Autonomic Management of Mobile Multimedia Services. XIII, 257 pages. 2006.

Vol. 4240: S.E. Nikoletseas, J.D.P. Rolim (Eds.), Algo-rithmic Aspects of Wireless Sensor Networks. X, 217 pages. 2006.

Vol. 4238: Y.-T. Kim, M. Takano (Eds.), Management of Convergence Networks and Services. XVIII, 605 pages. 2006.

Vol. 4235: T. Erlebach (Ed.), Combinatorial and Algo-rithmic Aspects of Networking. VIII, 135 pages. 2006.

Vol. 4217: P. Cuenca, L. Orozco-Barbosa (Eds.), Per-sonal Wireless Communications. XV, 532 pages. 2006.

Vol. 4195: D. Gaiti, G. Pujolle, E.S. Al-Shaer, K.L. Calvert, S. Dobson, G. Leduc, O. Martikainen (Eds.), Autonomic Networking. IX, 316 pages. 2006.

Vol. 4124: H. de Meer, J.P.G. Sterbenz (Eds.), Self-Organizing Systems. XIV, 261 pages. 2006.

Vol. 4104: T. Kunz, S.S. Ravi (Eds.), Ad-Hoc, Mobile, and Wireless Networks. XII, 474 pages. 2006.

Vol. 4074: M. Burmester, A. Yasinsac (Eds.), Secure Mobile Ad-hoc Networks and Sensors. X, 193 pages. 2006.

Vol. 4033: B. Stiller, P. Reichl, B. Tuffin (Eds.), Performability Has its Price. X, 103 pages. 2006.

Vol. 4026: P.B. Gibbons, T. Abdelzaher, J. Aspnes, R. Rao (Eds.), Distributed Computing in Sensor Systems. XIV, 566 pages. 2006.

Vol. 4003: Y. Koucheryavy, J. Harju, V.B. Iversen (Eds.), Next Generation Teletraffic and Wired/Wireless Advanced Networking. XVI, 582 pages. 2006.

Vol. 3996: A. Keller, J.-P. Martin-Flatin (Eds.), Self-Managed Networks, Systems, and Services. X, 185 pages. 2006.

Vol. 3976: F. Boavida, T. Plagemann, B. Stiller, C. Westphal, E. Monteiro (Eds.), NETWORKING 2006. Networking Technologies, Services, and Protocols; Performance of Computer and Communication Networks; Mobile and Wireless Communications Systems. XXVI, 1276 pages. 2006.

Vol. 3970: T. Braun, G. Carle, S. Fahmy, Y. Koucheryavy (Eds.), Wired/Wireless Internet Communications. XIV, 350 pages. 2006.

Vol. 3964: M.Ü. Uyar, A.Y. Duale, M.A. Fecko (Eds.), Testing of Communicating Systems. XI, 373 pages. 2006.

Vol. 3961: I. Chong, K. Kawahara (Eds.), Information Networking. XV, 998 pages. 2006.

Vol. 3912: G.J. Minden, K.L. Calvert, M. Solarski, M. Yamamoto (Eds.), Active Networks. VIII, 217 pages. 2007.

Vol. 3883: M. Cesana, L. Fratta (Eds.), Wireless Systems and Network Architectures in Next Generation Internet. IX, 281 pages. 2006.

Vol. 3868: K. Römer, H. Karl, F. Mattern (Eds.), Wireless Sensor Networks. XI, 342 pages. 2006.

Vol. 3854: I. Stavrakakis, M. Smirnov (Eds.), Autonomic Communication. XIII, 303 pages. 2006.

Vol. 3813: R. Molva, G. Tsudik, D. Westhoff (Eds.), Security and Privacy in Ad-hoc and Sensor Networks. VIII, 219 pages. 2005.

Vol. 3462: R. Boutaba, K.C. Almeroth, R. Puigjaner, S. Shen, J.P. Black (Eds.), NETWORKING 2005. XXX, 1483 pages. 2005.